Beltsville Symposia in Agricultural Research

A series of annual symposia held
at the Beltsville Agricultural Research Center
Agricultural Research Service
United States Department of Agriculture

XX Biotechnology's Role in the Genetic Improvement of Farm Animals

Beltsville Symposium XX Organizing Committee

H. D. Norman (co-chair), USDA-ARS

V. G. Pursel (co-chair), USDA-ARS

V. L. Hupfer (symposium secretary), USDA-ARS

M. R. Bakst (transportation), USDA-ARS

M. Beroza (Chair, Distinguished Scientist Award), Friends of Agricultural Research–Beltsville, Inc.

S. M. Hays (public relations), USDA-ARS

J. K. Lunney (finance), USDA-ARS

R. H. Miller (proceedings editor), USDA-ARS

C. E. Rexroad, Jr. (registration), USDA-ARS

Beltsville Symposia in Agricultural Research

XX Biotechnology's Role in the Genetic Improvement of Farm Animals

Edited by R. H. Miller, V. G. Pursel, and H. D. Norman

Invited papers presented at a symposium held
May 14–17, 1995, at the Beltsville Agricultural
Research Center, Beltsville, Maryland

Sponsored by Beltsville Agricultural Research Center,
Agricultural Research Service,
U. S. Department of Agriculture
and Friends of Agricultural Research–Beltsville

American Society of Animal Science
Savoy, Illinois

Library of Congress Cataloging-in-Publication Data

Biotechnology's role in the genetic improvement of farm animals : invited papers presented at a symposium held May 14-17, 1995, at the Beltsville Agricultural Research Center, Beltsville, Maryland / edited by R.H. Miller, V.G. Pursel, and H.D. Norman.
p. cm. -- (Beltsville symposia in agricultural research ; 20)
"Sponsored by Beltsville Agricultural Research Center, Agricultural Research Service, U.S. Department of Agriculture and Friends of Agricultural Research-Beltsville."
Includes bibliographical references and index.
ISBN 1-887458-01-8
1. Livestock--Breeding--Congresses. 2. Livestock--Genetic engineering--Congresses. 3. Livestock improvement--Congresses. 4. Animal biotechnology--Congresses. I. Miller, R. H., 1934- . II. Pursel, V. G., 1936- . III. Norman, H. D., 1942- . IV. American Society of Animal Science. V. Beltsville Agricultural Research Center. VI. Friends of Agricultural Research Beltsville, Inc. VII. Series.
SF105.B57 1996 96-9360
636.08'21--dc20 CIP

ISBN 1-887458-01-8

Designed by John W. Edwards

Published by the American Society of Animal Science, Savoy, Illinois, USA

Contents

Preface

R. H. Miller

USDA, ARS, Beltsville, MD 20705, U.S.A.

The twentieth Beltsville Symposium in Agricultural Research was organized to provide a comprehensive update on biotechnology with emphasis on how it may enhance or modify existing strategies for the genetic improvement of livestock and poultry. The Organizing Committee was gratified by the rapid acceptance of invitations to speak by many world-renowned specialists in their fields.

Session I, "Genetic Diversity," provides a resume of current efforts to document the world's array of livestock and poultry germplasm, identify unique populations that are in danger of being lost, and to make use of unusual genetic resources. The convention on Biological Diversity, which dealt primarily with plant resources, also considered animal germplasm. The implications of its provisions are discussed by Dr. Strauss. The Convention's provisions have been considered by United Nations bodies for implementation. Dr. Cunningham discusses how biotechnology can be used to identify unusual germplasm and thereby to make conservation efforts more efficient, and Dr. Keith Hammond presents the latest information on the worldwide efforts undertaken by the Food and Agriculture Organization. Dr. Roger Gerrits summarizes current U.S. and Canadian efforts to document animal germplasm resources.

Session II, "Mapping Technologies," covers an array of methods being applied to mapping genomes of several species, and also touches upon mapping with the specific goal of identifying genes involved in disease resistance. Methods discussed are microsatellites (Dr. Beattie), chromosome-specific libraries (Dr. Lunney), and expressed sequence tags and genetic regulatory elements (Dr. Lamont). Dr. Cockett discusses the state of the sheep genome map and efforts to make use of it. Dr. Womack discusses how the syntenic, physical, and linkage maps can be integrated to more fully understand the genome organization of domestic animals. Dr. Lewin presents a description of his work on resistance of cattle to the bovine leukemia virus and how his experiences may assist future researchers studying other genetic diseases of livestock.

Session III includes six presentations on the latest technologies for manipulating gametes and embryos of livestock. Dr. Day discusses the latest methods for the *in vitro* maturation and fertilization of pig oocytes and subsequent embryo transfer. Dr. Wilmut summarizes experiences with nuclear transfer in embryos and the factors that influence their survival and development. Dr. Johnson provides a comprehensive summary of progress in preselection of gender of embryos, emphasizing his methods for the identification and separation of X- and Y-bearing sperm of various species. Dr. Wall presents a discussion of modifying milk composition by the introduction of foreign genes expressed in the mammary gland. Dr. Stice summarizes methods and prospects for manipulating embryonic stem cells of various species. The session concludes with Dr. Wentworth's comprehensive presentation of his research on development of primordial germ cells of poultry and their potential usefulness in genetic improvement strategies.

Adoption of biotechnology in improving livestock inevitably depends on regulatory and consumer acceptance of the resulting animal products. Session IV was organized to consider how biotechnology is perceived and accepted by the general public and how regulatory agencies review proposals for their use ("Acceptance of Biotechnology: Political and Social Issues"). Dr. Bauer discusses how the Food and Drug Administration deals with public concerns about animal products derived from genetically modified livestock. Dr. Basu provides a similar discussion of how another regulatory agency, Food Safety and Inspection Service, USDA, deals with animal products appearing in the marketplace from species that have been altered via biotechnological methods. Dr. Woo surveys opinions and reactions of the consumers and public interest groups to the application of biotechnology to alter animals and their products. Finally, Dr. Zimbelman discusses how best to communicate about research with federal officials and public interest groups.

Among other developments, biotechnology has provided a virtually unlimited array of highly polymorphic DNA sequences that can be mapped to their location in the genome. These are potentially useful as markers if they are located in the immediate genetic neighborhood of important genes whose presence is not otherwise detectable. In Session V, "Marker Application," Dr. Weller provides an introduction to strategies for the detection of quantitative trait loci (QTL) and how markers for QTL can be used in selection. Dr. Andersson presents results on efforts to map specific genes affecting meat characteristics of swine. Dr. Dentine discusses how genetic markers are potentially useful in improving dairy cattle. Opportunities for use of markers in poultry are summarized by Dr. Fairfull. Dr. VanRaden concludes with a discussion of how genetic markers can be fitted into existing industry infrastructure, with particular reference to dairy cattle.

Acknowledgments

The editors would like to extend a special thanks to the following members of the organizing committee: Sandy M. Hays, Joan K. Lunney, Murray R. Bakst, Caird E. Rexroad, Jr., and Morton Beroza (Friends of Agricultural Research–Beltsville) and to the symposium secretary, Virginia L. Hupfer. A special thanks to other individuals who spent considerable time: Frank A. Ross, Jr. for data processing of registration information, Shirley A. Weese for helping with the overall effort, and Suzanne M. Hubbard for her preparation of the program/abstract book. In addition, we would like to acknowledge the following ARS employees or family who helped with the Beltsville Symposium XX: Valorie G. Akuffo, Helene C. Cecil, John R. Dobrinski, Lois J. Franklin, Edmond J. Lewis, Joel H. Megonigal, Jr., Shirley E. Moore, Kathryn E. Norman, Roslyn W. Norman, Anne M. Powell, Rex L. Powell, Leah L. Schulman, Patricia L. Seaman, William E. Shainline, Jr., Doris S. Stanley, Paul M. VanRaden, Leigh M. Walton, William P. Wergin, and Debra D. Wilhelm. The editors also gratefully acknowledge the assistance of John Edwards, Technical Editor for the American Society of Animal Science, for his thorough work in editing the manuscripts. The editors were also greatly aided by the assistance of the numerous scientists who provided anonymous peer review of the proceedings manuscripts.

Session I

Genetic Diversity

The Convention on Biological Diversity and Domestic Animal Diversity

Michael S. Strauss*, Donald L. Plucknett†, and John Hodges‡

**American Association for the Advancement of Science, Washington, D.C. 20005, U.S.A.; †Agricultural Research and Development International, Annandale, VA 22003, U.S.A.; and ‡European Association for Animal Production and Rare Breeds International, Mittersill, Austria A-5730*

Abstract The Convention on Biological Diversity was accepted internationally in 1992 by most states as a broadly based consensus affirming the need for global, regional, and national efforts to conserve biodiversity at all levels of plant, animal, and microbial species and the ecosystems of which they are part. The Convention has subsequently been ratified by more than 100 states. It calls for national strategies and programs for conservation action and offers financial and technological assistance to states without sufficient resources to implement such plans. In an approach that contrasts with some earlier concepts that viewed genetic resources as the common heritage of humanity, the Convention affirms the sovereignty of each state over the genetic resources within its borders. Agricultural biodiversity is a central component of global biodiversity, thus placing livestock or poultry resources within the mandate of the Convention. In calling for national conservation strategies and activities within the concept of sustainable use the Convention embraces livestock diversity. The Food and Agriculture Organization (FAO) effort to establish a global program for the management and conservation of domestic livestock and poultry genetic resources is reviewed in the light of the implications of the Convention for agricultural biodiversity.

Key Words: Convention on Biological Diversity, Conservation, Livestock, Animal Genetic Resources, Intellectual Property, Germplasm, Genetic Resources, Biological Diversity

Introduction

The loss of biological diversity is increasing. International concern for conserving biodiversity arose within broader concerns for the environment that intensified after the United Nations Conference on the Environment in Stockholm in 1972, which led to creation of the United Nations Environment Program (UNEP). Initially, international concern focused on the conservation of naturally occurring biodiversity. However, the United Nations Food and Agriculture Organization (FAO) identified the importance of domestic plant and animal biodiversity and, in the early 1980s, created the FAO Commission, the FAO Undertaking on Plant Genetic Resources, and a joint FAO/UNEP program for the Conservation of Animal Genetic Resources.

In 1987, UNEP convened intergovernmental negotiations for a Convention on Biodiversity with the aim of enrolling the commitment of all governments in the conservation of biodiversity through a legal framework with financial support for an agreed program. The initiatives of FAO in these negotiations led to recognition that agricultural biodiversity (plants and animals) is an important component that must be embraced within the Convention. This raised some difficult issues, including ownership rights, access, use, biotechnology and intellectual property rights. The negotiations were then no longer simply concerned with preservation of wilderness and natural ecosystems, but also embraced agroecosystems and their genetic resources. The negotiations were long and arduous. Nevertheless, after five years, the text of the Convention (UNEP, 1992) was signed by 157 presidents or heads of state at the United Nations Conference on Environment and Development (UNCED or the Earth Summit) in 1992. Governments must ratify the Convention in national legislation in order to become parties to it. Then, meeting as a Conference of the Parties (COP), these decide on its implementation.

The conservation of agricultural biodiversity is specifically recognized in the Preamble to the Convention as having critical importance for meeting the food, health, and other needs of the growing world population. This significant statement acknowledges that conservation is inextricably linked with the use of domestic plants and animals in sustainable development of agriculture. The Convention is thus committed, in principle, to the conservation of domestic livestock and poultry. The application of this principle to domestic animal diversity is one of the many tasks of the COP. This paper examines the present situation.

Livestock Diversity

Widespread concern about the loss of genetic diversity in domestic livestock is relatively recent. A significant step was the FAO/UNEP Intergovernmental Consultation held in 1980 which focused attention on the great risks facing livestock diversity in developing countries and called for the development of scientific and technical methods for livestock conservation. This Intergovernmental Consultation led in 1982 to a new program within FAO, supported by modest UNEP funding, resulting in an increasing number of surveys, training courses, studies and reports during the 1980s, and also enabled a number of governments of developing countries to start national data banks and subregional gene banks (reviewed by Hodges, 1990).

More progress was made in developed countries by some governments, for example France, and more particularly by NGOs, the most notable of which were the Rare Breeds Survival Trust (UK) and the American Minor Breeds Conservancy (now American Livestock Breeds Conservancy) in the United States. However, the general consensus from the 1980 FAO/UNEP Consultation onward has been that the greatest potential for loss of livestock diversity is in the developing world (FAO, 1984a,b, 1990, 1992; Fitzhugh and Strauss, 1992; National Research Council, 1993). Livestock breed surveys were started in Europe in 1982 by the European Association for Animal Production (EAAP), and a European Animal Genetic Resources Data Bank was set up in Germany in 1985, where FAO trained developing country scientists. The results of the EAAP pan-European surveys are reported by Simon and Buchenauer (1993). National livestock surveys, animal genetic data banks, and animal gene banks were established by some of the larger developing countries with FAO support. More recently FAO has published the World Watch List for Domestic Animal Diversity (Loftus and Scherf, 1993), which suggests that there are more than 4,000 breeds and strains of the 30 principal livestock and poultry species. This is probably a conservative estimate.

Thus, international concern at the national and international levels for livestock diversity has expanded over the 15 years since 1980. However, the activities do not match the more comprehensive and growing network of international activities on plant genetic resources. This plant program was led first by the International Board for Plant Genetic Resources (IBPGR) based during that period at FAO in Rome, and actively cooperating with its partners in the Consultative Group on International Agricultural Research (CGIAR). These included the International Rice Research Institute (IRRI) in the Philippines and the Centro Internacional de Mejoramiento de Maiz y Trigo (CIMMYT) in Mexico. The CGIAR has recently reorganized IBPGR as the International Plant Genetic Resources Institute (IPGRI) and given it full status as a CGIAR center. Plant genetic resources were also raised to high international profile by the creation in the 1980s of the FAO Commission and Undertaking on Plant Genetic Resources. On the animal side, two CGIAR centers, the International Laboratory for Research in Animal Diseases (ILRAD) and the International Livestock Center for Africa (ILCA), were concerned with livestock disease, development, and genetic resources issues in Africa. There has not been, however, a global system for livestock matching plant conservation activities (Plucknett et al., 1987). Since UNCED in 1992, the FAO Program on Conservation of Domestic Animal Diversity has been developed to address the needs of livestock conservation in developing countries (Hammond, 1994), although it still lacks the funds for implementation.

The Convention and the Conference of the Parties

The Convention came into force on December 29, 1993, 90 days after the ratification by the 30th government. The Conference of the Parties (COP) was then required to meet within a year. The First Conference of the Parties (COP-I) took place from November 28 to December 9, 1994, in Nassau, The Bahamas, by which time about 100 states had ratified the Convention. The United States has, subsequent to the UNCED meeting, signed the Convention but has not ratified it.

Overview of the Convention

The stated objectives of the Convention are conservation of the world's biological diversity; sustainable use of its component parts; and fair and equitable sharing of the benefits derived from its use, including appropriate access to genetic resources and by appropriate transfer of relevant technologies while taking account of rights. Biological diversity is defined by the Convention as the variability among living organisms in all the ecosystems of which they are part. The Convention thereby embraces diversity within and between species and the ecosystems, which are a dynamic complex of plant, animal and microorganism communities and their nonliving environment interacting as a functional unit. The Convention embraces biodiversity both *in situ* and *ex situ* and thereby extends its mandate from the population to the molecular levels. The Convention offers appropriate funding for conservation activities.

The components of the program for implementing the Convention that were agreed to at COP-I include:

National strategies for biodiversity conservation;
Projects integrating conservation and sustainable use;
Identification and monitoring of genetic resources;

In situ conservation;
Ex situ conservation;
Research and training for conservation;
Increasing public awareness and education on biodiversity;
Exchanging information to promote access, conservation, and sustainable use;
Providing for technical and scientific cooperation;
Enabling access and transfer of technology, biotechnology, and their products.

These were confirmed by conferees at the strategic level, but for many considerable further discussion will be needed for implementation.

Sustainability and Access in the Convention

The Convention states the areas of concern and associated activities in its articles. It calls for general measures to address the conservation and sustainable use of biological diversity (Article 6); identification and monitoring (Article 7); *in situ* (Article 8) and *ex situ* (Article 9) conservation; and a variety of activities to support these (Articles 10–14). Issues of access to genetic resources, transfer of technology, exchange of information, and scientific cooperation are covered (Articles 15–20). The financial resources of the Convention and a Mechanism for administering them are given in Articles 20–21. Other articles cover the administration of the Convention.

The language of the Convention is general and no specific statements are given on access, transfer, use and sharing of benefits from genetic resources and technology. Rather, the Convention calls upon parties to take the measures "as far as possible and as appropriate," to "agree mutual terms," to practice "informed consent," to seek that "national legislation on intellectual property rights are supportive of the objectives of the Convention." The Conference of the Parties is empowered to adopt protocols, establish subsidiary bodies, review progress, and seek further resolution of specific issues.

Within the next two decades world population is expected to exceed 6 billion. This increase in numbers is accompanied by increasing expectations and market demands for higher-quality foods. Thus, larger quantities of both crop and animal products are needed to satisfy the need for calories and quality protein in the human diet. While the primary focus of the Convention is biological diversity in naturally occurring species and ecosystems, the inclusion of agricultural biodiversity inevitably leads to some difficult issues. These include not only questions of access and intellectual property rights arising from the use of such diversity in biotechnology, but also the issue of how the genetic resources used in farming can be managed sustainably while supplying the ever-increasing need for food.

Common Heritage vs National Sovereignty

Significantly, the Convention departed from earlier arguments that biological diversity was the "common heritage" or patrimony of humanity, which meant that states had a global responsibility to preserve it for all (Strauss, 1994), a principle that had been important in gaining support for biodiversity conservation in general. That principle was not acceptable to many developing countries when the hard negotiations in agreeing on the Convention text were addressed. It was felt that reimbursement for genetic resources that originated within a nation's borders had to be recognized. The concerns and discussions were particularly sharp with regard to crop (plant) genetic resources.

Rather than common heritage, the Convention codifies the concept of national sovereignty over biological diversity. It acknowledges the value of this diversity to states and through programs it funds seeks to encourage states to conserve or sustainably develop their biodiversity. Thus, the conservation of indigenous livestock breeds, along with other biodiversity, is acknowledged within the Convention as a national concern and responsibility.

Development that depletes natural resources by causing, for example, desertification, salinization, pollution, or deforestation also affects and may deplete biological diversity. COP-I did not address poor development practices directly but indicated clearly that they envisioned the overall promotion of sustainable development as a primary task of the Commission on Sustainable Development, a new international body created by UNCED in 1992. COP-I sent a clear message to the Commission on Sustainable Development in which they sought a substantive relationship.

Financing of Convention Activities

The first Conference of the Parties (COP-I) in 1994 engaged in lengthy debate on some issues but on the whole was concerned with housekeeping matters. One was the Financial Mechanism for the substantial funds available under the Convention (about US$800 million in the first tranche), which, on an interim basis, the text of the Convention places in the hands of the Global Environmental Facility (GEF). The COP-I was deeply divided; developing countries, on the whole, wanted these funds within the administration of the Convention; the developed countries, who are the principal donors, wished it to remain in the GEF administration, where in fact it remained for the following year (1995).

Implications of the Convention

The Convention has had a significant influence on a variety of issues related to genetic resources conservation. Several of these are examined. Further reviews of the implications of the Convention are given by Dowdeswell (1994), Day et al. (1994), and Hodges (1995), who also reviews the results of the COP-I.

National Biodiversity Strategies

The Convention calls upon countries to develop national strategies, plans, and programs for the conservation and sustainable use of biological diversity. These are needed for information, setting priorities, and implementation. The Convention calls for conservation actions to be based upon these national biodiversity strategies. For livestock, this information will allow for the conservation of indigenous breeds of local or national importance. Development and implementation of these plans may be difficult for many developing countries. The Convention, therefore, allows for provision of financial support to develop such national strategies and their associated conservation projects in developing member states.

Legal and Policy Issues

The Convention provides for the sovereign rights of states over their genetic resources and authority to determine access to them. While most states have not yet adopted national strategies, some cases have occurred of states using the Convention to seek to limit the transfer or exchange of biological materials. However, the Convention is weak in offering a basis for legal claims and there is no clear agreed mechanism for items such as compensation. Rather, sovereign states are expected to respect the aims and values embodied in it. For example, there is recognition within the Convention of the rights of indigenous and local communities to an equitable share in the benefits arising from the use of the genetic material that has been traditionally in their care, a concept roughly equivalent to "farmers' rights" in the FAO Undertaking on Plant Genetic Resources.

Access and Exchange

The Convention gives parties the right to require researchers or collectors to obtain specific permission and to provide appropriate compensation when removing any material deemed part of the national biological diversity. This permission can be conditioned in a number of ways, at a state's discretion. It can require return of the material on request, set limits on transfer to a third party, decide whether to share material with other researchers, or direct compensation.

Under the Convention, the country of origin does not necessarily abdicate its right to compensation if collected material is passed to others. Thus, collectors must endeavor to monitor where materials have been distributed and may place limits of their own on the further transfer of germplasm. This is less problematic in livestock than in plant genetic resources, because regenerated samples of collected seeds can be passed through several gene banks or from one breeder to another. Further, because the most highly productive breeds and genotypes are in developed countries, whereas breeds in developing countries have genotypes with adaptive traits, livestock exchange may involve reciprocal access to this productive germplasm.

The Convention expects the collector and the party to have a clear agreement on the conditions of exchange and the rights to future access. It is possible, for example, that a state might retain essentially absolute rights to semen or embryos in an international facility, solely as insurance against loss. The capacity for such safety duplication is an important part of an international strategy, as events in the former Soviet Union and elsewhere, where loss of indigenous breeds has occurred, have demonstrated.

It is also possible to design the transfer conditions in a way that permits countries to provide reciprocal benefit to one another should the materials prove useful. Thus, an agreement might allow the evaluation, exchange, and use of the germplasm in breeding programs, and define a mechanism for compensation should the resulting animals be widely used in the industrialized world or become the source of an important gene or gene complex.

Biotechnology

The Convention calls not only for access to genetic materials but also for access to technologies and their products associated with genetic resources. This issue, with its implications for patents and intellectual property rights, is recognized within the Convention in principle and calls upon the parties to ensure that national and international legislation is framed appropriately to support and not to run counter to the objectives of the Convention. This issue was put forward as a principle reason for the United States not signing the Convention at the UNCED meeting and its lack of ratification of the subsequently signed document.

Rights and Innovations of Indigenous Peoples

The Convention recognizes the traditional knowledge of indigenous peoples and encourages the improved use of these resources while calling for the rights of the peoples to be incorporated in sharing of benefits.

Clearing House Mechanism

The Convention proposes a Clearing House Mechanism (CHM) to facilitate the exchange of technical and scientific information and to promote technical and scientific cooperation. The nature of this CHM was the basis for considerable debate by the COP-I (Putterman, 1994). While development of an independent brokerage mechanism for exchange was discussed, there remained significant disagreement, primarily divided between developed and developing countries. The former were seeking assurances of acceptance of intellectual property protection, and the latter wanting technology transfer "... based on concessional and favorable terms and not on commercial terms. ..." This difficulty highlights the challenge to provide through the Convention mutually agreed mechanisms to allow access to information on resources and technology, while preserving the benefits derived from their use. Further discussion of these issues was scheduled for COP-II.

Subsidiary Body

The Convention provides for the establishment of a Subsidiary Body on Scientific, Technical and Technological Advice (SBSTTA) with the aim of providing the parties with timely advice on the implementation of the Convention. The SBSTTA is multidisciplinary and is open to all Parties to participate. It reports directly to the COP. From a scientific and program point of view the SBSTTA is a highly important body. The COP-I convened the first meeting of the SBSTTA for September 1995 in readiness for COP-II in November 1995.

Implications of the Convention for Livestock

The relatively few species (but wide genetic diversity) that constitute most livestock are included in what the Convention is intended to address (Strauss, 1994). If the definition of livestock is extended to all animal species used by human society the number of species increases to many hundreds and includes not only mammal species but also fish, reptiles, and even insects (National Research Council, 1991). Thus, the Convention has implication for the conservation of domestic animal diversity.

The Convention calls for National Biodiversity Strategies and Plans. These place great importance on recognizing the country's diversity which includes livestock. A plethora of closely related breeds or indigenous types and resources insufficient to conserve them all can be problematic. In fact, where such breeds are closely related separate preservation may not only be economically difficult, but genetically unnecessary. Thus, this provision of the Convention implies an urgent need for adequate information to enable the setting of appropriate priorities. This compilation addresses the first challenge of the Convention: to assess the extent, diversity, and status of biological diversity at a national and international level. This requires the development of national inventories as well as appropriate databases.

Conservation of livestock carries with it the intent that the breeds or genotypes maintained are to be used to promote or improve animal production systems. With limited resources difficult decisions must be made as to which resources will receive national attention. A tension can result from the need to balance conservation of individual breeds and to make those resources available for improvement. The Convention recognizes all indigenous breeds as having potential value in such development. In this sense, indigenous breeds of livestock and poultry are significant parts of the biological diversity encompassed by the Convention.

Two broad approaches to conservation have been recognized in the reports of several studies (FAO, 1990; Fitzhugh and Strauss, 1992; National Research Council, 1993). One approach places importance on the development of improved breeding lines intended to focus on enhancing production and production systems. The other approach directs efforts to the preservation of unique indigenous breeds as the resource of genetic diversity. While the breeding lines of the former may be more immediately applicable to improvement programs, the distinct breeds held by the latter are more directly encompassed by the Convention.

One additional potential development is the extension of protection of the Convention to animal products other than breeding individuals. Such is the case where the Convention and its provisions of national sovereignty and compensation are cited to hinder collection of blood samples for use in taxonomic studies. While not of direct use in breeding or improvement, it could be argued that such samples, nonetheless, constitute genetic information about indigenous species. It is not hard to imagine such an interpretation being extended to encompass a range of animal products.

An important concern is the development of biotechnology, which has increased the value of diversity (that is of specific genes) and which has the potential both to enhance and to erode genetic diversity. Some techniques, such as artificial insemination, have existed for quite a while, but global economic conditions have led to the potential for increasingly widespread use of these technologies to promote what may be a narrowing set of genetic types. The capacity to apply powerful technologies to manipulate individual genetic traits was envisioned in the drafting of the Convention, which calls for equitable sharing of the benefits of such technologies but does not provide legally binding directives. The basis for the current FAO Undertaking on Plant Genetic is the transfer of plant genetic resources from developing countries to industrialized countries, where they are enhanced, and then sent back. For livestock, the primary loss is not removal of genetic resources, but replacement of indigenous breeds and strains by high-producing animals that originate in developed countries. Thus, the concern for livestock is less in controlling the dissemination of indigenous diversity and more in protecting it from highly productive and genetically narrow, but improved, types.

Under the Convention, any obligation to farmers in developing countries rests with the national government and national legislation (rather than with the international community, as is the case with the concept of "farmers' rights"). This obligation can be met by acts that collectively benefit agriculture and by actions that directly benefit farmers. At the same time, it would be logical to compensate individual farmers for maintaining particular indigenous varieties (Day et al., 1994). The Convention, therefore, could be seen to create an incentive to promote the maintenance of indigenous breeds. While such breeds may not have high value for breed improvement, the Convention confers upon them a significance as national resources and part of the national heritage.

Within the Convention's mandate is recognition of the rights and innovations of indigenous peoples and communities. This is an especially sensitive issue relating to access, particularly in its application to domestic livestock and poultry. Domestic livestock and poultry are unique when compared to other genetic resources, such as plants, in the sense that, typically, domestic animals are owned and maintained by individuals or communities. Further, the most threatened breeds in developing countries are often owned or under the stewardship of minority groups who are outside the mainstream of economic development. For these the indigenous breeds may be a central part of their culture and may be easily lost if, for a variety of reasons, their management ceases.

Conclusion

The conservation of and access to a diverse genetic resources of livestock and poultry is essential to developing national agricultures and to meeting growing needs for sustainable production of food. The Convention on Biological Diversity is an attempt by states to slow the erosion of biodiversity and to ensure that it is effectively conserved and used in a sustainable manner, while sharing the benefits equitably. The Convention alone, however, is not a solution. Achieving the goals it sets forth will require significant global cooperation in the implementation of its many provisions, particularly in the conservation of agricultural biodiversity, which is essential to ensure adequate food supplies for the growing human population.

Literature Cited

Day, P. R., M. S. Strauss, and J. H. Barton. 1994. Domestic animal resources and the Convention on Biological Diversity. In: Proc. 5th World Congr. on Genet. Appl. to Livest. Prod., August 7–12, 1994, Guelph, ON, Canada. 21:470.

Dowdeswell, E. 1994. Implications of the Biodiversity Convention on the Conservation and Sustainable Use of Domestic Animal Diversity. In: Prov. 5th World Congr. on Genet. Appl. to Livest. Prod., August 7–12, 1994. Guelph, ON, Canada. 21:462.

FAO. 1984a. Animal genetic resources conservation by management, data banks and training. FAO Animal Production and Health Paper No. 44/1, Food and Agriculture Organization, Rome, Italy.

FAO. 1984b. Animal genetic resources: Cryogenic storage of germplasm and molecular engineering. FAO Animal Production and Health Paper No. 44/2, Food and Agriculture Organization, Rome, Italy.

FAO. 1990. Animal genetic resources: A global programme for sustainable development. FAO Animal Production and Health Paper No. 80. Food and Agriculture Organization, Rome, Italy.

FAO. 1992. Expert Consultation on the Management of Global Animal Genetic Resources, Rome, Italy, April 7–10, 1992. Food and Agriculture Organization, Rome, Italy.

Fitzhugh, H., and M. S. Strauss. 1992. Management of Global Animal Genetic Resources: Organization, and Institutional Structure. pp 283–297. FAO Animal Production and Health Paper No. 104, Food and Agriculture Organization, Rome, Italy.

Hammond, K. 1994. Conservation of domestic animal diversity: Global overview. In: Proc. 5th World Congr. on Genet. Appl. to Livest. Prod., August 7–12, 1994, Guelph, ON, Canada. 21:423.

Hodges, J. 1990. Animal Genetic Resources: A Decade of Progress, 1980–1990. Food and Agriculture Organization, Rome, Italy.

Hodges, J. 1995. Convention on biological diversity: A description and appraisal of the Convention relative to domestic livestock and poultry following the First Conference of the Parties held in The Bahamas, 28 November to 9 December 1994. Available from the author at Lofererfeld 16, A-5730 Mittersill, Austria.

Loftus, R., and B. Scherf. 1993. World Watch List for Domestic Animal Diversity. Food and Agriculture Organization, Rome, Italy, and United Nations Environment Programme, Nairobi.

NRC. 1991. Microlivestock. National Academy Press, Washington, DC.

NRC. 1993. Managing Global Genetic Resources: Livestock. National Academy Press, Washington, DC.

Plucknett, D. L., N.J.H. Smith, J. T. Williams, and N. Murthi Anishetty. 1987. Gene Banks and the World's Food. Princeton Univ. Press, Princeton, NJ.

Putterman, D. 1994. Premium put on equity issues at Biodiversity Convention's First Conference of the Parties. Diversity 16:12.

Simon, D. L., and D. Buchenauer. 1993. Genetic diversity of European livestock breeds. European Association for Animal Production (EAAP) Publ. No. 66. Wageningen Pers, Wageningen, The Netherlands.

Strauss, M. S. 1994. Implications of the Convention on Biological Diversity. Management of Animal Genetic Resources and the Conservation of Domestic Animal Diversity. American Association for the Advancement of Science, Washington, DC.

United Nations Environment Programme (UNEP). 1992. Convention on Biological Diversity, Environmental Law and Institutions Programme Activity Centre, Nairobi.

Genetic Diversity in Domestic Animals: Strategies for Conservation and Development

Patrick Cunningham

Department of Genetics, Trinity College, Dublin 2, Ireland

Abstract Surprisingly few species of animals are used in today's agriculture. However, the great variety of contexts to which these species must respond calls for an enormous range of genetic diversity within each species. The present status of breed diversity of farm animals, and the factors that are tending to reduce it, are reviewed. Formal methods to assist decision making in the allocation of resources for conservation are discussed, and a method is proposed for relating costs and potential benefits in the preservation of individual breeds. Genetic diversity now can be documented at the DNA level. The use of mitochondrial and Y-chromosome sequences, together with microsatellite allele frequencies, permits measurement of genetic variability within and between populations with considerable precision, and facilitates measurement of gene flow between populations. Recent work on global diversity of cattle illustrates the utility of these biotechnological techniques.

Key Words: Genetic Diversity, Conservation Costs, Microsatellite, Mitochondria

Introduction

Genetic diversity exists in the animal world at three levels: between species, between breeds within species, and between individuals within breeds. This diversity underpins the capacity of livestock production systems to respond to a global range of physical and economic environments. Because of the increasing pace of economic change, breed diversity in domestic animals is being eroded rapidly, and there is growing concern that this erosion may weaken the capacity of livestock producers to respond to future needs, particularly in developing countries. There is therefore a need to document current animal genetic resources and to develop rational strategies for their conservation and development. Developments in molecular genetics in the past decade can contribute substantially toward these goals.

Diversity at the Species Level

At the present time, more than 1 million separate animal species are recognized; the vertebrates, with about 40,000 species, account for less than 5% of the total. New species are being named at the rate of more than 10,000 per year. It is likely that as many species of animals (mainly insects and roundworms) remain to be classified as are currently listed. It is also probable that at least as many species have become extinct as now exist. It is estimated, for example (Sibley and Ahlquist, 1985), that the 9,000 avian species at present in existence represent just 6% of all avian species that have at some time existed.

Most animal production is based on mammálian species. Approximately 4,000 species of mammals are known to exist, of which less than 20 are involved in animal agriculture. Among these, in turn, a few predominate (Table 1). Cattle are responsible for 90% of the world's milk and 32% of the world's meat. Pigs produce a further 40% of the world's meat; buffalo, sheep, goats, horses, asses, camels and poultry are also significant contributors to total output.

One may wonder why, with such a wealth of species available for domestication, world agriculture has been built on so few. The answer no doubt includes two elements: that the few chosen vary sufficiently in type and function to fulfill the main needs of agriculture, and that within each species selected there is sufficient variability to permit adaptation to a wide variety of climatic and environmental circumstances.

Are we likely to draw more species into the domestication process in the future? Probably not. The effectiveness with which modern breeding schemes can change the genetic nature of one of these species means that for any new set of production circumstances or needs, it is most likely that one of the existing domesticated species will provide the most cost-effective candidate.

However, there are a few cases in which domestication of new species has recently been undertaken. In the developed world, the percentage of personal expenditure devoted to food is commonly about 20%. This means that consumers have great latitude to pay a premium for novelty or other special features of particular food items. Thus venison, an unusual meat, commands a higher price than other meats. This demand, coupled with the special demand for antler velvet in the Far East, has led to the development of deer farming on a significant scale in New Zealand, and on a more modest scale in Europe. The advantage that deer have is not that they are more productive per unit of feed resource than other herbivores, or that they can exploit feed resources not used by cattle or sheep, but rather that the market pays a premium for their products.

Another significant domestication in recent years has been that of the Atlantic salmon. The total world catch of wild salmon is normally under 10,000 t per year, with the result that this high-quality fish had a substantial market premium over other fish. Techniques for rearing salmon in floating sea cages were developed in Norway in the early 1970s, and global production of farmed Atlantic salmon is now

Table 1. World numbers and contribution to world production of milk and meat, for different species[a]

Species	Numbers (millions)	Percentage of world production: Milk	Meat
Cattle	1,278	90	32
Buffalo	149	6	1
Sheep	1,110	2	4
Goat	592	2	2
Pig	871	—	40
Poultry	12,778	—	21

[a]Source: FAO Production Yearbook (1993).

about 250,000 t, or 25 times the wild catch. This process has involved not just the development of a husbandry system, but the screening of wild populations, and subsequent selection for both behavioral and performance traits important in production. This selection has gone a long way toward producing what could be called a domesticated species.

It is sometimes felt that the reason why more species are not cultivated in domesticated conditions is that the domestication process itself is a very long, and therefore expensive, one. The recent experience with deer and salmon suggests the contrary. This conclusion is further supported by the results of Belyaev (1979) on the domestication of silver foxes. His experiments, carried out over 18 generations of selection for tameness, demonstrated how rapidly domesticated characteristics can be established, and also show the physiological basis for these changes.

In summary, therefore, world agriculture depends on a handful of species. The reason why a wider range of species has not been brought into use is simply that the economic need for new species has not been present. If we need to harness the resources of additional species, it appears that we can do so in a relatively short time.

Diversity at the Breed Level

As the techniques of settled agriculture spread around the world, the main species of domesticated livestock were spread as part of that technology (Ammerman and Cavalli-Sforza, 1984). These species have since been subjected to diverse selection pressures appropriate to the great diversity of environments in which husbandry systems have evolved. For the main species, the wild ancestors of related species have almost ceased to exist. The animals we have today in husbandry systems are differentiated from these wild ancestors by from 1,000 to several thousand generations of selection in captivity. It can be argued that this process of domestication has not only greatly enlarged the genepool in each of these species, but has also greatly increased the genetic diversity it contains. Thus, we can see within cattle, sheep, or pig species an enormous range of local breeds, strains, or types. In the course of human history, these have been moved with the migration of peoples, crossed with other local strains, and selected continuously both by human intervention and by the forces of natural selection operating in the particular farming system of which they are a part.

Change in Farming Systems

Economic and demographic pressures cause changes in farming systems. These pressures are generally for intensification of systems. At a very broad level, this can be quantified by comparing annual rates of growth in the numbers of animals on the one hand, and of volume of production on the other.

Table 2 makes this comparison for cattle, buffalo, sheep and goats, and dairy cows. For these species it is clear that, over the past decade, ruminant meat output has increased annually more than three times faster than animal numbers. Milk production from dairy cows has increased at more than twice the growth in the number of cows.

These figures also incorporate great variability between systems and regions. In the developed world, ruminant animal numbers have generally declined in this period, while output has remained static for milk and risen slightly for meat. In sub-Saharan Africa, productivity of cattle has actually declined, that of sheep and goats has been static, and that of dairy cows has increased marginally. Latin America likewise showed modest changes in productivity. The most dramatic growth, both in output and in intensity, was in Asia. Although this region accounts for just 9% of the world's beef and veal production, and 30% of sheep and goat meat, it averaged annual growth rates of 7.8% and 6.3% in output of these commodities (against 1.2% and 1.8% annual growth in stock numbers). The most intense development was seen in the mixed farming systems with irrigation, indicating a rapid intensification of the interaction of livestock and crops in these systems.

Because of difficulties of enumeration, the process of intensification cannot be quantified in quite the same way for pigs and poultry. Asia accounts for 42% of the world's pig production and almost 90% of that in the developing countries. Annual output increased at an average of 7% in Asia, but at a rate of 17.8% per year in landless farming systems (which in turn accounted for 29% of output in the region). A similar pattern is evident in poultry production, with overall annual growth rates of 9.6% in Asia, rising to 17.3% for poultry meat, and 18.5% for eggs from the landless systems (which account for about half of total production in the region).

The components of these increases in productivity cannot readily be quantified. However, in different systems, they undoubtedly reflect more intensive feeding patterns, with increased offtake rates, higher slaughter weights and milk yields, and reduced losses from disease.

Impact on Genetic Constitution

Change in the farming system often involves changed breeding objectives for livestock. In traditional systems, number of livestock rather than output per head is often the main consideration. In these circumstances, traits related to survival in the face of nutritional, health, and climatic stress predominate. As farming systems become more market-oriented, volume and value of saleable product take over. In these

Table 2. Average annual rates of growth for the decade 1981–83 to 1991–93[a]

Species and product	Number of animals	Volume of production	Ratio in growth rates production/ animals
	Annual percentage of growth		
Cattle (meat)	.4	1.5	3.75
Buffalo (meat)	1.5	4.8	3.20
Sheep and goats (meat)	.7	2.4	3.43
Dairy cows (milk)	.4	.9	2.25

[a]Source: Agrostat.

circumstances, selection goals often change from multi-purpose use to much narrower targets. They shift to such traits as prolificacy, early maturity, individual growth rate and milk yield, and aspects of milk composition and carcass quality.

Selection for such traits has transformed the genetic constitution of many developed breeds. For example, in many non-market traditional cattle systems, cows first calve at 4 to 5 years and thereafter at 2-year intervals, producing perhaps 500 liters of milk per year. In modern highly developed dairy production systems, cows first calve at 2 years, and thereafter at yearly intervals, and produce up to 10,000 liters of milk per year. The genetic change in the animal has made possible the intensification of the system, while the evolution of the system of feed supply, management and healthcare has made possible the support for high-producing genotypes. The net result is a dramatic improvement in efficiency of resource use.

In addition to the direct effect of selection on the genetic constitution of livestock populations, changes in breeding objectives can also promote interpopulation gene flow. Thus, in West Africa demand for beef animals of higher body weight has led to substantial cross-breeding of larger *Bos indicus* on the smaller *Bos taurus* populations. In China, local breeds of pigs are being crossed with imported breeds at unprecedented levels. Such patterns of crossbreeding, over a few generations, can effectively lead to total replacement of the domestic pool of genetic diversity. In some cases, this can happen very rapidly, for example in some poultry systems, where domestic strains are simply bypassed by the importation of totally new genotypes.

The second route through which changes in farming systems can affect the genetic constitution of populations is through changes in breeding structure. In parts of Asia, the move from draft power to mechanical cultivation has meant that some cattle and buffalo populations are no longer dominated by working animals. The move to a market economy may lead to increased and selective offtake of animals from some pastoral systems, leading to changes in sex ratios and age groups. The introduction of hatcheries in poultry, artificial insemination in cattle, and hybrid breeding companies in pigs can all promote direct changes in the genetic constitution of populations and accelerate interpopulation gene flow.

Changes in breeding structure, reflecting changes in the farming system, can also have a major impact on population size. The most dramatic effect is through competition between populations or breeds. This can lead to the rapid growth to dominance of more specialized breeds and a parallel rapid decline in less competitive ones. It can also have an effect on population size in the system as a whole. With expanding demand, both animal numbers and animal productivity tend to rise. However, as can be seen in North America and Europe, limited growth in demand accompanied by rapid increase in animal productivity has produced substantial declines in dairy cow populations. Reduction in population size can affect the genetic constitution, mainly through reduction in diversity at the breed level, but also through the effects of inbreeding within small populations. Increasing population size is also a major direct factor in the interaction between livestock and the environment.

The Importance of Population Size

The principal effect of reduced population size is the associated decline in intralocus genetic variability within individuals (heterozygosity). This in turn is measured as the inbreeding coefficient. Many studies have shown that functional traits deteriorate at between 3% and 5% for every 10% increase in the inbreeding coefficient. In a random mating population of N individuals, with equal numbers of males and females and discrete generations, the increase in inbreeding per generation is a simple function of the population size: $1/(2N + 1)$. However, domestic livestock populations do not mate randomly, and they usually have many more breeding females than males. In these circumstances, the "effective population size" N_e is usually much lower than the number of animals in the population: $N_e = 4MF/(M + F)$, where M and F are the numbers of males and females, respectively. In a population with a large number of breeding females and a small number of males (as with artificial insemination, for example) this formula means that the effective population size is approximately four times the number of males.

The impact of small population size on genetic diversity within the population is also affected by a number of other factors. These include:

Generation turnover: long generation intervals, as in horses, can delay the negative effects of small population size and inbreeding.

Current and prospective changes in population size: if effective population numbers have been declining and are likely to continue the downward trend, then calculations based on current size are likely to underestimate the risk.

Changes in herd structure: in some small populations, the total number of herds may be quite small. The smaller the number, the greater the risk to the breed arising from the termination of any one herd.

The extent of cross-breeding: although cross-breeding can remedy the effects of inbreeding, it also means that the genetic diversity within the population is being reduced.

Translating all of these factors into guidelines for action to conserve genetic diversity at the breed level is not easy. Working groups established by the European Union, FAO, and EAAP have independently addressed this question and have produced different consensus threshold population sizes, below which breeds would be regarded as endangered. For cattle, for example, these numbers have varied from 1,000 to 5,000 breeding females.

A recent report on the status of breed diversity in Europe (Simon and Buchenauer, 1993) has proposed a classification of endangerment of breeds on the basis of arbitrary limits on acceptable inbreeding over a 50-year period. Because of the different generation intervals, this corresponds to different population sizes in each species. The five classifications are shown in Table 3.

With these definitions, populations of cattle, sheep and goat, pigs, and horses fall into the "normal" category with minimum effective population sizes of 139, 200, 303, and 111, respectively. These correspond to minima of at least 1,000 females and between 30 and 80 males for the different species.

The Present Status of Breed Diversity

Global domestic animal diversity at the breed level comprises some 4,000 different breeds or landraces. This is approximately equal to the total number of mammalian species currently in existence. Because of the rapid changes in farming systems, the

Table 3. Definition of breed status[a]

Status	Max. acceptable inbreeding, %	
	Over 50 years	Per generation
Normal	<5%	.36
Potentially endangered	5–14	1.07
Minimally endangered	15–25	1.79
Endangered	26–40	2.86
Critical	40+	

[a]Source: Simon and Buchenauer, 1993.

local breeds and types are coming under increasing pressure. Concern about the erosion of genetic diversity at the breed level has underpinned the establishment of a global program on animal genetic resources in FAO. The first initiative under this program was to produce a World Watch List of breeds at risk (Loftus and Scherf, 1993). Working with rather general criteria, a first, and necessarily incomplete, list has been compiled. This includes a total of 2,719 breeds across the seven main species. These represent some 68% of the total. Of those breeds with adequate data for assessment, 27% were classified as endangered. This is believed to substantially understate the degree of risk of breed loss, because undocumented breeds are often likely to be more vulnerable, and because a further 20 or so domesticated animal species have not been covered. The net conclusion is that there are likely to be at least 1,000 breeds in danger of extinction.

In Europe, a recent inventory (Simon and Buchenauer, 1993) of breed diversity showed that less than half of the total complement of 864 separate breeds was judged to be "normal." Some 158 or 30% were evaluated as "critical."

These figures illustrate the reality that the erosion of biodiversity at the breed level is not simply a concern for the distant future but is an active, ongoing process.

Rational Choices

Not every breed (or species) can survive indefinitely. Since resources for the preservation of diversity will always have some limits, choices must be made. At the broadest level we must consider which breeds (or species) to preserve. At the level of the individual breed we must ask a related but separate question: how much investment should be made in its conservation?

A theoretical framework for dealing with the first of these questions has recently been developed (Weitzman, 1992; Crozier and Kusmierski, 1994). The objective is to develop a rational basis for decision making aimed at maximizing a "diversity function." This function is derived from a matrix of pair-wise genetic distances between all the breeds (or species) in the set. It also requires some *a priori* estimate of extinction probabilities, which in turn would normally be a function of effective population size. It is then possible to calculate an "expected present discounted diversity" for the group of breeds (or species) as a function of the extinction probabilities, genetic distances, and the discount rate used. While this cannot be given a strict economic value, it does provide a quantified measure of expected results from a particular conservation program and could therefore help in guiding rational choice. It should be noted that valid estimates of genetic distances provide the foundation for this measure of diversity, and this serves to emphasize the importance of good DNA-level information in planning conservation programs.

The preferred path to the conservation of a particular breed or landrace is that it should continue to function in an economically viable farming system. The costs and benefits of such a strategy must first be assessed by the farmer. Where the economic balance is strongly against conservation, it will usually not be possible with subsidy schemes to alter the farmer's choice. The search for methods to ensure *in situ* conservation therefore should focus on methods to ensure the viability of the farming system, rather than just the animal population.

Calculating costs and returns from such interventions is extremely difficult. This particularly applies to the valuation in economic terms of the benefits of breed retention. The probability of widespread recourse to the genotypes conserved at some point in the future and the scale of the benefits this would bring are impossible to quantify. However, the problem can be approached from the other end, by asking what minimum level of ultimate benefit would justify a continuing support now for a small population.

If A = annual cost of input support, B = ultimate benefit (in year n) from survival of a breed, and r = an appropriate discount rate, then the total discounted cost to year n is

$$\sum_{i=1}^{n} A/(1 + r)^i \quad = \quad A/r \text{ if n is large.}$$

The discounted value of the benefit is $B/(1 + r)^n$. If the costs and returns are equal, the ratio of ultimate return to annual cost is $B/A = (1 + r)^n/r$. This ratio is a function of the discount rate and period considered.

Table 4 shows the break-even ratios of ultimate return to annual cost for different discount rates and periods. It can be seen that very long-term support programs, particularly at high discount rates, require an enormous ultimate payoff if the costs are to be recovered. However, the normal time horizon for such programs is more like a human generation, and discount rates of under 5% are commonly applied in cases like this. In these circumstances, a fairly modest technical benefit, which could be applied to a wider population and perhaps taken advantage of over an extended period, would recoup the costs. As a rough guide, one could say that an expected benefit of 100 to 200 times the annual cost could repay a 50-year investment in the conservation of a breed.

Table 4. Ratio of ultimate benefit after n years to annual support cost required for investment to break even

Discount rate	Years to return: 25	50	75	100
.025	74	137	255	472
.050	68	229	777	2,630
.075	81	496	3,024	18,441
.100	106	1,174	12,719	137,806

Utility of Molecular Data

A recently completed cycle of studies in cattle (Bradley et al., 1994; Loftus et al., 1994; MacHugh et al., 1994) illustrates the power of molecular information in clarifying the genetic origins of populations and the strategic options for conservation. Thirteen breeds were selected for study: Angus, Charolais, Friesian, Hereford, Jersey, and Simmental from Europe and N'dama from West Africa, all representing *Bos taurus* types; and White Fulani from West Africa, Butana and Kenana from East Africa, and Hariana, Sahiwal, and Tharparker from India, all representing *Bos indicus* types. A total of 349 animals were involved.

Three kinds of genetic variation were investigated: that of the maternally inherited and swiftly evolving mitochondrial chromosome (mtDNA), that of highly polymorphic autosomal microsatellite markers, and that detected by a chromosome Y-specific probe. mtDNA, because of its clonal inheritance and high variation, has been the subject of many productive intraspecific genetic studies . In this study, a 380-bp segment of the most variable region, the D loop, was sequenced by polymerase chain reaction amplification of a 1.6-kb fragment containing the control region, using oligonucleotide primers from published sequences and whole DNA isolate as template. Nuclear DNA polymorphism was examined using eight microsatellite markers. These loci show high variability in the length of short, PCR-amplified units of DNA. Polymorphism is due to the presence of tandem repeats of very short motifs that are known to vary in copy number. Y chromosomal inheritance was examined using Southern blotting and a DNA probe.

Sequence analysis of the complete D-loop region showed two major lineages, differing at approximately 5% of sites sequenced. Surprisingly, these lineages did not follow the *taurus-indicus* division. Instead, they grouped all of the African and European breeds together in one group and the Asian breeds in another. The extent of divergence between these two major lineages suggests (from a molecular clock estimation) that they diverged between 200,000 and 1 million years ago. This is clearly not consistent with the theory that *B. indicus* represent artificially selected *B. taurus*, but instead indicates that Asian cattle were domesticated separately from European and African. Also, the presence in Africa, without exception, of only variants from the taurine lineage implies a crossbred history for the zebu of that continent. The East African animals have previously been considered as typical *indicus*.

Eight simple sequence microsatellite loci were assayed by PCR for length variation in DNA from 349 animals representing the 13 breeds of cattle involved in the study. Genetic distances were calculated, which clearly distinguished between *Bos taurus* and *Bos indicus*. However, unlike within the mtDNA phylogeny, the African zebu cattle grouped with the Asians, whereas the taurine N'Dama formed the first branch off the European clade.

The crossbred nature of the African cattle, implied by the mtDNA results, is clearly established by the microsatellite data. Four of the loci tested possessed alleles that were predominant in the Asian breeds but absent in the Europeans. African zebu possessed, in a decreasing cline from the most zebu in the East, a mixture of *taurus* and *indicus* alleles. The Nigerian N'Dama population also, although predominantly taurine, showed some evidence of zebu crossbreeding. The Y-specific probe yielded distinct zebu and taurine types in 65 European and African cattle tested. There was a marked tendency for the *indicus* variant to introgress substantially into morphologically *taurus* populations where some crossbreeding had occurred.

The most widely accepted account of the domestication of the wild ox, the aurochs, places the primary events in the emerging civilizations of the middle East approximately 10,000 years B.P. (Epstein, 1971). *Bos indicus* is given as having developed as a later variant from herded *Bos taurus* on the Eastern fringe of this area. However, the mtDNA data described here represent strong molecular evidence for the separate domestication of the two cattle sub-species.

The microsatellite data provide evidence on recent gene flow in the African populations and permit a quantification of the zebu contribution to the genetic makeup of each African breed studied. Also, the marked introgression of the *indicus* Y-specific variant into morphologically taurine populations suggests that zebu gene flow may have been primarily through male transmission.

Implications

More than a quarter of the animal genetic resources at breed level are at risk in developed countries, and a higher proportion in developing countries. All cannot, or need not, be sustained indefinitely. Reasoned choice for conservation between breeds, and between alternative conservation strategies for any particular breed, requires considerable knowledge of the genetic structure of breeds, their place in production systems, and also of the genetic relationships between them. For most breeds in developing countries, these kinds of information are not available. However, developments on two fronts are contributing substantially to improved knowledge. The first is the range of programs, coordinated by FAO, to document global animal genetic resources. The second is the rapid advance in and use of methods of DNA analysis that facilitate the study of genetic structure within and between populations.

Acknowledgments

I would like to thank my colleagues Dan Bradley, Paul Sharp, Ronan Loftus and David McHugh. This work was in part supported by research grants under the STD Programme of the European Union.

Literature Cited

Ammerman, A. J., and L. L. Cavalli-Sforza. 1984. The Neolithic Transition and the Genetics of Populations in Europe. Princeton Univ. Press, Princeton, NJ.

Belyaev, D. K. 1979. Destabilising selection as a factor in domestication. J. Hered., 70:301.

Bradley, D. G., D. E. MacHugh, R. T. Loftus, R. S. Sow, C. H. Hoste, and E. P. Cunningham. 1994. Zebu-taurine variation in Y chromosomal DNA: A sensitive assay for genetic introgression in West African trypanotolerant cattle populations. Anim. Genet. 25:7.

Crozier, R. H., and R. M. Kusmierski. 1994. Genetic distances and the setting of conservation priorities. In: (Ed.) V. Loeschche,J. Tomiuk, and S. K. Jain, Conservation Genetics. Basel, Switzerland.

Epstein. 1971. The Origin of the Domestic Animals of Africa. Africana Publishing Corporation, New York.

Loftus, R. T., D. E. MacHugh, D. G. Bradley, P. M. Sharp, and P. Cunningham. 1994. Evidence for two independent domestications of cattle. Proc. Natl. Acad. Sci. USA 91:2757.

Loftus, R., and B. Scherf (Ed.) 1993. World Watch List for Animal Diversity. Food and Agriculture Organization, Rome, Italy.

MacHugh, D. E., R. T. Loftus, D. G. Bradley, P. M. Sharp, and P. Cunningham. 1994. Microsatellite DNA variation within and among European cattle breeds. Proc. R. Soc. Lond. Ser. B 256:25.

Sibley, C. G., and J. E. Ahlquist. 1985. Reconstructing bird phylogeny by comparing DNA's. Sci. Am. p 68.

Simon, D., and D. Buchenauer. 1993. Genetic diversity of European livestock breeds. EAAP Publ. No. 66. Wageningen Pers, Wageningen, The Netherlands.

Weitzman, M. L. 1992. On diversity. Q. J. Econ. 107:363.

The FAO Global Program for the Management of Farm Animal Genetic Resources

Keith Hammond and H. W. Leitch

Animal Genetic Resources Group, Food and Agriculture Organization of the United Nations, Rome, Italy

Abstract The more than 170 member governments of the Food and Agriculture Organization of the United Nations (FAO) have resolved to design, help fund, coordinate, and facilitate the implementation and maintenance of a Global Program for the Management of Farm Animal Genetic Resources aimed at 1) documenting, monitoring, and better understanding the range of breeds of livestock developed during domestication within each of the more important species of animals used directly by humankind for food and for a range of other agriculture items, 2) better developing and using this range of genetic material to achieve highly productive and sustainable agriculture in each of the wide spectrum of production environments throughout the world, and 3) maintaining for possible future use those unique genetic resources of little current interest to farming communities. The program design must also provide for Agenda 21 and be in harmony with the Convention on Biological Diversity (UNEP, 1992), to maximize opportunities for success and minimize duplication of effort. The program is now being launched and comprises 1) a country-driven global structure, 2) an intergovernmental mechanism to enable governments to effectively monitor progress and develop international policy particularly concerning access to and maintenance of genetic resources, 3) a technical program of work partitioned into seven elements with a range of activities planned for each, and 4) a cadre of experts to help guide the program's development over the coming decades. All facets of animal breeding eventually will be involved. Success will require the active involvement of governments; non-governmental, including private sector, organizations; research and training institutions; and other international agencies. Each country is being invited to identify a National Focal Point and National Coordinator for the program and to facilitate and establish a within-country network. The rationale for the initiative, the nature of the program, and the range of opportunities for the involvement of groups will be described.

Key Words: Animal Genetic Resources, Global Management, Breeding, Conservation, Diversity

Introduction

At the current rate of population growth, during the second decade of the next century consumption of food and agriculture products will be equivalent to that in the last 10,000 years. Domestic animals are a crucial element in meeting future global food requirements. The more than 40 species that have been domesticated contribute directly and indirectly some 30 to 40% of the total value of food and agriculture production. Incorporating animal with plant species will commonly increase production and productivity of sustainable agriculture in most production environments. Animal genetic diversity allows farmers to select stocks or develop new breeds in response to changes in the environment, threats of disease, new knowledge of human nutrition requirements, changing market conditions and societal needs, all of which are largely unpredictable. Animal genetic diversity is critical for achieving food security for the rapidly growing population, not only with respect to the local or national situation, but also because countries are becoming increasingly interdependent for access to unique animal genetic resources. Furthermore, lifting trade barriers should create opportunities for developing countries to achieve more efficient and effective livestock sectors capable of increasing foreign exchange earnings and security for the world's majority. In addition, in developed countries consumer emphasis on product quality is increasing, markets are becoming more segmented, and efficiency of input resource use is being increasingly emphasized. These needs will intensify and should broaden efforts in the genetic development of animals for production, productivity, and product quality and help sustain primary national and regional agroecosystems.

The need for countries to develop and strengthen their capacity to benefit fully from their biological resources highlights the need for a global strategy for conservation of biological resources. The catalyst for such a global strategy was endorsement of the Convention on Biological Diversity (CBD) by representatives of 167 countries at the UN Conference on Environment and Development (UNCED) in Rio de Janeiro in June 1992. Following its ratification by 30 countries the CBD became law in December 1993. As of April 1995, 117 countries have ratified the CBD. The CBD specifically addresses agriculture and identifies the objectives of conserving diversity, using it in a sustainable manner, and sharing benefits arising from utilization of genetic resources.

In recognition of the importance of domestic animal genetic resources, and of the sizable proportion that are currently at risk of loss, and in keeping with FAO's mandate under the CBD, a special action program for the global management of domestic animal genetic resources was launched by FAO in 1992. This report addresses the rationale and the operation of FAO's Global Program for the Management of Farm Animal Genetic Resources. The range of opportunities for the involvement of groups is also described.

Achieving Sustainability

The contribution of livestock production to global food security is generally undervalued, particularly by developed country communities. Sansoucy et al. (1995) provides a review of contributions of livestock in achieving sustainable production systems. They also suggest that as human needs for food and agriculture products double over the coming decades, global food production from animal products will increase more rapidly than from plants in most developing countries with increasing purchasing power. Over this same period and beyond, the large between-region difference in human needs for food and agriculture and in production capacity will persist at the global level, with almost 75% of world agriculture remaining at the low- to medium-input levels. Indigenous livestock breeds often possess valuable traits such as disease resistance, high fertility, good maternal qualities, unique product qualities, longevity, and adaptation to harsh conditions and poor-quality feed, all desirable qualities for achieving sustainable agriculture under low-input conditions. In both developing and developed countries, food and agriculture production systems must increasingly utilize mixes of biological diversity, plant and animal, several species, and combinations of lines across species, with line development applying increasingly to particular primary production environments. With continual loss of agricultural land at alarming rates, the challenges of understanding, better measuring, and introducing more sustainable systems must be accepted.

National Strategies for Resource Use The fitness traits associated with the adaptation complex in a particular environment are generally much more difficult to measure and change than are the "production" traits. This seriously questions the common approach of introducing to low- to medium-input environments high-input/high-output exotic breeds developed primarily under comparatively benign environments and then relatively quickly expanding the exotic breeds, expecting them to rapidly adapt in arriving at sustainable systems. Instead, a superior strategy for these low to medium environments will aim to develop the production traits in one or more indigenous breeds that are already adapted and that local farmers are still prepared to use. Remember, much of the very large amount of low- to medium-input agriculture is likely to remain for much of the next century. Some medium- to even high-input pockets of agriculture are emerging in some developing countries, particularly around large cities, and are better able to utilize waste products. Although these situations pose land use and rural development policy issues, they may enable use of some more demanding and higher-producing genetic resources, particularly when disease, parasite, and climatic stresses are substantially *and* permanently reduced by other changes to the particular production environment. However, even in many of these countries low-input rural production from a range of species will remain important and will require highly superior animal adaptation to achieve sustainable agriculture. It is also of interest that in these adapted types, under the low-input and otherwise high-stress environments, the relative variability between animals for the production traits is often very high, suggesting that rapid rates of improvement in these traits under these stresses will be feasible if an effective breeding strategy can be designed and maintained.

Genetic Type by Production Environment Differences We have little understanding of the important ranking differences between most production environments of the genetic types developed under low-, medium-, and high-input levels. Included among the low level category are the combination high-stress and scavenger production environments, so relevant because of their continuing importance, particularly scavenger poultry and goats in the rural communities in much of the developing world. The high selection intensities that have ruled for very long periods in these environments surely have resulted in types that are more likely to differ between different low environments, and also from lines developed under more benign high-input conditions. Studies to characterize this re-ranking of genetic types among environments are not easy to design and conduct well; however, they are very important and are likely to be rewarding relatively quickly, particularly for the prolific species with short generation intervals.

A Sinking Ark?

In 1990 FAO initiated, with the support of the United Nations Environmental Programme (UNEP) and the European Association of Animal Production, an inventory and basic description of the breeds of the domestic livestock species. Global surveys were commenced for the more important species and breed population data were also recorded where available. The first very preliminary analysis of this data involving just 1,400 breeds and seven species resulted in the first edition of the World Watch List for Domestic Animal Diversity (FAO/UNEP, 1993), suggesting that about 30% of animal genetic resources globally were at very high risk of loss.

Tables 1 and 2 summarize the analysis of FAO's Global Databank for Animal Genetic Resources by species and region. The Global Databank now lists 3,882 breeds for 28 domestic species. The global analysis is published in the second edition of World Watch List for Domestic Animal Diversity (FAO/UNEP, 1995). The information contained in this second edition shows that globally 30% of breeds are classified as endangered and critical based on population size. The FAO defines endangered populations as those having fewer than 1,000 breeding females and fewer than 20 breeding males; critical populations are defined as those having fewer than 100 breeding females and fewer than 5 breeding males. Of the breeds listed under these two categories, 36% are managed either through a conservation program or maintained by an institute. Presumably, the risk of loss of these actively managed or maintained breeds is far lower than the risk of loss of breed populations outside such management programs. Of the total number of breeds with population data identified globally (2,924), 19% are categorized as endangered or critical and lack a breed conservation management program. As such, there is a very high risk of loss of these animal genetic resources.

Rushing to One Genetic Resource per Species There are several factors that place breeds at risk of loss and threaten domestic animal diversity. These factors are listed in Table 3. By far the greatest cause of genetic erosion is the growing trend for global reliance on one or a few modern breeds suited for the high input-output needs of industrial agriculture. This trend is of paramount concern; a cursory examination of the research literature suggest that about 50% of the total variation at the quantitative level is between-breed, and the remainder is common to all breeds. Hence, a move to one breed would eliminate half this variation in the species, in addition to jeopardizing readily available gene combinations in other remaining unique genetic resources. Marked increases in world population size and food demand necessitate our reliance on utilizing already formed genetic types rather than on establishing new genetic packages, at least for the adaptation complex in each primary production environment and the species with long generation intervals, such as cattle.

Animal breeding increasingly promotes and otherwise facilitates the universal use of these very few "superior" breeds, via international sale and advertising of stock and semen, international genetic evaluation, international assistance, and perhaps in future even via international patenting of generally important biological procedures. Abrupt changes in breed composition continue to occur in developing countries, with a small number of exotic breeds being rapidly introduced and expanded as straight- and crossbreds, and with concurrent dramatic reduction in population numbers for many of the indigenous breed resources.

The Impact on Longer-Term Food Security These very few genetic lines will not achieve high levels of production, productivity, and sustainability for each of the broad range of production environments and combination of needs. To rely on these few lines globally, or even on continual (systematic) crossbreeding incorporating one or more of these lines, will substantially reduce food security over time in many situations.

Table 1. Breeds of domestic animals at high risk by species[a]

Species	Total no. breeds on file	No. breeds with population data and % of total	No. of breeds categorized as critical or endangered and % of which are maintained	No. of breeds at high risk of loss (%)[b]
Mammalian				
As	77	24 (31%)	9 (0%)	9 (38%)
Buffalo	72	55 (76)	2 (0)	2 (4)
Cattle[c]	787	582 (74)	135 (41)	80 (14)
Goat	351	267 (76)	44 (16)	37 (14)
Horse	384	277 (72)	120 (20)	96 (35)
Pig	353	265 (75)	69 (25)	52 (20)
Sheep	920	656 (71)	119 (29)	85 (13)
Yak	6	6 (100)	0	0 (0)
Dromedary	50	40 (80)	2 (0)	2 (5)
Bactrian Camel	7	7 (100)	1 (0)	1 (14)
Llama	3	3 (100)	0	0
Alpaca	4	4 (100)	0	0
Guanaco	2	2 (100)	0	0
Vicuna	3	3 (100)	0	0
Total	3,019	2,191 (89%)	501 (39%)	364 (17%)
Avian				
Chicken	606	512 (85)	334 (61)	131 (26)
Turkey	31	29 (94)	11 (9)	10 (35)
Domestic Duck	62	54 (87)	29 (38)	18 (33)
Muscovy Duck	14	13 (93)	5 (40)	3 (23)
Domestic Goose	59	51 (86)	28 (11)	25 (49)
Guinea Fowl	22	17 (77)	4 (0)	4 (24)
Quail	24	23 (96)	16 (100)	0
Pigeon	19	16 (84)	2 (0)	2 (13)
Pheasant	8	7 (88)	0	0
Partridge	11	4 (36)	0	0
Ostrich	7	7 (100)	3 (33)	2 (29)
Total	863	733 (85%)	372 (48%)	195 (27%)
Global	3,882	2,924 (75%)	873 (36%)	559 (19%)

[a]Source: Adapted from FAO/UNEP (1995).
[b]At risk determined based on breeds with population data having < 1,000 breeding females or < 20 breeding males and for which there is no conservation program in place.
[c]Includes mithan and banteng.

Conservation Imperatives for Animal Genetic Resources

With the projection that more than 1,000 of the probably 4,000 breeds of Earth's domestic animal species are currently at high risk of extinction, and with so little known about most, scarce international funds cannot at this point be concentrated on a small number of breed rescue projects. Emphasis must be on implementing a sound global management infrastructure and broad technical program, which has the potential to help many countries design and implement national action strategies, as required under the CBD. With these considerations firmly in mind, the imperatives for conservation of domestic animal genetic diversity can be stated:

1. *Identify* and *understand* those unique genetic resources that collectively constitute the global gene pools for each of the important animal species domesticated and used to provide food and agriculture.
2. *Develop* and properly *utilize* the associated diversity to increase production and productivity, achieve sustainable agricultural systems, and meet demands for specific product types. Hence, the effective use of breeds is also an essential component of conservation, and perhaps the most cost-effective, a further reason for enabling the development and use of more breeds. Development and use of genetic resources must be integral to an effective conservation effort for the domestic animal species.
3. *Monitor* particularly those resources that are currently represented by small populations of animals, or which are otherwise being displaced by one or other breed replacement strategies.
4. *Preserve* the unique resources for which sufficient current demand cannot be engendered.
5. *Train* and *involve* people in management of these resources, including their best use and development, and in the maintenance of diversity.
6. *Communicate* to the world community the importance of our domestic animal genetic resources and of the associated diversity, its current exposure to loss, and its irreplaceability.

Table 2. Breeds of domestic animals at risk by region[a]

Region	Total no. breeds on file	No. breeds with population data and % of total	No. of breeds categorized as critical or endangered and % of which are maintained	No. of breeds at high risk of loss (%)[b]
Africa	396	239 (60%)	27 (0 %)	27 (11%)
Asia and Pacific	996	710 (71)	105 (8)	97 (15)
Europe and former USSR	1,688	1,501 (89)	638 (44)	358 (24)
Latin Am. and Caribbean	220	143 (65)	29 (7)	27 (20)
Near East	378	214 (57)	15 (40)	9 (7)
North America	204	117 (57)	59 (31)	41 (50)
Global	3,882	2,924 (75%)	873 (36%)	559 (19%)

[a]Source: Adapted from FAO/UNEP (1995).
[b]At risk determined based on breeds with population data having < 1,000 breeding females or < 20 breeding males and for which there is no conservation program in place.

Table 3. List of causes for risk of loss or extinction of breeds

Reason	Description
Aid	Lack of incentive to develop and use breeds, giving preference to those few developed for use in high-input, high-output, relatively benign environments.
Product	Undue emphasis placed on a specific product or trait leading to the rapid dissemination of one breed of animal to exclusion and loss of others.
Cross-breeding	Indiscriminate cross-breeding that can quickly lead to the loss of original breeds.
Storage	Failure of the cryopreservation equipment and inadequate supply of liquid nitrogen to store samples of semen, ova or embryos; or inadequate maintenance of animal populations for breeds not currently in use.
Technology	Introduction of new machinery to replace animal draught and transport, resulting in permanent change of farming system. Poorly interpreted international genetic evaluation.
Biotechnology	Artificial insemination and embryo transfer leading to rapid replacement of indigenous breeds.
Violence	Wars and other forms of socio-political instability.
Disaster	Natural disasters such as floods, drought, or famine.

Working Definitions Clear terminology is necessary in the conservation effort, to advance understanding, facilitate education and training, communicate successfully with the wider public, and realize a common purpose in application. Terms must accommodate all practical situations. In the case of domestic animals, these must include all genetic resources and diversity associated with each species, both resources currently in use and those not in use, the common and the rare, the long-developed and the new, the commercial lines and the research stocks. The conservation literature includes a number of terms that are not well understood in the domestic animal context. Interpretation of these terms to form working definitions for domestic animal diversity (DAD) is required if meaningful involvement in conservation of DAD by the necessary range of people is to be realized. A minimum set of working definitions is presented in Table 4.

Lessons from Plant Genetic Resource Conservation? Much of the global effort to date concerning agrobiodiversity has focused on the conservation of plant genetic resources. In fact, the first genebanks for plants were established in the 1960s and very large numbers of resources are catalogued, stored, and readily available for regeneration. More than 100,000 accessions per year are now made to these genebanks across the world for breeding purposes. Is there anything that can be learned from these conservation efforts that would be of benefit to animal genetic conservation?

Although the underlying principles of genetics and gene action are similar for plants and animals, there are several differentiating characteristics between them (Table 5). Animal resources tend to be more mobile, comparatively high cost per unit, have very low fecundity, and "seed" needs to be deep-frozen to survive. Many animal diseases spread rapidly and have serious impact both within and across animal species, including *Homo sapiens*. These primary differences lead to a suite of related considerations that will affect resource management strategies, such as the relatively small total number of animal genetic resources and extremely small number of these that have been exposed to date to modern development technology, the high impact of some biotechnologies on dispersal ability of animal resources, the few remaining wild relatives of domestic animal species, the low level of knowledge of animal biology compared with the wide range of environments over which sustainable production is being sought, the high impact of animal disease protocols on international transport and access, and the low level of genebanking of animal resources at risk. These substantial differences indicate that policy, legal, and technical uniqueness must be addressed to achieve effective management of animal genetic resources.

Table 4. Working definitions proposed for conservation

Animal Genetic Resources (AnGR) At the breed level, the genetically unique breed populations formed throughout all domestication processes within each animal species used for the production of food and agriculture, together with their immediate wild relatives (here "**breed**" is accepted as a cultural rather than a technical term, i.e., to emphasize ownership, and also includes strains and research lines).

Domestic Animal Diversity (DAD) The genetic variation or genetic diversity existing among the species, breeds, and individuals for all animal species that have been domesticated and their immediate wild relatives.

Conservation (of domestic animal diversity) The sum total of all operations involved in the management of animal genetic resources, such that these resources are *best used and developed* to meet *immediate and short-term* requirements for food and agriculture, and to ensure the diversity they harbor remains available to meet possible *longer-term* needs.

Conservation (in general) The management of human use of the biosphere so that it may yield the greatest sustainable benefit to *present* generations while maintaining its potential to meet the needs and aspirations of *future* generations. Thus conservation is positive, embracing preservation, maintenance, sustainable utilization, restoration, and enhancement of the natural environment (IUCN/UNEP/WWF/FAO/UNESCO, 1980).

***In situ* Conservation** Primarily the active breeding of animal populations for food production and agriculture, such that diversity is both best utilized in the short term and maintained for the longer term. Operations pertaining to *in situ* conservation include performance recording schemes and development (breeding) programs. *In situ* conservation also includes ecosystem management and use for the sustainable production of food and agriculture. For wild relatives *in situ* conservation, generally called *in situ* preservation, is the maintenance of live populations of animals in their adaptive environment or as close to it as practically possible.

***Ex situ* Conservation** In the context of conservation of domestic animal diversity, *ex situ* conservation means storage. It involves the preservation as animals of a sample of a breed in a situation removed from its normal production environment or habitat, and/or the collection and cryopreservation of resources in the form of living semen, ova, embryos, or tissues, which can be used to regenerate animals. While other methods of genetic manipulation, such as the use of various recombinant DNA techniques, may represent useful means of studying or improving breeds, but do not constitute *ex situ* conservation, and may not serve conservation objectives. At present the technical capacity to regenerate whole organisms from isolated DNA does not exist.

The FAO's Global Management Program: Structure and Work Elements

The FAO is the sole intergovernmental organization with a broad international mandate for improving agriculture and food production for current and future populations, with particular emphasis on developing countries. Given the global nature of animal genetic resources (AnGR), few institutions have the capacity to coordinate the geographic, species, technical, and intergovernmental issues that are necessarily involved in developing a successful global program of management for the domestic animal sector. Based on the conservation imperatives, and in accordance with the CBD and Agenda 21 and the oversight provided by the International Commission on Sustainable Development, FAO's Global Program for the Management of Farm Animal Genetic Resources is about to be launched. The planned global strategy involves four components: 1) an intergovernmental support mechanism for enabling direct government involvement and ensuring continuity of policy advice, 2) a geographically and country-based global structure, to assist and coordinate national actions, 3) a technical program of activities grouped under seven elements, and 4) a cadre of experts from throughout the world to guide the program.

The Global Structure Countries possess different subsets of the global total of breeds forming each domestic animal species. While the Convention on Biological Diversity clearly accepts each country's sovereignty over its genetic resources, countries are becoming increasingly interdependent in their seeking to access unique animal genetic resources from elsewhere. Therefore, effective conservation programs by nations provide the foundation for a successful global program of management for each species. The FAO's management program provides the structure for achieving country-based emphasis combined with the necessary regional and global coordination of policy and effort (Figure 1). This essential primary global infrastructure will lead to the early implementation of the necessary range of activities to assist countries that have not done so to design, implement, and maintain comprehensive national strategies for the management of their animal genetic resources. Scarce financial resources are concentrated on initiating the key infrastructure required. Some aspects of the program are already being implemented, although complete implementation will take some years and will depend on strong collaborative support for the program.

National Focal Point for each country will comprise a coordinating institution strongly linked to the regional focus (Figure 1) and a national technical coordinator nominated and supported by the Member Government. This coordinator will be the point of contact for the country's involvement in the FAO AnGR Program and will assist in establishing and maintaining the essential in-country network. To date, 50 countries have established focal points, 38 in the European Community and 12 in Asia.

Regional Focal Point in each major genetic storehouse region of the world will help develop effective national coordinators, design and implement effective regional networks as integral components of the global structure, help achieve early and wide introduction of national strategies, and trigger a range of most effective projects covering the conservation complex for domestic animals. Regional Focal Points will need to be established with the assistance of extra-budgetary funds. Regional focal

Table 5. Principle differences of sectors involved in utilization and conservation of genetic resources for food and agriculture[a]

Aspect	Terrestrial animals	Plants		Fish	
		Crops	Forest	Fisheries	Aquaculture
No. of species used for food	Less than 15 incl. poultry supply most	>100 but 4 supply about 60% and 10-12 about 75%	Several hundred for food and thousands for other uses	More than 1,000	200-400
Mobility of organism	Mobile	Immobile	Immobile	Mobile	Mobile
Potential of many diseases to spread and impact within and over species	High	Low	Moderate	Low	Low
Fecundity	Low, medium with biotechnologies	High	High	High	High
Cost of producing unit	High	Low	Medium to low	Low	Low
Level of domestication	High	High	Medium to low	None	Low
No. of varieties and breeds in each species	Few to moderate	Many	Few	Very few	Few/moderate
State of knowledge about genetic resources	Medium/low	High	Medium to low	Low	Medium
Location of resource	Farms, pastures, some nature	Farms	Nature, some farms	Nature	Farms and nature
Common/private property resource	Private	Private	Private and some public	Common	Private
Wild relatives exist	Few	Some	All	All	All
Dispersal ability, impact of biotechnologies on dispersal ability	Low, high	Medium low	Low, high	High, low	Medium to high, low
Ease to storage/transport of live material	Moderate to difficult	Very easy	Easy to difficult	Moderate	Moderate
Biosafety record (genetically modified organisms)	Limited, but appears good	Extensive and good	Moderate and good	Non-existent	Extremely limited and inconclusive
Genebanking-seed	Sperm and embryos possible, not oocytes; continuous freezing and strict hygiene essential	Storage of seeds and embryos easy to difficult; freezing not required		Sperm only Continuous freezing and strict hygiene essential	

[a]Source: FAO (1995).

points are planned for Asia and the Pacific, Europe, the Americas and Caribbean, Africa, and the Near East and Mediterranean. The regional focal point for Asia and the Pacific has now been initiated in Bangkok through funding from Japan.

Global Focus for the Program is being established at FAO headquarters to facilitate, communicate, and coordinate the global effort, involve the range of governmental, non-governmental, and intergovernmental parties essential for the Program's success, maintain the Global Databank for AnGR (centered on global surveys) and the Early Warning System (the World Watch List for Domestic Animal Diversity), provide the secretariat for the intergovernmental mechanism for AnGR, coordinate development of the necessary guidelines and procedures and the assistance to countries in establishing their management strategies for AnGR, and seek the essential extra-budgetary funding for the Program.

Work Elements

The structure must be accompanied by a cost-effective program of work if the global management is to be effective in achieving the imperatives over time, both intergovernmental and technical (Figure 2).

Global Information System for Domestic Animal Diversity A key feature in supporting the global operational structure is the Global Information System (DAD-IS) under development. The DAD-IS will form the information axis for all aspects of the Program and will utilize the Internet to enable low-cost, effective, ongoing communication and information transfer. A prototype of the first stage of DAD-IS

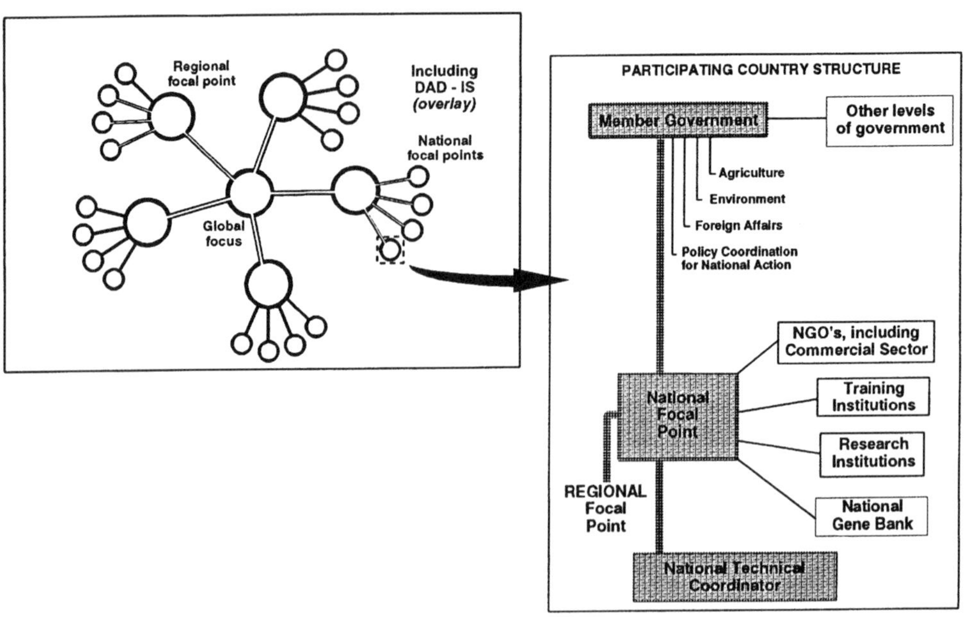

Figure 1. Structure for FAO's Global Program for Management of Farm Animal Genetic Resources.

Table 6. List of species included in Global Project for Measurement of Domestic Animal Genetic Diversity (MoDAD)

Cattle[a]	Ass	Chicken
Sheep	Horse	Domestic goose
Goat	Camel[b]	Duck
Pig	Lamoid[c]	Turkey
Buffalo	Rabbit	

[a]Includes all *Bos* as well as yak, banteng, and mithan.
[b]Includes dromedary and bactrian.
[c]Includes llama, alpaca, guanaco, and vicuna.

will be field-tested in late 1995. DAD-IS will serve:

To accommodate the range of essential and unique databases for the Program covering the global inventory and description and monitoring of resources for both species and genebanks, functionaries and for Project MoDAD (see below), and to link other specific genetic databases under development;

To provide The Global Early Warning System for AnGR and to facilitate continuous updating and ongoing access to this database;

To lower the cost and increase the amount and effectiveness of training and education in animal conservation genetics and procedures, through a system of shared expertise and information;

To serve as a central and reliable source of aids for experimental design and data analysis, in order to increase cost-effectiveness of, and capacity for, research;

To provide the global bibliography for AnGR;

To assist in management of the Program, and execution of activities including the effective networking in project development and management; and

To facilitate active involvement of the world community in the Global Program.

Characterization Establishing the magnitude of existing animal genetic diversity and reliable rates of loss are cornerstones for the Program. The enormity of this task is highlighted by the fact that globally there are some 4,500 unique breeds of more than 40 species; at least some of these breeds are found in most countries and for the most part there is no systematic monitoring in place or baseline information available. By mid-year survey data contained in FAO's Global Databank for AnGR will include basic descriptive data on roughly 85% of these known breeds representing 28 species. The population data provided through surveys enable the monitoring of breeds at risk of extinction. This information was first summarized in FAO's World Watch List for DAD (FAO/UNEP, 1993). Information in the Databank will be regularly updated to incorporate missing data and as a mechanism to breed genetic resources at risk and rates of loss. Each of these activities, together with information on the status of *ex situ* collections, forms The Global Early Warning System for AnGR.

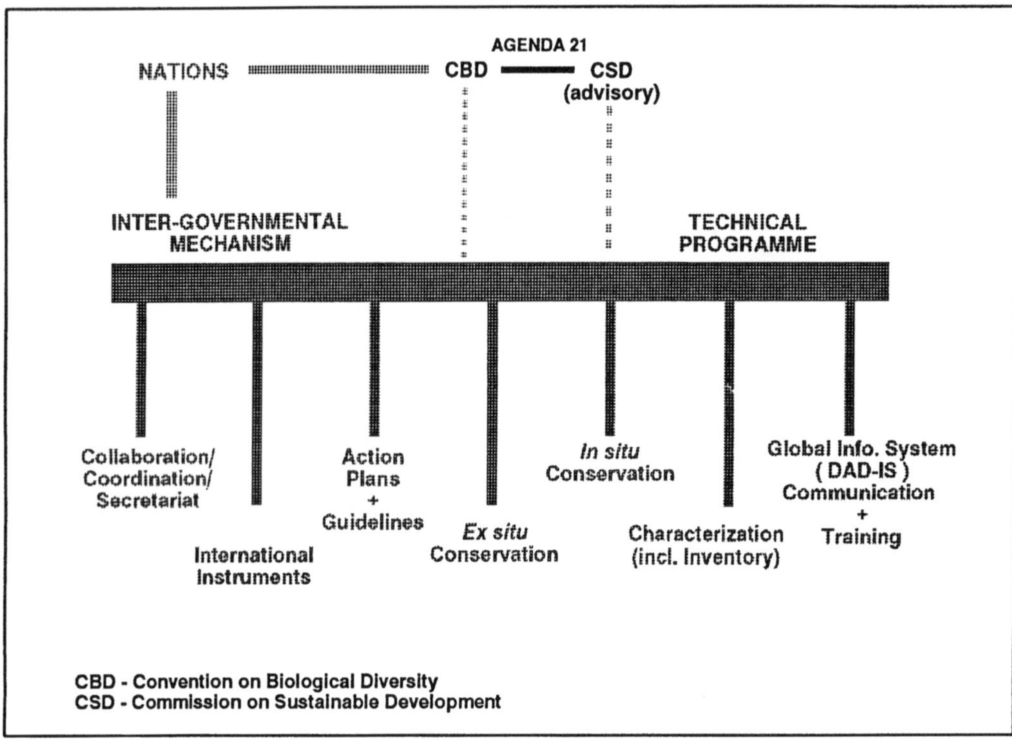

Figure 2. Intergovernmental and technical work elements for the FAO Global Program for Management of Farm Animal Genetic Resources.

Characterizing Animal Genetic Diversity Comprehensive genetic evaluation at the breed level to cover all breeds for both current and future production potential for all primary production environments is not feasible or required. However, knowledge of the amount of the breed level variation in each species and of the size of each breed's contribution to this will assist priority setting for overall AnGR management. In order to better understand this relative uniqueness of animal genetic resources, a global research project is planned in genetic distancing (Barker et al., 1993). This project, referred to as the Global Project for the Measurement of Domestic Animal Genetic Diversity (MoDAD), will utilize microsatellite technology and initially focus on the analysis of genetic variation within some 14 species (Table 6) accounting for above 90% of food and agriculture production globally. The primary objective of MoDAD is to substantially increase the cost-effectiveness of the total Global Program by establishing the comparative uniqueness of genetic resources in each species to aid rationalization of the total management task for countries and globally. This will be achieved primarily by directly utilizing the results to objectively reduce the number of breeds that will need to be maintained. Secondary objectives of MoDAD are to establish global repositories for both AnGR microsatellite data and DNA, for enabling more effective research and use in capacity building; and to identify breed

Table 7. Summary of state-of-the-art in cryopreservation for the most important domestic species. Status of technology identified as: routine technique available (**RT**), technique needs further development prior to commercial use (**PND**), no technique yet available (**NP**), or current status of technology unknown (**SU**)

Species	Semen	Ova	Embryos
Cattle	RT	PND	RT
Buffalo	RT	SU	PND
Goats	RT	SU	RT
Sheep	RT	SU	RT
Pigs	PND	SU	NP
Horses	RT	SU	RT
Asses	RT	SU	SU
Lamoids	RT	SU	SU
Camels	RT	SU	SU
Chickens	PND[ab]	NP	PND[c]
Turkeys	PND[ab]	NP	NP
Ducks	SU	NP	NP
Goose	SU	NP	NP

Sources: Thibier et al. (unpublished date); [a]Etches et al. (1995); [b]Reedy et al. (1995); and [c]Kino et al. (1995).

combinations that have the highest potential for heterosis for improvement in traits related to adaptation. Project MoDAD will benefit all nations. A detailed formulation document for the project for use in assisting countries in becoming involved is available from FAO.

Mechanisms for Conserving Animal Genetic Resources Conservation is not an end in itself but a means of ensuring that animal genetic resources are better understood and available and more effectively used by present and future generations. Once genetic resources have been identified and characterized, there are two basic conservation activities that may then be defined as either *in situ* or *ex situ*.

In Situ Conservation Generation and loss of alleles is a dynamic process that should be maintained at about equilibrium through sound management. The strategy for the global *in situ* activity emphasizes "wise use" of indigenous animal genetic resources by establishing and implementing breeding goals and strategies for sustainable production systems. Effective development of more of these adapted resources to meet the requirements is all-important and will form the focus of FAO's *in situ* emphasis. Because *in situ* involves the maintenance of live populations of animals in their adaptive environment, animal populations continue to evolve and be developed for more effective use. Infrastructure for animal recording and breeding is well established for developed countries. But infrastructure that is appropriate to developing country systems remains scarce. Modalities for the simplified animal recording, genetic development, and dissemination are needed for each species and a range of national livestock structures for developing countries.

Ex Situ Conservation *Ex situ* conservation includes cryogenic preservation and the maintenance as animal populations of breeds of domesticated species in farm parks, zoos and locations away from the environment in which they are being developed; in effect this is storage of AnGR for which farmers in a country are no longer interested in using. The biggest shortcoming of *ex situ* genebanks is that, once stored, animal genetic resources are removed from the evolutionary process they undergo in nature; unless a concerted effort is made, the level of knowledge about them is also frozen. Most *ex situ* conservation is likely to utilize cryopreservation, at least for the larger species. No selection pressure is imposed from changing natural conditions. Cryopreservation is technologically demanding and not yet developed for storing both male and female gametes of all species of interest. A range of quarantine issues must be overcome before international storage and access can be effective. Table 7 provides a summary of the current state of the art for cryopreservation of semen, ova, and embryos for several livestock species.

The Global Programs's *ex situ* conservation strategy is still being developed but is based primarily on the use of live animal populations wherever possible, backed up by cryopreservation where technology exists or can be developed, combining within-country genebanks with global gene repositories of last resort, in keeping with the Convention on Biological Diversity. Interested governments, NGOs, research institutions, and private enterprises would also be encouraged to maintain *in vivo* samples of breeds at risk, with national inventories of these being established and maintained current, so that the genetic resources are directly available for use and study.

In situ and *ex situ* conservation are complementary, not mutually exclusive; their application for a particular AnGR depends on farmers' current use of it and comparative uniqueness. Further, frozen germplasm can play an important role in the support of *in situ* breeding strategies. For example, the use of artificial insemination (AI) in *in situ* conservation of populations may enable much greater male selection differentials and dissemination than would be practical via natural mating using live adult males. The use of AI in a back-crossing system enables efficient regeneration of a population, and alternate use of male breeds in reciprocal crossbreeding systems may also be advantageous.

Global Action Plan and Guidelines The Program provides assistance to countries in the development and implementation of comprehensive and practical guidelines for designing National Action Strategies for the management of animal genetic resources to also harmonize with the provisions of the CBD. The Global Action Strategy will be further developed by integrating all national action plans and will be continually updated as knowledge, technology, the negotiation of the CBD by member countries, and implementation of policies progress.

A further early activity in the Program involves sub-regional project identification missions being carried out to establish a global portfolio of the most effective conservation activities, ready for formulation, funding, and execution to add to the basic framework now being implemented. The specially designed missions are being trust-funded by various donors and focus on the following regions: China, Sub-Saharan Africa (francophone and anglophone), Central and eastern Europe, former USSR, Latin America and the Caribbean (split into two sub-regions), Near East and the Mediterranean, and the Indian sub-continent. The first four of the intended nine missions are complete, with the remainder still to be funded.

Table 8. Education needs in conservation genetics[a]

Global aspects on animal genetic diversity, including evolution and history of domestic species and breeds, the breed and population concepts, animal populations in various parts of the world and present development trends.

Livestock production systems in various regions of the world, and the prospects and constraints as regards different animal populations in relation to environmental and socio-economic conditions.

Molecular and qualitative genetics, including, e.g., gene regulation and expression, additive and non-additive effects, gene mapping, genetic markers, inbreeding, effective population size, gene frequencies, and the effects of migration, mutation, selection and genetic drift.

Quantitative genetics and selection theory, including the makeup of genetic, phenotypic and environmental variances and covariances, genetic parameters and breeding values and their estimation, selection and selection response, breeding methods and their long-term effects on the genetic variance and inbreeding.

The genetics of production and functional traits in different species and breeds, including reproduction and disease resistance and their relationships with production under variable environmental conditions.

Breeding programs for utilization of animal genetic resources, including breeding objectives, recording and organization of trait information, use of artificial insemination, multiple ovulation and embryo transfer (MOET) and nucleus-herd breeding systems, optimization and organization of breeding schemes for pure and cross-breeding, breed "competition" and the influence of market forces, and, finally, the short- and long-term economic and biologic consequences of the breeding programs.

Characterization and documentation of animal populations with regard to distribution and trait characteristics in relation to defined environments, organization and utilization of data banks, descriptors to monitor population changes, and genetic distances between breeds.

Methods for preservation of animal genetic resources, *in situ* and *ex situ*. *In situ* preservation: in particular breeding strategies for small animal populations under no or little genetic change. *Ex situ* preservation: cryopreservation (i.e., storage of frozen semen, embryos, cell culture, or DNA, including methods of collection, sample sizes needed, and record keeping).

[a]Source: Malmfors et al. (1994).

International Instruments The global program for plant genetic resources operates with technical input from FAO's intergovernmental Commission on Plant Genetic Resources. No such mechanism is in place to facilitate intergovernmental input to the management of AnGR. Because many of the technical logistics, and the policy and legal issues, are fundamentally similar for both plants and animals, and because of the need to promote the development and maintenance of more sustainable agriculture systems, plans are underway to expand the Commission on Plant Genetic Resources to include domestic animals (and eventually forestry and fish genetic resources) under the title of the Commission on Genetic Resources for Agriculture and Food.

Intellectual Property Rights A main feature of the CBD relates to the sovereignty that nations have over their biological resources. National sovereign rights issues are entwined with access and intellectual property rights (IPR) issues, which have serious implications for access to biological resources and the sharing of benefits derived from their use, the transfer of technologies, and compensation for rural and indigenous people. Because of the growing reliance among countries for genetic resources, it has been suggested that access to genetic resources used in agriculture and food could be agreed on a multilateral basis. The broadened Commission will provide the logical mechanism for such intergovernmental negotiation of policy.

A Global Management Entity Successful management of AnGR involves a range of technical development, coordination, facilitation, and operational activities for countries and for non-governmental and international parties involved. The FAO is now establishing a management entity for this Global Program, with the mandate to help countries requesting assistance and coordinate activities regionally and globally; develop uniform basic procedures, guidelines, and protocols; maintain The Global Early Warning System for AnGR; bolster capacity building and upgrade training; increase awareness and communicate issues globally; and serve as the Secretariat for the global intergovernmental mechanism for AnGR.

Strengthening Scientific and Technological Capacity

A New Paradigm for Animal Breeding Traditional animal breeding has dwelt on quantitative genetic principles and theory. At the other extreme, conservation has been mainly focused on storage of genetic material. There can be no denying the enormous success of traditional animal breeding programs in terms of improving production potential of modern livestock, particularly in developed countries (Smith, 1984a,b; Gibson and Smith, 1989). However, socio-political changes underway are emphasizing the need to also account for the longer-term costs and benefits and for sustainability of production systems, including the breeding programs involved. This broadens the domain that genetic development must address. The term "conservation genetics" perhaps better describes the new paradigm; the objective is now to wisely manage all of a country's genetic resources in terms of better understanding, developing, using, and maintaining agrobiodiversity. Education and training programs relating to management of animal genetic resources must address these needs. Malmfors et al. (1994) has identified areas where university teaching need focus. A summary is provided in Table 8.

Capacity-Building Imperatives The ability of countries to implement effective national action strategies for conservation of animal genetic resources will largely depend on human resources and institutional capacities. Countries will need to determine their training and technological and research needs. Countries may seek guidance and require support to complement local initiatives in order to strengthen their capacity and technical competence. The FAO's Program will address these needs as part of, and in support of, the technical elements of the Program.

Implications

The outcomes being sought by FAO's Global Program for Management of Farm Animal Genetic Resources are inventory control and early warning; establishing the comparative uniqueness of the breeds that constitute each species; more effective development and use of a greater range of unique resources; a much higher level of public awareness; major reduction in the number of breeds at high risk of loss; maintenance of and ready access to the more unique AnGR; and helping countries to meet their obligations under the CBD. DAD represents a resource that is critical for achieving food security for the rapidly growing human population, not only with respect to the local or national situation, but also because countries are becoming increasingly interdependent for unique animal genetic resources. The country-based global structure of the Program is designed to recognize and emphasize responsibility and activities within countries, to involve countries and other essential parties, to maximize opportunities for conservation action and fully align with the CBD.

The Program introduces some changes of emphasis to and a broadening of our long-accepted approach to animal breeding principles and practice, to harmonize with the decisions of a large section of the world community. Much remains to be done if the vast majority of largely unique genetic resources in developing countries in particular are not to be lost under the increasingly intense pressures now being applied. Countries need assistance now.

This FAO Program has the potential to be effective for all species of domestic animals upon which humankind must rely for much of its existence. However, the Program's implementation is heavily dependent on funding from developed countries. The primary reason being offered by a wide spectrum of developed countries for being unable to currently provide substantial support is budgetary constraint. The easiest short-term decision is to delay, but in so doing it must also be accepted that these domestic animal genetic resources are irreplaceable, and "sharing of benefits" will inevitably also mean the granting of access to those providing assistance. There are several ways in which interested groups may become involved, including:

- Learn more about your country's program from your national focal point or international government representative;
- Help advance the way genetics and animal breeding is taught, at all academic and applied levels;
- Help address the new research and development needs;
- Help develop DAD-IS, and particularly its training and research modules;
- Help communicate the issues to the public and commercial business to rethink the overemphasis on the development and sale of few AnGR;
- Help your country to identify, better understand, and monitor its AnGR;
- Help establish national genebanks for each species;
- Help develop more of the world's AnGR in each of the species used for food and agriculture;
- Help develop the range of detailed guidelines required to assist countries to upgrade their management of AnGR;
- Contribute to the development and maintenance of your country's national action strategy for management of AnGR of each species; and
- Contribute to funding of your national program and the global program to conserve AnGR.

Literature Cited

Barker, J.S.F., D. G. Bradley, R. Fries, W. G. Hill, M. Nei, and R. K. Wayne. 1993. An integrated global programme to establish the genetic relationships among the breeds of each domestic animal species. FAO Animal Production and Health Paper, Rome, Italy.

Etches, R. J. 1995. Reproduction in Poultry. CAB Press, Wallingford, Oxon, U.K.

FAO/UNEP. 1993. World Watch List for Domestic Animal Diversity (1st Ed.). R. Loftus and B. Scherf (Ed.). FAO, Rome, Italy.

FAO/UNEP. 1995. World Watch List for Domestic Animal Diversity (2nd Ed.). B. Scherf (Ed.). FAO, Rome, Italy.

Gibson, J. P., and C. Smith. 1989. The incorporation of biotechnologies into animal breeding strategies. In: M. Moo-Young, L. A. Babiuk, and J. P. Phillips (Ed.) Animal Biotechnology: Comprehensive Biotechnology, First Supplement. pp 203-231. Pergamon, Elmsford, NY.

IUCN/UNEP/WWF/FAO/UNESCO. 1980. World Conservation Strategy. Living Resources Conservation for Sustainable Development. IUCN, Switzerland.

Kino, K., B. Pain, S. P. Leibo, M. Cochran, M. E. Clark, and R. J. Etches. 1995. Production of chicken chimeras from injection of frozen thawed blastodermal cells. Proc. Poultry Science Annual Meeting, Edmonton, Alberta, Canada.

Malmfors, B., J. Philipsson, and M. Haile-Mariam. 1994. Education and training in the conservation of domestic animal diversity - student needs and field experience. Symposium on Conservation of Domestic Animal Diversity. In: Proc. 5th World Congr. Genet. Appl. Livest. Prod. XXI:485.

Reedy, S. E., S. P. Leibo, M. E. Clark,and R. J. Etches. 1995. Beyond freezing semen. In: M. R. Bakst, and G. J. Wishart (Ed.) Proc. First Int. Symp. on the Artificial Insemination of Poultry. Poultry Science Assoc., Savoy, IL.

Sansoucy, R., M. A. Jabbar, S. Ehui, and H. Fitzhugh. 1995. The contribution of livestock to food security and sustainable development. Roundtable on Livestock Development Strategies for Low Income Countries (In press).

Smith, C. 1984a. Estimated costs of genetic conservation of farm animals. In: Animal Genetic Resources Conservation and Management, Data Banks and Training. FAO Animal Production and Health Paper no. 44/1. pp 21-30. FAO, Rome, Italy.

Smith, C. 1984b. Genetic aspects of conservation in farm livestock. Livest. Prod. Sci. 11:37.

UNEP. 1992. Convention on Biological Diversity Final Text. UN Department of Public Information, New York.

Animal Germplasm Programs in North America

R. J. Gerrits* and J.N.B. Shrestha†

***National Program Staff, USDA-ARS, Beltsville, MD 20705, U.S.A. and †Centre for Food and Animal Research, Agriculture and Agri-Food Canada, Ottawa, ON K1A OC6, Canada**

Abstract The importance of biological diversity in all life forms, including animals, has received international attention in recent years. National programs to protect animal genetic resources were initiated in the United States and Canada in 1990. The Centre for Food and Animal Research (CFAR) was designated to provide the leadership for the conservation of animal genetic resources in Canada. The Canadian program was developed as a working partnership among countries, governments, universities, non-governmental organizations, industries, conservation groups, veterinary practitioners, and private individuals. Their objectives were to establish an inventory of Canada's animal genetic resources, a gene bank for preservation of germplasm resources, and a program of research for the long-term conservation of animal genetic resources and guidelines to ensure use by the Canadian animal industry. Guidelines for the U.S. National Animal Germplasm Program (NAGP) were outlined in the 1990 Farm Bill. The program is administered by the Secretary of the United States Department of Agriculture through the Agricultural Research Service. The NAGP was established by a national committee composed of federal and state agricultural experiment stations and industry scientists. The principal objectives of the NAGP are to characterize, preserve, and develop systems to utilize animal germplasm and to identify (through a gene mapping program) genes or groups of genes associated with desirable traits and disease resistance in animals. The United States and Canada are establishing germplasm database systems on animals and repositories for the long-term preservation of unique, diverse, and valuable animal germplasm.

Key Words: Genetic Resources, Germplasm, Database, Gene Mapping

Introduction

The genetic diversity of animals has been used effectively by scientists and producers and led to improvements in production efficiency of animals and the quality of animal products. Animal products will continue to provide a significant portion of the nutrients to human diets in the world's ever-increasing population. Conservation of genetic diversity is necessary to provide optimum access to desirable genes and gene complexes that will contribute to the future quantity and quality of our food and fiber supply. Research on the evaluation, conservation, identification, and utilization of unique genetic stocks and genes of livestock and poultry should provide the foundation for effective utilization of germplasm on a global basis. In the last two decades, significant advances have been made using techniques of molecular biology (i.e., gene mapping and identification of genes). This new technology, when used in conjunction with traditional animal breeding programs, has potential to significantly increase the current annual rate of genetic progress in animals.

The Animal Germplasm Program in Canada

In 1990, the first National Workshop on Conservation of Animal Germplasm served as a basis to establish the national program. It was determined that the Centre for Food and Animal Research (CFAR) of the Research Branch was the best qualified to provide the leadership for the program because of their expertise in genetics and cryopreservation. The national program was endorsed by all organizations in Canada affiliated with animal genetics, reproduction, and research (Shrestha, 1990).

In 1992, the Government of Canada developed and funded a "Green Plan" to ensure that the quality of the environment in Canada was maintained. Funds from the Green Plan were made available to CFAR and were supplemented with those provided by Agriculture and Agri-Food Canada to initiate the conservation of animal genetics program.

In response to Canada's Green Plan, the National Workshop on Conservation of Animal Germplasm, and the Convention on Biodiversity, the concept of a National Network on Animal Germplasm Conservation with a working partnership among CFAR and interested animal industry groups was initiated (Chiperzak and Shrestha, 1992).

The partnership proposal called for the government to provide leadership and coordination with universities and industry to conduct related research, share cryopreservation activities, and establish an updated Inventory and Database of Canada's Domestic Animal Genetic Resources. The non-governmental conservancy groups are to continue to play a vital role in live animal conservation, public awareness programs, and participate in cryoconservation of animal genetic resources. The foundations and trusts are to acquire funds for the continual support of all aspects of conservation activities.

A Canadian Steering Committee on Animal Germplasm was established to provide leadership in the development of strategies, programs, and priorities relevant to a sound national system. In addition, a Canadian Animal Germplasm Technical Experts Board capable of addressing scientific, legal, and ethical concerns was established. This committee established guidelines for conservation and set priorities in the following areas: identification of specific populations at risk, minimizing inbreeding, gene bank strategy, disease control for conservation, ownership, access and use of germplasm inventory information, and support of conservancy efforts (Lister and Ho, 1995).

A program was established at CFAR to catalogue Canada's animal genetic resources by developing a database that would be capable of networking with similar data banks in other organizations and countries to identify species or breeds at risk of extinction. This activity was to be coupled with gene banks participating in conserving live animals and poultry, semen, embryos, nucleic acids, and stem cells to preserve unique, diverse, and valuable germplasm.

An inventory of genetic resources maintained by breeders, researchers, veterinarians, hobbyists, institutions, cooperatives, breeding companies, universities, and provincial and federal governments is being established. A recently conducted survey of domestic animal genetic resources showed that Canada has 58 breeds of horses, 54 breeds of cattle, 41 breeds of sheep, 14 breeds of goats, and 14 breeds of swine (Shrestha, 1994). Many of these breeds, maintained in small populations, are subject to diseases and are at risk due to dramatic changes in the economics of the industry.

The U.S. Program

The current U.S. Program on National Genetic Resources was established according to Subtitle C of the 1990 Farm Bill (Public Law 101-624, 1990). An official U.S. National Animal Germplasm Program (NAGP) was initiated in 1990. The program is administered by the Secretary of the United States Department of Agriculture through the Agricultural Research Service (ARS). The program was established by a national committee composed of state and federal scientists, and included representatives from private industry and commodity groups and non-profit groups involved in conservation (National Animal Genetic Resources, 1990). The NAGP has been designed to maintain and enhance the collection, preservation, and dissemination of genetic material of importance to U.S. food and agriculture production. The primary species to be included in the program include beef and dairy cattle, swine, poultry, sheep, goats, and aquatic forms. The activities that the proposed program will include are identifying and evaluating promising and unique sources of germplasm, establishing repositories for germplasm preservation, forming a data bank for storage and retrieval of germplasm information, and fostering increase of stored germplasm (U.S. National Genetic Resources Program, 1992). The program will include research on characterization, selection, collection, processing, and culture and cryopreservation of sperm cells, embryos, and embryonic stem cells. The program of research also includes research on the mapping of the genome of each species. National Coordinating Committees, established for each species, will develop germplasm policies, identify program and research needs for each species, and implement the program.

Database A national animal germplasm database has been established and is part of the USDA-ARS Genetic Resources Information Network system (GRIN) at Beltsville, Maryland. The NAGP databases will contain descriptors and characteristics of the world's animal genetic resources; DNA sequences or mapping information on relevant animal genomes will also be available because the databases of NAGP will be relational. The long-range plans call for a national germplasm database system that includes all animal germplasm and genome information. At present, the database includes breeds of cattle (55), sheep (41), swine (17), horses (55), and asses (7). We have established a core database for poultry and will soon be able to include the data in the GRIN system. We plan to hold a workshop to initiate the program on aquaculture within the next year. The data currently contained in the GRIN database on animals are the result of a cooperative agreement between USDA-ARS and the American Livestock Breeds Conservancy. The database is on the World Wide Web and is publicly accessible through the Internet.

Central and Satellite Gene Bank Repositories A central gene bank repository is being developed at Beltsville, Maryland. The gene bank will meet the projected needs of the program for about 10 years and will permit the fulfillment of the initial objectives of the program. The objectives of the gene bank repository component of the program call for providing long-term storage space for sperm, embryos, oocytes, stem cells, cell lines, and DNA from designated genetics covering a wide range of domestic animal species and aquaculture. The ARS also has satellite repositories in Clay Center, Nebraska and at Greenport, New York to hold transient samples for disease research. The ARS also plans to establish a satellite repository for poultry at the East Lansing, Michigan laboratory. A national committee on cryopreservation will work with the species committees to establish the priorities for genetic material that will be maintained in the repositories.

Gene Mapping The development of a successful National Animal Germplasm Program to conserve, improve, and effectively utilize the genetic diversity of animals will require the development of gene maps for all species. Animal genome research is a worldwide effort. For each livestock species, genetic maps of polymorphic loci must be developed with sufficient resolution to permit the location, definition, and use of genes affecting economically important traits. Essential related objectives include the analysis of the fine structure of candidate genes and gene families, the definition of the genetic basis of quantitative trait loci to be used to implement marker-assisted selection, and development of new experimental technologies for utilization of genome information. Efficient use of carefully designed biological resources and development of a broadly based data management system are essential for this effort. In 1993, a 5-year National Animal Genome Program was initiated as part of the National Genetic Resources Program. The Cooperative State Research Education and Extension Service (CSREES) was designated to serve as leader of this effort. The project is supported by state funds, and scientists at land-grant universities were selected by CSREES to serve as coordinators for cattle, poultry, swine, and sheep.

Recent Progress on Gene Mapping Genetic linkage maps for cattle (Bishop et al., 1994 and S. M. Kappes et al., personal communication) (1,200 microsatellite-linked markers) and swine (Rohrer et al., 1994, 1996) (1,042 microsatellite-linked markers) have been developed by ARS scientists at the U.S. Meat Animal Research Center (USMARC), Clay Center, Nebraska. A sheep map containing over 500 microsatellite markers has been developed jointly by New Zealand and USDA scientists at USMARC (Crawford et al., 1995). Scientists at ARS and Michigan State University scientists at East Lansing, Michigan have developed a linkage map and database (chick base) for chickens (Levin et al., 1994). To date, over 250 microsatellite and 500 total markers have been mapped. The ARS scientists at USMARC have developed a relational interactive database that contains raw genotypic data on individual animals within pedigrees and displays the genetic linkage maps for cattle, sheep, and swine (Keele et al., 1994). The database developed for cattle, sheep, and swine is on the World Wide Web and is publicly accessible through the Internet.

Literature Cited

Bishop, M. D., S. M. Kappes, J. W. Keele, R. T. Stone, G. A. Hawkins, R. Fries., S. Solinas-Toldos, M. D. Grosz, S.L.F. Sunden, J. Yoo, L. V. Cundiff, and C. W. Beattie. 1994. A genetic linkage map for cattle. Genetics 136:619.

Chiperzak, J., and J.N.B. Shrestha. 1992. The conservation and maintenance of domestic animal genetic resources in Canada. In: L. Anderson and I. Bodo (Ed.) Genetic Conservation of Domestic Livestock. Vol. II. C.A.B. International, Wallingford, U.K.

Crawford, A. M., K. G. Dodds, C. A. Pierson, A. J. Ede, G. W. Montgomery, H. G. Garmonsway, A. E. Beattie, K. Davies, J. F. Maddox, S. M. Kappes, R. T. Stone, T. C. Nguyen, J. M. Penty, E. A. Lord, J. E. Broom, J. Buitkamp, W. Schwaiger, J. T. Epplen, P. Matthew, M. E. Matthews,

D. J. Hulme, K. J. Beh, R. A. McGraw, and C. W. Beattie. 1995. An autosomal genetic linkage map of the sheep genome. Genetics 140:703.

Keele, J. W., J. Wray, D. Behrens, G. A. Rohrer, S.L.F. Sunden, S. M. Kappes, M. D. Bishop, R. T. Stone, L. Alexander, and C. W. Beattie 1994. A conceptual model for animal genomic research. J. Comp. Biol. 1:63.

Levin, I., L. Santangelo, H. Cheng, L. Crittenden, and J. Dodgson. 1994. An autosomal genetic linkage map of the chicken. J. Hered. 85:79.

Lister, E. E., and S. K. Ho. 1995. Canadian farm animal germplasm conservation: A plan for the future. Research Branch, Agriculture Canada, Ottawa, ON, Canada.

National Animal Genetic Resources. 1990. ESCOP, ARS, and CSRS Committee Report. December 1990.

Public Law 101-624, Subtitle C—National Genetic Resources Program. United States Farm Bill. November 1990.

Rohrer, G. A., J. Alexander, Z. Hu, T.P.L. Smith, J. W. Keele, and C. W. Beattie. 1996. A comprehensive map of the porcine genome. Genome Res. (In press).

Rohrer, G. A., L. J. Alexander, J. W. Keele, T. P. Smith, and C. W. Beattie. 1994. A microsatellite linkage map of the porcine genome. Genetics 136:231.

Shrestha, J.N.B. 1994. Domestic animal genetic resources in Canada. Proc. of Third Global Conference on Conservation of Domestic Animal Genetic Resources (In press).

Shrestha, J.N.B. 1990. Proceedings of the first national workshop on conservation of animal germplasm. Research Branch, Agriculture Canada, Ottawa, ON, Canada.

U.S. National Genetic Resources Program. USDA-ARS Initial Report to Congress. April 1992.

Session II

Mapping Technologies

Development of Detailed Microsatellite Linkage Maps in Livestock

Craig W. Beattie

USDA, ARS, Roman L. Hruska U.S. Meat Animal Research Service, Clay Center, NE 68933-0166, U.S.A.

Abstract The development of comprehensive genetic linkage maps for livestock is contingent upon the integration of existing low-resolution linkage maps and improving overall coverage through physical assignment of additional polymorphic (framework) markers to individual chromosomes. Chromosomal segments currently bereft of physically assigned polymorphic markers limit our ability to define the extent of genome size and reliably estimate physical to genetic distance(s). A strategy to improve coverage that incorporates map integration, comparative linkage mapping between closely related livestock species, and directed assignment of polymorphic markers would aid in the timely completion of such maps to identify economically relevant loci. Such a strategy might include the physical assignment of previously linked, highly polymorphic, short repetitive elements such as microsatellites through the use of 1) Direct In situ Single Copy-Polymerase Chain Reaction (DISC-PCR), 2) iterative screening of large insert (cosmid, genomic lambda, bacterial (BAC) or yeast artificial chromosome (YAC) DNA libraries for previously linked microsatellites and subsequent assignment using fluorescent in situ hybridization (FISH), or 3) the use of microdissected, microcloned chromosomal regions to provide a large number of polymorphic loci within a specific chromosomal region. When placed in the context of closely related species, such as ruminant bovids, where significant functional and structural conservation for microsatellite sequences exists, framework and high-resolution comprehensive linkage maps for multiple species can be developed.

Key Words: Genetic Linkage, Microcloning, Microdissection, Mapping, Livestock

Introduction

Extensive genetic linkage maps for humans and mice crucial to identification of disease-related and other loci are now available (Weissenbach et al., 1992; Buetow et al., 1994; Dietrich et al., 1994; Murray et al., 1994). Each consists of thousands of polymorphic marker loci integrated into framework and comprehensive maps (Matise et al., 1994; Murray et al., 1994). Among the most useful of these markers are microsatellites (ms), abundant, repetitive elements that display multiple alleles due to variation in the number (read length) of repeat units following amplification by the polymerase chain reaction (PCR). Microsatellites are uniformly distributed throughout the genome of numerous species, including humans (Weber and May, 1989), and represent one form of a class of repetitive elements that includes minisatellites, large tandem repetitive elements (VNTRs). In contrast to minisatellites, ms arise from defined (single) loci and segregate as codominant markers in outbred populations, making their genotypic profile easier to interpret and transferable to a population outside of the one used to develop the polymorphism. These properties also set them apart from random amplified polymorphic DNAs (RAPDs) generated by mutation or insertion:deletion events that change the interval between primer pairs generated from arbitrary (sequence) oligonucleotides.

The process of map integration can take several forms. As with map construction, polymorphic markers from a variety of published and unpublished sources can be genotyped across several pedigrees that comprise single or multiple populations by one or more laboratories, errors identified, and a "consensus" marker order(s) developed. Alternatively, extant marker genotypes from several laboratories, again typed over a well-described population (e.g., the Centre d'Etude du Polymorphisme Humain [CEPH] families [Murray et al., 1994]), can be integrated using computer programs based on readily available algorithms that allow ordering of multiple loci (Ott, 1988; Green et al., 1989; Ott and Donis-Keller, 1994). The use of PCR-based anonymous markers such as ms as sequence tagged sites (STS), where specific primer pairs provide unique reference points on genetic and physical maps, allows direct transfer of mapping reagents between laboratories as information files containing primer sequences (Olson et al., 1989; Murray et al., 1994).

This strategy has worked well in developing the high resolution (< 5 cM or 5 megabase [Mb] intervals) human comprehensive map either by constructing "meiotic bins" (of marker loci) on each chromosome, where meiotic bins are initially defined by loci at relatively uniform intervals on a skeletal linkage map (Murray et al., 1994), or development of an expert system computer program to automate the extensive, repetitive human-computer interaction essential to resolving marker location and order irrespective of marker or DNA source (Matise et al., 1994). For each meiotic event within a (family) reference panel the most likely parenteral chromosomal assignment of each allele (phase) of each genotype is used to identify the bin in which each recombination event could be uniquely assigned (Murray et al., 1994). Final marker order within a bin or interval is determined by permutation of all possible orders of loci assigned to each bin and selected essentially on the basis of the one having the fewest number of recombination events within the previously identified subset of individuals showing a recombination event for that interval (Murray et al., 1994). Overall map distance, in centimorgans (cM; 1 cM $\cong 1 \times 10^6$ base pair [bp]), for all ordered loci on the map are determined using maximum likelihood methods (Green et al., 1989).

The novel algorithms in the expert system called "MultiMap" sequentially build a map using primary genotype data and continuously revalidate the map as marker loci are added in a non-random manner (Matise et al., 1994). This approach, termed mapping by locus content, defines the content of a marker locus as a function of several parameters that rank its potential usefulness in map construction (Matise et al., 1994). Importantly, each strategy seeks to limit the amount of map inflation due to errors arising from incorrect genotyping or data entry errors. Although overall map length can be reduced as errors are corrected, individual maps (irrespective of species) are commonly shorter due to differences in marker number (Murray et al., 1994). Framework markers or polymorphic markers ordered with high local statistical support, that are preferably physically assigned, provide the frame on which all other markers are localized, creating a comprehensive map (Matise et al., 1994; Murray et al., 1994). While this approach of physically assigning markers with high local support can establish or confirm an order determined by linkage analysis, local order may conflict as marker density increases. This can be partially overcome by physically assigning previously linked polymorphic loci to chromosomal regions that are relatively "marker poor," particularly in those species where marker density remains relatively sparse.

In mice, two high-resolution comprehensive linkage maps are currently available. One consists primarily of ms loci and is anchored to the mouse gene (linkage) map through simple sequence length polymorphisms (SSLPs; ms; Type II markers; O'Brien, 1991). This map was constructed by genotyping all SSLPs over a single (C57BL/6J-ob × CAST/Ei) F_2 interspecific cross and currently has over 4,000 markers ordered (Dietrich et al., 1994). The map provides markers useful for the analysis of inter- or intraspecific genetic crosses. A second gene map currently contains over 1,500 polymorphic markers and consists entirely of restriction fragment length polymorphisms (RFLPs)(Copeland et al., 1993) generated with restriction endonucleases that cleave the genome at sequence-specific sites. The sites appear near or within chromosomal regions coding for specific genes (Type I marker; O'Brien, 1991) when restricted genomic DNA is hybridized to a cDNA probe for the gene in question. Since genes within a region are identified, a comparison of the correspondence between the human and murine genomes can be made. The two complementary maps have been integrated by genotyping SSLP markers relatively well spaced across the mouse genome onto a subset of progeny used to develop the gene map (Copeland et al., 1993). Total length of the murine genome accounted for by the gene and SSLP maps is reasonably close, 1,350 cM vs 1,450 cM, respectively, but regions of recombinational suppression or transmission ratio distortion may preclude perfect alignment of the two maps (Copeland et al., 1993). Obviously, the generation and typing of virtually all SSLPs across a single cross by one group (Dietrich et al., 1994) simplified map construction, marker ordering, and the integration of the gene and SSLP maps.

In plants, RFLPs seem to provide the most useful approach to map construction because individual marker heterozygosity is relatively high in inter- and intraspecific crosses and the degree of conservation of coding sequences between species (e.g., rice, maize, and wheat) is high compared to similar crosses in mammals (Ahn et al., 1992, 1994; Van Deynze et al., 1994; Kurata et al., 1994). This has permitted the rapid exchange of markers and helped align species maps (Devos et al., 1994).

Integration and Comparison of Linkage Maps in Livestock

Several low-resolution (10 to 20 cM) linkage maps are available for cattle (Barendse et al., 1994; Bishop et al., 1994), swine (Ellegren et al., 1994b; Rohrer et al., 1994; Archibald et al., 1995), sheep (Crawford et al., 1995), and chicken (Burt et al., 1995). Although continuously updated, the total number of linked markers for all livestock species combined is currently less than 50% of those linked on either the human (Murray et al., 1994) or mouse (Copeland et al., 1993; Dietrich et al., 1994) maps. Linkage maps for each species have evolved in a manner analogous to that in humans and mice. In cattle, several groups placed a wide variety of type II markers across a single international bovine reference panel, the IBRP, and genotypes were subsequently integrated into a single linkage map (Barendse et al., 1994). A parallel effort by the USDA placed additional Type II markers as well as those typed across the IBRP on a large U.S. reference population, the USDA, ARS, U.S. Meat Animal Research Center (MARC) panel (Bishop et al., 1994). The three initial linkage maps developed for swine took a similar approach to that in cattle, each with its own set of SSLP markers, with limited genotyping of markers from any one group across an individual DNA reference panel (Rohrer et al., 1994). Initial attempts to integrate linkage groups assigned to single chromosomes by different groups using different

DNA reference panels have focused on chromosome 23 in cattle (Beever et al., 1996) and chromosomes 6 and 7 in swine (Paszek et al., 1995; Rohrer et al., unpublished data); chromosomes thought to potentially harbor economic trait loci (ETLs). Each effort has yielded a low-resolution "consensus" map for the respective chromosome, but coverage is neither complete nor the number of framework markers extensive.

An initial sheep linkage map was constructed using an integrated approach to mapping where Type I and II markers were placed across an international panel (international mapping flock, IMF), largely by one group, and markers were ordered by the same group (Crawford et al., 1995). A significant advantage to this effort was the conservation of ms usefulness across species; as approximately 25% of all ms markers linked in the sheep genome were derived from the bovine sequence (Crawford et al., 1995). A recent update of the sheep linkage map, which now comprises >500 linked markers, has increased the percentage of bovine ms sequences to approximately 75% of all linked markers (M. de Gortari, unpublished data). Marker integration on the updated sheep map incorporated the use of the international DNA panel and the MARC sheep map reference population to increase meioses, genotyping of all new and selected previously typed markers across all informative progeny, and on-line integration of all available genotypic information.

An initial comparison of the cattle and sheep linkage maps suggests that apart from regions of known chromosomal translocation and recombination, marker order within individual linkage groups seems to be well conserved, as is genetic interval. This clearly suggests that a comparative linkage map can be constructed for these closely related species in relatively short order. Should ms previously shown to be informative in sheep and cattle be typed across a subset of families used to construct a (RFLP) gene map for either species, the two could be integrated relatively easily. A comparative linkage map, particularly if integrated with a gene map, would facilitate defining economic trait loci (ETLs) in both species and the subsequent use of marker-assisted selection (MAS) for incorporating those traits. Because the economic cost(s) of developing resource families to identify ETLs are likely to be significant, the impact of such a map could also be significant. Extension of this approach to other closely related ruminants might also be possible; ms sequences developed from cattle, sheep, and goats have been reported to be informative across each species (Moore et al., 1992, 1994; Kemp et al., 1993; Vaiman et al., 1994).

The poultry genome presents its own set of challenges and opportunities. The current linkage map incorporates approximately 500 markers over a small (1.2×10^9 bp) but extremely complex genome (n = 39) that includes numerous micro chromosomes (Burt et al., 1995). A variety of Type II markers is genotyped across two reference populations and a significant number of ms are typed across both populations. Incorporating a comparative and directed (linkage) mapping strategy that includes additional economically important poultry species may hasten the overall development of MAS in these species and allow rapid characterization of those genetic disorders resulting from single gene mutations.

Directed Mapping Strategies

Although remarkable progress has recently been made in genetic linkage mapping in cattle, swine, and sheep (Barendse et al., 1994; Bishop et al., 1994; Rohrer et al., 1994; Ellegren et al., 1994b; Archibald et al., 1995; Crawford et al., 1995), economic reality suggests that the current mapping effort within each species will not link more than 25 to 30% of the approximately 6,000 markers currently linked on either the human or mouse maps in the near future. Relatively few new ms markers are being developed from small insert libraries of randomly cloned cattle, swine, or sheep DNA. Construction of consensus linkage maps should improve resolution within individual linkage groups. However, because only indirect estimates of genomic size are available for any of the major livestock species and the informative telomeric and centromeric markers necessary for a more rigorous analysis are not yet available, consensus maps may not immediately improve overall genomic coverage (Beattie, 1994).

Integration of informative, physically anchored loci into a comprehensive linkage map significantly improves estimates of chromosomal coverage and resolution by improving linkage group orientation, verifying sequential order, and detecting differential recombination rates within the genome. Unfortunately, marker-deficient regions (gaps) in the current linkage maps for cattle, swine, and sheep and paucity of polymorphic loci physically anchored to individual chromosomes have slowed development of high-resolution linkage maps for these species. A strategy to overcome this limitation might include the physical assignment of previously linked, highly polymorphic, short repetitive elements such as ms through the use of 1) direct in situ single copy polymerase chain reaction (DISC-PCR), 2) iterative screening of large insert (cosmid, genomic lambda, bacterial [BAC], or yeast artificial chromosome [YAC]) DNA libraries for previously linked ms and subsequent assignment using fluorescent in situ hybridization (FISH), or 3) the use of microdissected, microcloned chromosomal regions to provide a large number of polymorphic loci within a specific chromosomal region.

The importance of increasing chromosomal coverage and reducing interval length is found in even a cursory examination of the human linkage map. With > 5,800 mapped loci, and markers at an average density of .7 cM, the map still contains large intervals ranging from 7.5 to 25 cM and a lack of physically assigned informative loci at chromosomal centromeres and telomeres (Murray et al., 1994).

One way to estimate genome size and reveal (local) differential recombination rates is to compare the relationship between physical and genetic distance using anchored markers (Murray et al., 1994; Rohrer et al., 1994). This requires that linked polymorphic markers be physically anchored by in situ methods at relatively even intervals across an entire genome to ensure maximal coverage. At present, this is not the case for any livestock linkage map. Framework markers reduce conflicts between linkage data and physical map points within species (Murray et al., 1994). For example, Ellegren et al. (1994a) recently reported that the rate of genetic recombination in relation to physical DNA content is 1 cM per 2 to 4 Mb of physical distance over at least 80% of swine chromosome 1. Many human chromosomes show a much higher recombination rate than anticipated near their telomeres (Henke et al., 1993); conversely, loci near centromeres separated by 5 cM on the genetic map are sometimes separated by 15 to 25% of the chromosome's length due to a low recombination rate (Murray et al., 1994). Genetic/physical correlations are not well defined throughout any livestock genome, so the perceived distance a given point lies from a gene of interest can be erroneous due to the marked variation in recombination rate per unit of physical distance (Murray et al., 1994).

Direct In situ Single Copy PCR

Direct in situ single copy-PCR (DISC-PCR) offers the opportunity to directly assign ms to a location on a metaphase chromosomal spread in karyotypes such as the pig where chromosomal architecture allows discrimination of marker location in the absence of concomitant R, Q, or G banding. While the similarity in chromosomal size and acrocentric nature of the bovine autosomal karyotype currently does not lend itself to DISC-PCR, the method can be extremely useful in swine for anchoring previously linked ms that flank a large marker interval, or define the end of a linkage group, particularly if a cosmid or genomic lambda clone containing the ms is not available to assign the ms by fluorescence in situ hybridization (FISH) (Troyer et al., 1994a,b, 1995). Since coding sequences can also be assigned in a similar manner (Troyer et al., 1994b) gene location is placed into the context of a linkage or comparative map.

Iterative Screening

High stringency, iterative, PCR screening of genomic lambda, cosmid, BAC, or YAC libraries using primer pairs that flank a previously linked ms can also be used to obtain sufficient numbers of large insert DNA clones suitable for assignment by FISH such that each chromosome, irrespective of species, is covered by framework markers at relatively even intervals (T. P. Smith, unpublished data). This strategy allows a rapid estimation of actual chromosomal coverage by a linkage group, particularly when incorporated with DISC-PCR. Repeated screening of closely linked STS also offers the possibility of constructing contigs or an ordered group of continuous pieces of DNA that would allow mapping of a locus accounting for a major portion of genetic variation for a trait.

Microdissection

Finally, physical dissection of metaphase chromosomes seems to be the most straightforward approach to directly reach a defined chromosomal region for detailed molecular genetic analysis (Johnson, 1990), and thus is likely to prove a powerful method in a directed strategy of assigning polymorphic markers to specific regions within a livestock genome (Schmutz et al., 1994). Removal of targeted regions of interest from metaphase chromosomes using ultra-fine glass needles and micromanipulators (Kao, 1987) has found wide use in human (Hozier and Davis, 1992) and mouse (Bahary et al., 1993; Zhang et al., 1994) genome projects but has not seen wide use in the livestock genome effort to date (Ponce de León and Robl, 1992; Schmutz et al., 1994). When combined with the use of degenerate oligonucleotide primers and a low annealing temperature, the procedure is species-independent and is a simple and rapid method for representative amplification of regionally dissected DNA from multiple evenly dispersed sites (Viersbach et al., 1994). It obviates the use of fluorescent cytometric photometry to isolate whole chromosomes, which is impractical in the bovid with its acrocentric karyotype, reduces contamination from other chromosomes (Ellegren and Basu, 1995), and allows any recognizable position in the genome to be cloned and screened for informative Type II markers. DNA fragments and libraries obtained from microdissected chromosomes can also be used to develop high-resolution linkage maps of genomes by providing a large number of ms for a defined area and a set of probes that, in turn, can be used to isolate secondary cosmid or YAC clones (Djabali et al., 1991). The latter, in turn, can be used to generate additional informative markers, such as ms (Seki et al., 1994) and fluorescent probes to "paint" metaphase chromosomes and identify regions of similarity even in distantly related species (Lyons et al., 1994; Scherthan et al., 1994; Rettenberger et al., 1995).

Discussion

The continued development of genetic maps in major livestock species depends upon a realistic approach to map construction and integration. Scarce resources dictate the sharing of laboratory reagents, perhaps as data files from the extant livestock databases in Australia, Europe, and the United States containing all information relevant to a STS. This is particularly true as we attempt to increase marker resolution, since economic reality suggests the number of STS generated from random cloned, small insert DNA libraries is not likely to increase significantly beyond 1,500 markers. Consensus maps of individual chromosomes, while useful for map integration in a limited sense, are relegated to a single chromosome and offer only a discrete number of framework markers and relatively slow publication rate. Future collaborative efforts at map integration might consider genotyping a variety of polymorphic markers, irrespective of source, that are distributed relatively evenly over the whole genome on a single DNA reference panel of an individual species. Should the physical location of a significant number of these polymorphic loci be known, development of truly comprehensive maps would be hastened.

Conclusion

A strategy that incorporates directed and comparative linkage mapping will help merge the genetic and physical linkage maps of closely related livestock species into relatively high-resolution comprehensive maps. Directed addition of highly polymorphic microsatellite markers will also improve marker coverage and in turn allow a more robust estimate of genome length. Taken together, the strategy provides a basis for the continued development and integration of the linkage maps essential to a systematic whole genome approach to dissection of single and multigenic economically important traits in livestock.

Acknowledgments

The author is indebted to S. Kappes, G. Rohrer, and J. Keele for helpful discussion and timely criticism of the manuscript; T. Smith, D. Troyer, F. A. Ponce de León, M. de Gortari, A. Crawford and K. Dodds for the use of unpublished observations; and S. Kluver for manuscript preparation.

Literature Cited

Ahn, S., J. A. Anderson, M. E. Sorrells, and S. D. Tanksley. 1994. Homeologous relationships of rice, wheat, and maize chromosomes. Mol. Gen. Genet. 241:483.

Ahn, S., C. N. Bollich, and S. D. Tanksley. 1992. RFLP tagging of a gene for aroma in rice. Theor. Appl. Genet. 84:825.

Archibald, A. L., C. S. Haley, J. F. Brown, S. Couperwhite, H. A. McQueen, D. Nicholson, W. Coppieters, A. Van de Weghe, A. Stratil, A. K. Winterø, M. Fredholm, N. J. Larsen, V. H. Nielsen, D. Milan, N. Woloszyn, A. Robic, M. Dalens, J. Riquet, J. Gellin, J.-C. Caritez, G. Burgaud, L. Ollivier, J.-P. Bidanel, M. Vaiman, C. Renard, H. Geldermann, R. Davoli, D. Ruyter, E.J.M. Verstege, M.A.M. Groenen, W. Davies, B. Heyheim, A. Keiserud, L. Andersson, H. Ellegren, M. Johansson, L. Marklund, J. R. Miller, D. V. Anderson Dear, E. Signer, A. J. Jeffreys, C. Moran, P. Le Tissier, Muladno, M. F. Rothschild, C. K. Tuggle, D. Vaske, J. Helm, H.-C. Liu, A. Rahman, T.-P. Yu, R. G. Larson, and C. B. Schmitz. 1995. The PiGMaP consortium linkage map of the pig (Sus scrofa). Mamm. Genome 6:157.

Bahary, N., D. A. Siegel, J. Walsh, Y. Zhang, L. Leopold, R. Leibel, R. Roenca, and J. M. Friedman. 1993. Microdissection of proximal mouse chromosome 6: Identification of RFLPs tightly linked to the ob mutation. Mamm. Genome 4:511.

Barendse, W., S. M. Armitage, L. M. Kossarek, A. Shalom, B. W. Kirkpatrick, A. M. Ryan, D. Clayton, L. Li, H. L. Niebergs, N. Zhang, W. M. Grosse, J. Weiss, P. Creighton, F. McCarthy, M. Ron, A. J. Teale, R. Fries, R. A. McGraw, S. S. Moore, M. Georges, M. Soller, J. E. Womack, and D.J.S. Hetzel. 1994. A genetic linkage map of the bovine genome. Nature Genet. 6:227.

Beattie, C. W. 1994. Livestock genome maps. Trends Genet. 10:334.

Beever, J. E., H. A. Lewin, W. Barendse, L. Andersson, S. M. Armitage, C. W. Beattie, B. M. Burns, S. K. Davis, S. M. Kappes, B. W. Kirkpatrick, R. Z. Ma, R. A. McGraw, R. T. Stone, and J. F. Taylor. 1996. Report of the first workshop on the genetic map of bovine chromosome 23. Anim. Genet. (In press).

Bishop, M. D., S. M. Kappes, J. W. Keele, R. T. Stone, G. A. Hawkins, R. Fries, S. Solinas-Toldo, M. D. Grosz, S.L.F. Sunden, J. Yoo, L. V. Cundiff, and C. W. Beattie. 1994. A genetic linkage map for cattle. Genetics 136:619.

Buetow, K. H., J. L. Weber, S. Ludwigsen, T. Scherpbier-Heddema, G. M. Duyk, V. C. Sheffield, Z. Wang, and J. C. Murray. 1994. Integrated human genome-wide maps constructed using the CEPH reference panel. Nature Genet. 6:388.

Burt, D. W., N. Bumstead, J. J. Bitgood, F. A. Ponce de León, and L. B. Crittenden. 1995. Chicken genome mapping: A new era in avian genetics. Trends Genet. 11:190.

Copeland, N. G., N. A. Jenkins, D. J. Gilbert, J. T. Eppig, L. J. Maltais, J. C. Miller, W. F. Dietrich, A. Weaver, S. E. Lincoln, R. G. Steen, L. D. Stein, J. H. Nadeau, and E. S. Lander. 1993. A genetic linkage map of the mouse: Current applications and future prospects. Science (Wash DC) 262:57.

Crawford, A. M., K. G. Dodds, C. A. Pierson, A. J. Ede, G. W. Montgomery, H. G. Garmonsway, A. E. Beattie, K. Davies, J. F. Maddox, S. M. Kappes, R. T. Stone, T. C. Nguyen, J. M. Penty, E. A. Lord, J. E. Broom, J. Buitkamp, W. Schwaiger, J. T. Epplen, P. Matthew, M. E. Matthews, D. J. Hulme, K. J. Beh, R. A. McGraw, and C. W. Beattie. 1995. An autosomal genetic linkage map of the sheep genome. Genetics 140:703.

Devos, K. M., S. Chao, Q. Y. Li, M. C. Simonetti, and M. D. Gale. 1994. Relationship between chromosome 9 of maize and wheat homologous group 7 chromosomes. Genetics 138:1287.

Dietrich, W. F., J. C. Miller, R. G. Steen, M. Merchant, D. Damron, R. Nahf, A. Gross, D. C. Joyce, M. Wessel, R. D. Dredge, A. Marquis, L. D. Stein, N. Goodman, D. C. Page, and E. S.

Lander. 1994. A genetic map of the mouse with 4,006 simple sequence length polymorphisms. Nature Genet. 7:220.
Djabali, M., C. Nguyen, I. Biunno, B. A. Oostra, M. G. Mattei, J. E. Ikeda, and B. R. Jordan. 1991. Laser microdissection of the fragile X region: Identification of cosmid clones and of conserved sequences in this region. Genomics 10:1053.
Ellegren, H., and T. Basu. 1995. Filling the gaps in the porcine linkage map: Isolation of microsatellites from chromosome 18 using flow sorting and SINE-PCR. Cytogenet. Cell Genet. 71:370.
Ellegren, H., B. P. Chowdhary, M. Fredholm, B. Hoyheim, M. Johansson, P. B. Nielsen, P. D. Thomsen, and L. Andersson. 1994a. A physically anchored linkage map of pig chromosome 1 uncovers sex and position-specific recombination rates. Genomics 24:342.
Ellegren, H., B. P Chowdhary, M. Johansson, L. Marklund, M. Fredholm, I. Gustavsson, and L. Andersson. 1994b. A primary linkage map of the porcine genome reveals a low rate of genetic recombination. Genetics 137:1089.
Green, P., K. Falls, and S. Crooks. 1989. Documentation of CRI-MAP, Version 2.4. Washington Univ. School of Medicine, St. Louis.
Henke, A., C. Fischer, and G. A. Rappold. 1993. Genetic map of the human pseudoautosomal region reveals a high rate of recombination in female meiosis at the Xp telomere. Genomics 18:478.
Hozier, J. C., and L. M. Davis. 1992. Cytogenetic approaches to genome mapping. Anal. Biochem. 200:205.
Johnson, D. H. 1990. Molecular cloning of DNA from specific chromosomal regions by microdissection and sequence-independent amplification of DNA. Genomics 6:243.
Kao, F. 1987. Chromosome microdissection and microcloning in human molecular genetics. Somatic Cell Mol. Genet. 13:375.
Kemp, S. J., L. Brezinsky, and A. J. Teale. 1993. A panel of bovine, ovine, and caprine polymorphic microsatellites. Anim. Genet. 24:363.
Kurata, N., G. Moore, Y. Nagumara, T. Foote, M. Yano, Y. Minobe, and M. Gale. 1994. Conservation of genome structure between rice and wheat. Bio/Technology 12:276.
Lyons, L. A., M. M. Raymond, and S. J. O'Brien. 1994. Comparative genomics: The next generation. Anim. Biotechnol. 5:103.
Matise, T. C., M. Perlin, and A. Chakravarti. 1994. Automated construction of genetic linkage maps using an expert system (MultiMap): A human genome linkage map. Nature Genet. 6:381.
Moore, S. S., W. Barendse, K. T. Berger, S. M. Armitage, and D.J.S. Hetzel. 1992. Bovine and ovine DNA microsatellites from the EMBL and Genbank databases. Anim. Genet. 23:463.
Moore, S. S., K. Byrne, K. T. Berger, W. Barendse, F. McCarthy, J. E. Womack, and D.J.S. Hetzel. 1994. Characterization of 65 bovine microsatellites. Mamm. Genome 5:84.
Murray, J. C., K. H. Buetow, J. L. Weber, S. Luwigsen, T. Scherpbier-Heddema, F. Manion, J. Quillen, V. C. Sheffield, S. Sunden, G. M. Duyk, J. Weissenbach, G. Gyapay, C. Dib, J. Morrissette, G. M. Lathrop, A. Vignal, R. White, N. Matsunami, S. Gerken, R. Melis, H. Albertsen, R. Plaetke, S. Odelberg, D. Ward, J. Dausset, D. Cohen, and H. Cann. 1994. A comprehensive human linkage map with centimorgan density. Science (Wash DC) 265:2049.
O'Brien, S. J. 1991. Mammalian genome mapping. Curr. Opin. Genet. Dev. 1:105.
Olson, M. V., L. Hood, C. Cantor, and D. Botstein. 1989. A common language for physical mapping of the human genome. Science (Wash DC) 245:1434.
Ott, J. 1988. Analysis of Human Genetic Linkage. The John Hopkins Univ. Press, Baltimore.
Ott, J., and H. Donis-Keller. 1994. Statistical methods in genetic mapping. Genomics 22:496.
Paszek, A. A., L. B. Schook, C. F. Louis, J. R. Mickelson, G. H. Flickinger, J. Murtaugh, J. R. Mendiola, M. A. Janzen, C. W. Beattie, G. A. Rohrer, L. J. Alexander, L. Andersson, H. Ellegren, M. Johansson, P. Mariani, L. Marklund, B. Høyheim, W. Davies, M. Fredholm, A. L. Archibald, and C. S. Haley. 1995. First international workshop on porcine chromosome 6. Anim. Genet. 26:377.
Ponce de León, F. A., and J. Robl. 1992. Microisolation and microcloning of the bovine X chromosome. Proc. ISAG Conf., Interlaken, Switzerland. p 62. Univ. of Berne, Berne, Switzerland.
Rettenberger, G., C. Klett, U. Zechner, J. Kunz, W. Vogel, and H. Hameister. 1995. Visualization of the conservation of synteny between humans and pigs by heterologous chromosomal painting. Genomics 26:372.
Rohrer, G. A., L. J. Alexander, J. W. Keele, T. P. Smith, and C. W. Beattie. 1994. A microsatellite

linkage map of the porcine genome. Genetics 136:231.

Scherthan, H., T. Cremer, U. Arnason, H. U. Weier, A. Limadefaria, and L. Fronicke. 1994. Comparative chromosome painting discloses homologous segments in distantly related mammals. Nature Genet. 4:342.

Schmutz, S. M., T. G. Berryere, J. S. Moker, T. D. Thue, and D. C. Winkelman. 1994. Gene mapping from a bovine 1;29 DNA library prepared with chromosome microdissection. Mamm. Genome 5:138.

Seki, N., M. Yamauchi, T. Saito, Y. Harada, and T. Hori. 1994. Sequence tagged sites of microclones obtained by microdissection of a human chromosomal region 11q23 and isolation of yeast artificial chromosomes. Jpn. J. Hum. Genet. 39:249.

Troyer, D. L., D. W. Goad, H. Xie, G. A. Rohrer, L. J. Alexander, and C. W. Beattie. 1994a. Use of direct in situ single copy (DISC) PCR to physically map 5 porcine microsatellites. Cytogenet. Cell Genet. 67:199.

Troyer, D. L., H. Xie, and D. W. Goad. 1995. Use of DISC-PCR to map a porcine microsatellite. Anim. Biotechnol. 6:51.

Troyer, D. L., H. Xie, D. W. Goad, and D. Z. Skinner. 1994b. Use of a new technique to map the porcine α interferon gene to chromosome 1. Mamm. Genome 5:112.

Vaiman, D., M. Imam-Ghali, K. Moazami-Goudarzi, G. Guérin, C. Grohs, H. Levéziel, and N. Saïdi-Mehtar. 1994. Conservation of a syntenic group of microsatellite loci between cattle and sheep. Mamm. Genome 5:310.

Van Deynze, A. E., J. Dubcovsky, K. S. Gill, J. C. Nelson, M. E. Sorrells, J. Dvorak, B. S. Gill, E. S. Lagudah, S. R. McCouch, and R. Appels 1994. Molecular-genetic maps for group 1 chromosomes of Triticeae species and their relation to chromosomes in rice and oats. Genome 38:45.

Viersbach, R., G. Schwanitz, and M. M. Nothen. 1994. Delineation of marker chromosome by reverse chromosome painting using only a small number of DOP-PCR amplified microdissected chromosomes. Hum. Genet. 93:663.

Weber, J. L. and P. E. May. 1989. Abundant class of human DNA polymorphisms which can be typed using the polymerase chain reaction. Am. J. Hum. Genet. 44:388.

Weissenbach, J., G. Gyapay, C. Dib, A. Vignal, J. Morissette, P. Millasseau, G. Vaysseix, and M. Lathrop. 1992. A second-generation linkage map of the human genome. Nature (Lond.) 359:794.

Zhang, Y. Y., R. Proenca, M. Maffei, M. Barone, L. Leopold, and J. M. Friedman. 1994. Positional cloning of the mouse obese gene and its human homologue. Nature (Lond.) 372:425.

Chromosome Specific Libraries and Their Use in Swine Genome Mapping

J. K. Lunney[*,1], D. R. Grimm[*], R. Holley-Shanks[*], T. Goldman[*], F. A. Ponce de Leon[†], S. Ambady[†], J. Robl[†], J. Mendiola[‡], L. Buoen[‡], and C. Louis[‡]

[*]Immunology and Disease Resistance Laboratory, ARS, USDA, Beltsville, MD 20705, U.S.A.; [†]Department of Veterinary and Animal Science, Paige Laboratory, University of Massachusetts, Amherst, MA 01003-6410, U.S.A.; and [‡]Department of Veterinary Pathobiology, University of Minnesota, St. Paul, MN 55108, U.S.A.

Abstract Swine chromosome-specific libraries have been prepared by our laboratories using both flow cytometric and microdissection techniques. Such libraries have been successfully used in swine and other species to expand the number of genomic markers available, particularly for chromosomes, or areas of chromosomes, that are linked to important traits. Our target has been to produce more swine chromosome 6 microsatellite markers so that we can identify unique loci near the porcine stress syndrome defect that determine favorable swine carcass traits. For flow cytometry, chromosomes were prepared from activated lymphoblasts and sorted based on their staining with Hoechst 33258 and chromomycin A3. Libraries were produced from size-selected DNA fragments amplified from fewer than 1,000 sorted chromosomes. Chromosomes have also been isolated by microdissection techniques and then PCR-amplified using specifically designed primers. Purity of these populations was verified by fluorescent in situ hybridization (FISH) analyses of painting probes produced from the isolated products and ultimately by the reliability of the isolated chromosomes to result in genomic markers for the specified chromosome. The swine libraries were screened with dinucleotide repeat probes and, after sequencing, PCR primers designed for each marker, with the result that many chromosome 6 microsatellite-specific markers have been identified and mapped. The development of chromosome-enriched libraries has enabled many new markers to be isolated that can be used to identify the actual genes that influence production traits.

Key Words: Microdissected Chromosomes, Flow-Sorted Chromosomes, Chromosome-Enriched Libraries, Swine Gene Mapping, Porcine Stress Syndrome, Microsatellites

Introduction

The last decade has seen dramatic advances in the definition of the genome maps in many species. The Genethon human genome map has a 2.9 cM average (Gyapay et al., 1994) and the Cooperative Human Linkage Center map has a .7 cM average marker spacing (Murray et al., 1994). The mouse genome map reported by Dietrich et al. (1994) has an average marker spacing of .35 cM. Developments in livestock genome maps have been no less impressive, as discussed in other articles in this volume. The First Pig Gene Mapping Workshop in 1992 reported only 120 markers in 25 linkage groups (Andersson et al., 1993). Two years later, four laboratories have

---

[1]Author to whom correspondence should be addressed.

reported significant developments. As discussed in this volume by Beattie et al. (1995), Rohrer et al. (1994) mapped 376 microsatellites and 7 restriction fragment length polymorphisms (RFLPs) to 24 linkage groups; linkage groups were assigned definitively to 13 autosomes and the X chromosome. Ellegren et al. (1994) developed a comprehensive linkage map with 68 microsatellite markers and 60 anchored reference markers with an average spacing of 11 cM. Archibald et al. (1995) and Yerle et al. (1995) reported on the progress of the international PiGMAP effort with linkage and physical mapping data on 240 genes and microsatellites. Results in cattle and chickens have been equally impressive, as reported in other articles in this volume (Beattie et al., 1996; Lamont et al., 1996; Lewin et al., 1996).

The increased number of genetic markers on the linkage maps of these species has created greater opportunities for studies searching for genes of interest. Determining the linkage of genetic markers with genes that affect quantitative traits in livestock breeding populations is one area that has benefited from the improved genetic maps. Genes associated with production traits are excellent candidates for linkage relationships with quantitative trait loci (QTL) (Soller, 1991, 1994), as discussed in many articles in this volume. The linkage of genetic markers with QTL has actually been studied for over 40 years, but several recent advances have been documented. In swine, Rothschild et al. (1995) recently identified alleles of the estrogen receptor gene that positively influence litter size. Andersson et al. (1994a,b, 1996) have used gene mapping strategies with a limited number of pig genetic markers to identify multiple loci on pig chromosome 4 where allelic variance contributes to differences in fat deposition and growth.

Importance of Swine Chromosome 6

Our research has been aimed at a specific chromosome, swine chromosome 6. Numerous researchers have reported the relationship between porcine stress syndrome (PSS), which maps to this chromosome, and swine carcass traits. In PSS, certain breeds of pigs exhibit an increased likelihood of sudden death in response to stressors such as transport, restraint, or fighting. Inheritance of PSS is known to affect a range of important production traits; animals exhibit superior lean meat content, ham content, loin eye muscle area, and decreased backfat depth compared to other breeds (Webb et al., 1982; Jensen and Barton-Gade, 1985; Zhang et al., 1992). A number of studies have indicated that while PSS animals grow slightly more slowly than normal pigs, they have superior feed conversion (Carlson et al., 1980) and rank favorably on indices commonly used to compare breeding stock. Furthermore, we and others have shown that the heterozygous individual is intermediate to the PSS and normal animals for a number of production and carcass traits (Vecchionacce, 1981; Webb et al., 1985; Rempel et al., 1995), resulting in improved carcass yield with no stress losses, or losses due to the pale, soft, exudative (PSE) meat characteristic of PSS pigs (Harrison, 1972; Eikelenboom and Minkema, 1974; Jensen and Barton-Gade, 1985).

An essentially identical syndrome in humans has been termed malignant hyperthermia. Some of these families, and all cases of PSS identified to date, have been shown to result from a mutation in a single gene (MacLennan et al., 1990; Fujii et al., 1991; Gillard et al., 1991; Otsu et al., 1991; MacLennan and Phillips, 1992; Mickelson et al., 1992). Each results from a single base pair mutation, Arg^{615} to

Cys ^{615} in the 5039 amino acid skeletal muscle isoform of the sarcoplasmic reticulum calcium release channel (ryr1) that is the receptor for the plant alkaloid ryanodine. In the pig, the *ryr1* gene has been physically mapped to chromosome 6q11→q21 by in situ hybridization (Davies et al., 1988; Harbitz et al., 1990; Yang et al., 1992). The genomic organization of the porcine and human *ryr1* locus has been probed using contigs of cosmids, yeast artificial chromosomes (YACs) (Burke et al., 1987), and bacterial artificial chromosomes (BACs) (Leeb et al., 1993; Rouquier et al., 1993, 1994).

The nature of the gene(s) responsible for the positive carcass quality traits of PSS swine remains to be identified. One proposal is that the various effects of the gene responsible for PSS on swine performance arise from two or more different genes very close together on swine chromosome 6, one coding for the gene responsible for PSS and other(s) giving rise to favorable carcass traits. Moreover, chromosome 6 is syntenic with a major portion of human chromosome 19 that is known to be particularly gene-rich (GC-rich and high abundance of CpG islands; Larsen et al., 1992), so it likely contains a number of additional genes that affect important carcass and production traits in swine, as already indicated by Clamp et al. (1993a,b). We and others are interested in the development of more markers in this region of the chromosome to identify such performance genes (QTL) that may be linked to the *ryr1* gene. However, as indicated by Andersson et al. (1995), there are undoubtedly other genes that also contribute to muscle and adipose tissue growth and development that are neither linked to the PSS allele of the *ryr1* nor are located on chromosome 6, and it will require additional markers to identify those genes.

The mapping of the PSS gene in swine to chromosome 6 caused an increased amount of interest in developing the physical and linkage maps of this chromosome. To aid this effort, the first swine chromosome 6 workshop was organized. Ten markers were exchanged by four independent laboratories and mapped on 583 animals derived from four reference populations; this resulted in the assignment and ordering of 52 markers onto this important chromosome (Figure 1) (Paszek et al., 1995). The framework map consisted of 10 highly polymorphic markers with an average marker distance of 13.5 cM; the comprehensive map included 52 markers with an average distance of 3.2 cM. While this would indicate that most markers on chromosome 6 are less than 5 cM apart, there remain large (26.8 cM in the sex-averaged map) unmapped intervals of the chromosome 6p arm (Paszek et al., 1995). Thus, targeted approaches to prepare more markers are still necessary.

Chromosome-Enriched Library Production Using Flow-Sorted Chromosomes

Chromosome flow-sorting and microdissection techniques have been used to isolate individual chromosomes to prepare whole chromosome paints for fluorescent in situ hybridization (FISH) analyses as well as chromosome-enriched libraries for marker development for individual chromosomes (Telenius et al., 1992; Kao, 1993; VanDevanter et al., 1994) Chromosome-enriched DNA libraries have been constructed for humans, rodents, and swine using a range of strategies (Disteche et al., 1982; Van Dilla et al., 1986; Fuscoe et al., 1989; Baron et al., 1990; Miller et al., 1992; Davies et al., 1994; Frengen et al., 1994; Zilstra et al., 1994). In the pig, every chromosome has been sorted and chromosome identification confirmed by several independent techniques and by FISH using non-specific polymerase chain reaction (PCR) amplification of sorted chromosomal material (Langford et al., 1993; Yerle et al., 1993). The assignment of all the chromosomes to their exact peak has been verified for each chromosome by DNA content, by comparative flow karyotyping of chromosomes from male and female pigs, by flow karyotypes of chromosomes from pigs with known translocations, and by PCR-based mapping of specific genes (Schmitz et

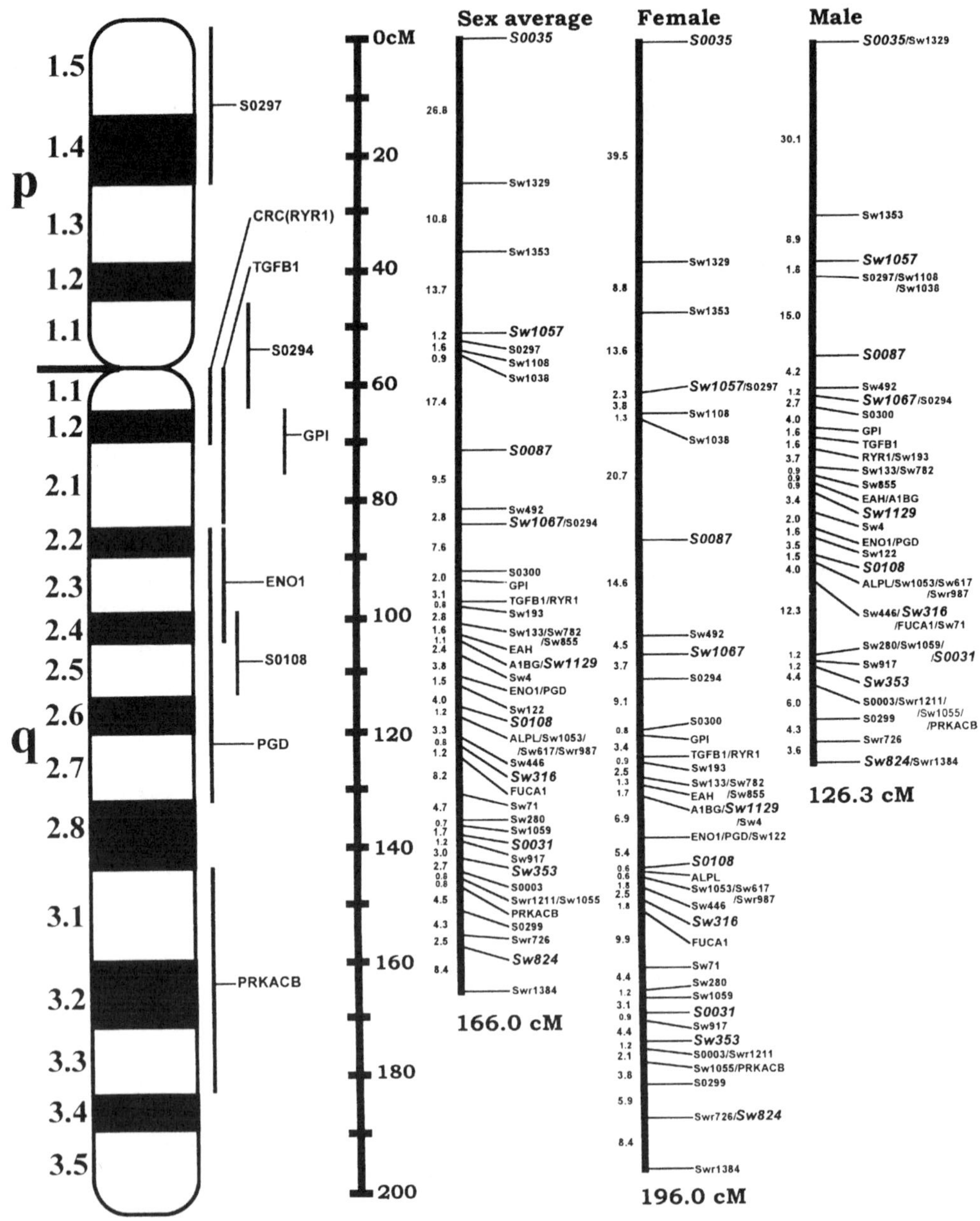

Figure 1. Comprehensive cytogenetic and linkage map of swine chromosome 6. Physical locations of markers are based on literature citations. Markers given in larger, italicized font are selected framework map markers (reprinted with permission from Paszek et al., 1995).

Table 1. Summary of libraries screened[a]

Library	Clones screened	Positive clones[b]	Clones sequenced	Clones with true repeats[c]	Primers synthesized[d]	Genotypes completed	Linkage completed
Plasmid	5,000	6	6	6	4	2	2
Lambda	95,000	89	89	54	28	22	9
Scraped	60,000	65	30	25	22	15	8
Total	160,000	160	125	85	54	39	19

[a]A summary of results as of October 1995, from libraries produced from flow-sorted chromosomes (Plasmid and Lambda) and microdissected chromosomes (Scraped).

[b]Clones that were positive for dinucleotide repeats following two rounds of hybridization to labeled poly-d(CA)$_n$ probe.

[c]Clones that contained uninterrupted repeats that had enough dinucleotides (>8) repeats for mapping work.

[d]Primers were synthesized because: 1) repeat containing clones were unique when compared to US MARC database; 2) repeat located in a positon that allowed for flanking primers to be designed.

al., 1992a,b). Results from these studies have confirmed that non-specific PCR primers (degenerate oligonucleotide-primed PCR [DOP-PCR, Telenius et al., 1992] or PHARM-PCR [Yerle et al., 1993]) on DNA from 300 to 500 sorted chromosomes (<5 minutes sorting time) yielded sufficient material to produce chromosome-specific paints when hybridization was blocked with repetitive pig DNA. The biotin-dUTP-labeled probes also confirmed that the full length of each chromosome was amplified, thus supporting such non-specific techniques for producing random PCR amplified single chromosome libraries.

We have used these approaches to prepare chromosome-specific paints and libraries for swine chromosome 6. Unlike the National Laboratory Gene Library Project that sponsored the development of chromosome-specific libraries in the 1980s, the object of our efforts is to produce microsatellite markers, not RFLP markers (Fuscoe et al., 1989). Therefore, their approach of preparing cosmid libraries using flow-sorted chromosomes isolated from human x hamster somatic cell hybrids has been greatly modified. Similar approaches for mouse chromosome 19 necessitated the collection of 700,000 chromosomes to produce a small (3 x 10^4 clones) library of which only 5/9 clones tested were chromosome-specific (Baron et al., 1990). We prepared chromosomes from metaphase spreads of blastogenic porcine peripheral blood lymphocytes using the polyamine isolation protocol (Cram et al., 1990). The chromosomes were stained for bivariate flow cytometry by incubating in chromomycin A3, followed by Hoechst 33258.

Dual color flow cytometry was performed on the Coulter Epics V FCM in collaboration with Glenn Welch and Larry Johnson (Germplasm and Gamete Physiology Laboratory, USDA, ARS, Beltsville, MD). The chromosomes are sorted according to the flow cytometric karyotype that results from the characteristic bivariate fluorescence pattern of each chromosome. As described above, all the chromosomes have been successfully sorted and isolated. Our resulting bivariate flow karyotype is shown in Figure 2.

For our flow-sorted chromosome 6 libraries, limited numbers of swine chromosome 6 were isolated by flow cytometry. DNA was extracted from the flow-sorted chromosomes in minimal volumes of extraction buffer containing Proteinase K and the DNA precipitated in the presence of a carrier. *Sau*3A digestions were then performed in a small volume followed by ligation of the entire DNA digest into the

*Bam*HI site of pBluescript SK II+ (Stratagene). The cloned DNA was amplified by PCR using the T7 and T3 primers, precipitated, and then restriction digested with *Xba*I and *Pst*I. Amplified fragments (300 to 700 bp) were size-selected on agarose gels and then used to construct swine chromosome 6 enriched libraries by ligation into pBluescript SK II+ or Lambda ZAP Express (Stratagene). The library in pBluescript SK II+ had only 52% recombinants, while the Lambda library proved to be more efficient, containing 95% recombinants and a titer approximately 20-fold greater (see Table 1). Marker identification from these libraries will be discussed below.

Chromosome-Enriched Library Production Using Microdissected Chromosomes

In order to pinpoint the production of new markers for each chromosome in even greater detail, scientists have developed techniques for microdissection of individual chromosomes or of chromosome regions. Indeed, chromosome region-specific libraries produced by microdissection have already been generated for Drosophila

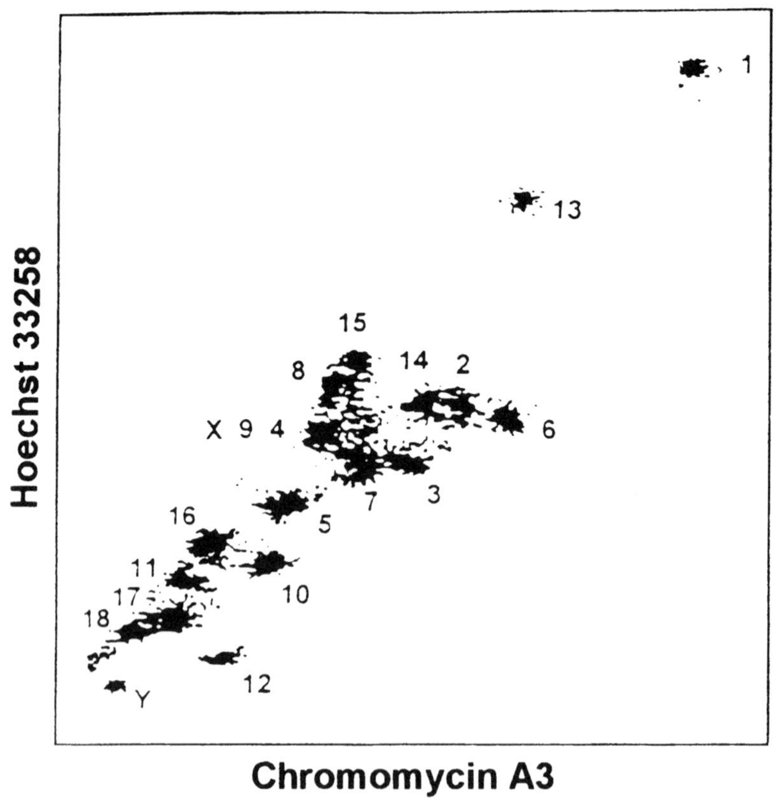

Figure 2. Bivariate flow karyotype of the male swine chromosomes isolated from PHA-stimulated peripheral blood lymphocytes. The linear fluorescence intensities were quantified using a dual-beam flow cytometer after staining with Hoechst 33258 and Chromomycin A3. The locations of each of the separated chromosomes are indicated.

(Scalenghe et al., 1981; Pirrota et al., 1983), mice (Fisher et al., 1985; Brockdorff et al., 1987), and humans (Kaiser et al., 1987; Kao, 1987; Richa and Lo, 1989; Ludecke et al., 1989; Davis et al., 1990). We have generated whole chromosome-specific libraries for bovine (Ambady and Ponce de Leon, 1994; Ponce de Leon et al., 1996), chicken (Li et al., 1992; Ponce de Leon et al., 1996), and swine (Ambady et al., unpublished data).

Our work is an adaptation of the strategy developed by Saunders et al. (1989). Specifically, metaphases were prepared on coverslips and stored at -20°C. Microisolation of chromosomes was carried out with the aid of a Nikon inverted microscope using scraping needles controlled by hydraulic micromanipulators. Ten copies of chromosome 6 were scraped from the coverslip containing the metaphase spreads and the DNA purified by down-scaling the standard purification protocol using Proteinase K digestion and phenol/chloroform extractions in an oil chamber in 1 nL hanging droplets. All necessary manipulation of drops was done with a mechanical micromanipulator. Purified DNA was digested to completion with *Sau*3A I to generate small PCR-amplifiable chromosomal fragments. An adaptor, developed and tested in our laboratory (Ponce de Leon et al., 1996), was ligated to the chromosomal inserts; the ligated adaptor sequences were then used as priming sites for PCR amplification of the chromosomal inserts. Amplified chromosomal inserts were then ligated to the *Bam*H I site of Lambda Zap Express to generate the chromosome-enriched library. The resultant swine chromosome 6 library contains about 4 chromosome equivalents and has been screened for clones containing microsatellite sequences to be developed as genetic markers.

Chromosome-Enriched Library Purity Analysis

Confirmation of chromosome origin of the amplified chromosomal fragments was performed by FISH using either DOP-PCR amplified material for the flow-sorted chromosomes or biotinylated material resulting from PCR with adaptor-specific primers for the microdissected chromosome (Telenius et al., 1992; Ponce de Leon et al., 1996). The FISH analyses were performed on swine metaphase chromosome spreads using the protocol outlined by Lichter et al. (1990). The hybridization reaction mix contained competitor DNA (swine genomic DNA and salmon sperm DNA digested to 300 bp) and the biotinylated painting probe. Binding was detected with avidin-conjugated FITC. For both the microdissected and flow-sorted chromosome 6 painting probes, positive signals were observed over the entire length of chromosome 6. Therefore, the purity of our preparations has been demonstrated to the level of resolution afforded by FISH.

Use of Swine Chromosome 6 Libraries for Microsatellite Marker Development

For several mammalian species, microsatellites containing $(CA)_n$ dinucleotide repeats with high polymorphism information content are known to be widely dispersed throughout the genome. Therefore, the chromosome 6 enriched libraries were screened first for these repeats. Blots were developed by autoradiography and positive clones were confirmed by a secondary round of screening before sequencing.

Template DNA was then amplified for cycle sequencing with the T7 or T3 primers according to the standard Applied Biosystems protocol. The purified extension products were then sequenced using the Applied Biosystems Model 373A DNA sequencer. Sequences were checked for uniqueness with respect to the current swine linkage map, and new cloned sequences were used to generate primer pairs with the PRIMER 0.5 program (Lincoln et al., 1991).

Screening of the plasmid and lambda flow-sorted chromosome 6 enriched libraries resulted in the isolation of 95 unique $(CA)_n$ microsatellite repeats, as seen in Table 1. To enhance marker development, the stringency of both hybridization and wash steps was increased during each round of screening; this resulted in a twofold increase in the efficiency of isolation of positive clones. To date, all 95 clones have been sequenced and we have found 32 unique clones that contain $(CA)_n$ repeats suitable for amplification by PCR. These unique cloned repeat sequences were used to generate primer pairs and are currently being used to genotype the University of Illinois swine reference family (Clamp et al., 1993a) and the genotypes subjected to linkage analyses using CRI-MAP 2.4 (Green et al., 1990). Our efforts have resulted in the addition of five more markers to the swine chromosome 6 linkage map. Genotypes of another 12 markers are being scored for linkage assignment. The data are summarized in Table 1. Attempts to screen the libraries with other repeat probes (AT_{10}, TCC_7, GTG_7, and $GGAT_5$) have proven unsuccessful in isolating clones containing these repeats.

The microdissected library is currently being screened with a $(GT)_{10}$ oligonucleotide probe, and 15 $(CA)_n$ repeat-containing clones have been isolated; 50 additional clones that were positive after screening are being further characterized. The results from screening this library are also shown in Table 1. PCR primer pairs have been synthesized for the 15 repeat containing clones, and 12 of these primer pairs produced multiple alleles when screened in unrelated swine. These microsatellites are currently being genotyped on the University of Illinois swine reference family. To date, 8 of these new markers have been added to the chromosome 6 linkage map as a result of linkage analysis.

Taken together, our data show that we have been successful in adding 13 new microsatellite markers to the swine chromosome 6 linkage map as a result of using chromosome-specific approaches to generating our libraries. The initial results indicate that the microdissection was a more effective method of constructing a chromosome 6 enriched library, since much smaller numbers of chromosomes were used (only 10 copies of chromosome 6 as opposed to approximately 1,000 flow sorted copies) and the microdissected libraries constructed were analyzed by FISH to select the purest one to screen. Despite our success in using the whole chromosome approach to constructing our libraries, we will specifically target the PSS region of chromosome 6 for future marker development. Therefore, we plan to produce a chromosome 6 PSS region-specific library using the microdissection of banded chromosomes described by Ludecke et al. (1989). This will provide us with the necessary genomic material to generate more markers directly in the PSS region.

Future of Region-Specific Marker Development

Chromosome-enriched libraries have been very useful in developing markers to add fine detail to the swine genome linkage map, but as smaller areas of the genome are pursued, our ability to define important genes using linkage procedures becomes limited by the size of the mapping population being analyzed, that is, our number of genetic breakpoints (bins) is limited (Cox et al., 1994). For example, in the pig, with a genome estimated to be between 1,500 and 2,000 cM (Ellegren et al., 1994; Rohrer et al., 1994) with approximately 80,000 genes, the average distance between adjacent genes would be .025 cM (approximately 14 kb). This problem is further complicated by the fact that genes are not evenly distributed over the genome (Antonarakis, 1994). To overcome this limitation of break points, new approaches had to be developed.

The current linkage maps with an average marker spacing of about 10 cM will provide enough power to perform QTL analysis and potentially localize loci of interest to intervals of this size. However, once a QTL is detected, in order to clone the actual genes and to differentiate between closely linked genes, the current average marker spacing is too large. A 10 cM distance on average represents approximately 10 million base pairs, so the regions between existing markers will require more markers internal to those to allow for molecular techniques to be performed to isolate and distinguish between the many genes in that region. In order to fulfill this requirement, the approaches described above have been used to increase marker density on individual chromosomes. However, in order to develop more markers within smaller, defined areas of the chromosome, techniques have been developed to target the isolation of new markers to specific regions of the chromosomes. The *ryr1* region of chromosome 6 is one example of this since there have been no markers isolated beside the *ryr1* gene itself that show linkage to the QTL of this region. Increased marker coverage is therefore necessary to differentiate the known QTL from the *ryr1* gene.

There are several techniques that have been described by other laboratories for the development of new markers within small, targeted regions of a chromosome. As described above, the flow sorting or microdissection of whole chromosomes has been used to target new marker development to specific chromosomes, but several techniques have now become useful in targeting new markers to chromosomal regions. The microdissection of chromosomal regions as opposed to whole chromosomes is another way to develop new markers within specific regions of a chromosome. Coincidence cloning has been used to isolate probes within genomic regions common to two independent sources of complex DNA (Frengen et al., 1994). Genetically directed representational difference analysis (GDRDA) has been used to identify new markers from regions of the genome that contain genotypic differences between two DNA samples (Lisitsyn et al., 1994). This procedure can therefore be used to isolate markers from regions controlling phenotypic traits. Finally, the arraying of clones from large insert libraries is also useful in the development of region-specific markers. Yeast artificial chromosomes (YACs) were the original large insert libraries of choice (Burke et al., 1987), but now there are also bacterial artificial chromosome libraries (BACs) and P1 artificial chromosome libraries (PACs) that have been developed as alternative approaches (Pierce and Sternberg et al., 1992; Shizuya et al., 1992). We plan to use microdissection of the *ryr1* region of chromosome 6 to isolate DNA from that region to be used in the construction of a *ryr1* region-specific library.

The *ryr1* region of chromosome 6 is one example of how important these targeted approaches to marker development will be as the genes that affect performance traits are localized and cloned. All of these techniques will continue to be essential as smaller regions of the chromosome are searched for the genes that affect economically important traits. They will be useful in the generation of more markers in the regions where QTL are detected, and increased marker coverage in small regions of the chromosome will be important for the eventual cloning of the actual genes that code for QTL.

Summary

The development of specific-chromosome isolation methods and the use of chromosome-specific libraries have effectively generated more chromosome-specific polymorphic markers, thus enhancing the construction of linkage maps for individual chromosomes. Production of new swine microsatellites that are chromosome-specific will create a linkage map with greater marker density; moreover, techniques such as microdissection will allow specific regions of each swine chromosome to be targeted. This will be helpful in increasing the resolution of the genome linkage map in regions of each chromosome that are suspected to have associated QTLs, but as of now are in need of more markers. The PSS region of swine chromosome 6 is a good example of such a region. The majority of the markers that have been mapped to chromosome 6 in the pig are within one linkage group (L6) which maps to 6p11→q25 (Andersson et al., 1993; Rohrer et al., 1994). In order to perform fine mapping studies to isolate genes associated with quantitative traits, high-resolution chromosome maps are needed. The development of more chromosome 6 genetic markers will allow detailed mapping studies to be performed in resource families of pigs characterized for their carcass traits as well as their inheritance of the PSS allele of the ryanodine receptor. Overall, the development of chromosome-enriched libraries has enabled swine genome mappers to quickly develop many new microsatellite markers so that they are better able to start to identify the actual genes that influence production traits.

Literature Cited

Ambady, S. and F. A. Ponce de Leon. 1994. Microisolation and cloning of bovine chromosome 23. Anim. Genet. 25:37.

Andersson, L., C. S. Haley, I. Edfors-Lilja, H. Ellegren, M. Johansson, and L. Marklund. 1994a. Mapping trait loci by crossbreeding genetically divergent populations of domestic animals. Anim. Biotechnol. 5:225.

Andersson, L., C. S. Haley, H. Ellegren, S. A. Knott, M. Johansson, K. Andersson, L. Andersson-Eklund, I. Edfors-Lilja, M. Fredholm, I. Hansson, J. Hakansson, and K. Lundstrom. 1994b. Genetic mapping of quantitative trait loci for growth and fatness in pigs. Science (Wash DC) 263:1771.

Andersson, L. A., A. L. Archibald, J. Gellin, and L. B. Schook. 1993. First pig gene mapping workshop (PGM1). Anim. Genet. 24:205.

Andersson, L. 1996. Mapping genes influencing meat characteristics in pigs. In: R. H. Miller, V. G. Pursel, and H. D. Norman (Ed.) Beltsville Symposia in Agricultural Research XX: Biotechnology's Role in the Genetic Improvement of Farm Animals. pp 276–281. American Society of Animal Science, Savoy, IL.

Antonarakis, S. E. 1994. Genome linkage scanning: Systematic or intelligent. Nature Genet. 8:211.

Archibald, A. L., C. S. Haley, J. F. Brown, S. Couperwhite, and H. A. McQueen, et al. 1995. The PiGMaP Consortium linkage map of the pig (*Sus scrofa*). Mamm. Genome 6:157.

Baron, B., P. Metezeau, H. Kiefer-Gachelin, and M. E. Goldberg. 1990. Construction and characterization of a DNA library from mouse chromosome 19 purified by flow cytometry. Biol. Cell 69:1.

Beattie, C. W. 1996. Development of detailed microsatellite linkage maps in livestock. In: R. H. Miller, V. G. Pursel, and H. D. Norman (Ed.) Beltsville Symposia in Agricultural Research XX: Biotechnology's Role in the Genetic Improvement of Farm Animals. pp 51–60. American Society of Animal Science, Savoy, IL.

Brockdorff, N., E.M.C. Fisher, J. S. Cavanna, M. F. Lyon, and S. D. Brown. 1987. Construction of a detailed molecular map of the mouse X chromosome by microcloning and interspecific crosses. EMBO J. 6:3291.

Burke, D. T., G. F. Carle, and M. V. Olson. 1987. Cloning of large segments of exogenous DNA into yeast by means of artificial chromosome vectors. Science (Wash DC) 236:806.

Carlson J. P., L. L. Christian, D. L. Kuhlers, and B. A. Rasmusen. 1980. Influence of the porcine stress syndrome on production and carcass traits. J. Anim. Sci. 50:21.

Clamp, P. A., J. E. Beever, R. L. Fernando, D. G. McClaren, and L. B. Schook. 1993a. Detection of linkage between genetic markers and genes affecting growth and carcass traits in pigs. J. Anim. Sci. 70:2695.

Clamp, P. A., R. Feltes, D. Shalhevet, J. E. Beever, E. Atac, and L. B. Schook. 1993b. Linkage relationships between ALPL, ENO1, GPI, PGD, and TGF 1 on porcine chromosome 6. Genomics 17:324.

Cox, D. R., E. R. Green, E. S. Lander, D. Cohen, and R. M. Myers. 1994. Assessing mapping progress in the human genome project. Science (Wash DC) 265:2031.

Cram, L. S., M. Campbell, J. J. Fawcett, and L. J. Deaven. 1990. Polyamine buffer for bivariate human flow cytogenetic analysis and sorting. Methods Cell Biol. 33:377.

Davies, W., I. Harbitz, R. Fries, G. Stranzinger, and J. G. Hauge. 1988. Porcine malignant hyperthermia carrier detection and chromosomal assignment using a linked probe. Anim. Genet. 19:203.

Davies, W., B. Hoyheim, B. Chaput, A. L. Archibald, and G. Frelat. 1994. Characterization of microsatellites from flow-sorted porcine chromosome 13. Mamm. Genome 5:707.

Davis, L. M., G. Senger, H.-J. Ludecke, U. Claussen, B. Horsthemke, S. S. Zhang, B. Metzroth, K. Hohenfellner, B. Zabel, and T. Shows. 1990. Somatic cell hybrid and long-range physical mapping of 11p13 microdissected genomic clones. Proc. Natl. Acad. Sci. USA 87:7005.

Dietrich, W., J. C. Miller, R. G. Steen, and M. Merchant, D. Damron, R. Nahf, A. Gross, D. C. Joyce, M. Wessel, R. D. Dredge, A. Marquis, L. D. Stein, N. Goodman, D. C. Page, and E. S. Lander. 1994. A genetic linkage map of the mouse with 4006 simple sequence length polymorphisms. Nature Genet. 7:220.

Disteche, C. M., L. M. Kunkel, A. Lojewski, S. H. Orkin, M. Eisenhard, E. Sahar, B. Travis, and S. A. Latt. 1982. Isolation of mouse X-chromosome specific DNA from an X-enriched lambda phage library derived from flow sorted chromosomes. Cytometry 2:282.

Eikelenboom, G., E. Minkema. 1974. Prediction of PSE muscle with a halothane-induced porcine malignant hyperthermia syndrome. Neth. J. Vet. Sci. 99:421.

Ellegren, H., B. P. Chowdhary, M. Johansson, L. Marklund, M. Fredholm, I. Gustavsson, and L. Andersson. 1994. A primary linkage map of the porcine genome reveals a low rate of genetic recombination. Genetics 137:1089.

Fisher, E.M.C., J. S. Cavanna, and S.D.M. Brown. 1985. Microdissection and microcloning of the mouse X chromosome. Proc. Natl. Acad. Sci. USA 82:5846.

Frengen, E., P. D. Thomsen, A. Schmitz, G. Frelat, and W. Davies. 1994. Isolation of region-specific probes from pig chromosome 6 by coincidence cloning. Mamm. Genome 5:497.

Fujii, J., K. Otsu, F. Zorzato, S. deLeon, V. K. Khanna, J. E. Weiler, P. J. O'Brien, and D. H. MacLennan. 1991. Identification of a mutation in porcine ryanodine receptor associated with malignant hyperthermia. Science (Wash DC) 253:448.

Fuscoe, J. C., J. S. McNinch, C. C. Collins, and M. A. Van Dilla. 1989. Human chromosome-specific DNA libraries: Construction and purity analysis. Cytogenet. Cell Genet. 50:211.

Gillard, E. F., K. Otsu, J. Fujii, V. K. Khanna, S. deLeon, J. Derdemezi, B. A. Britt, C. L. Duff, R. G. Worton, and D. H. MacLennan. 1991. A substitution of cysteine for arginine 614 in the ryanodine receptor is potentially causative of human malignant hyperthermia. Genomics 11:751.

Green, P., K. Falls, and S. Crook. 1990. Documentation for CRI-MAP, Version 2.4. Washington University School of Medicine, St. Louis, MO.

Gyapay, G., J. Morissette, A. Vignal, C. Dib, C. Fizames, P. Millasseau, S. Marc, G. Bernardi, M.

Lathrop, and J. Weissenbach. 1994. The 1993-94 Genethon human genetic linkage map. Nature Genet. 7:246.

Harbitz, I., B. Chowdhary, P. D. Thomsen, W. Davies, U. Kaufman, S. Kran, I. Gustavsson, K. Christensen, and J. G. Hauge. 1990. Assignment of the porcine calcium release channel gene, a candidate for the malignant hyperthermia locus to the 6p11→q21 segment of chromosome 6. Genomics 8:243.

Harrison, G. G. 1972. PSE pork, PSS and malignant-hyperthermia. J. S. Afr. Vet. Med. Assoc. 43:630.

Jensen P., and P. Barton-Gade. 1985. Performance and carcass characteristics of pigs with known genotypes for halothane susceptibility. Eur. Assoc. Anim. Prod. Publ. 33:80.

Kaiser, R., J. Weber, K.-H. Grzeschik, J. E. Edstrom, A. Driesel, S. Zengerling, M. Buchwald, L. C. Tsui, and K. Olek. 1987. Microdissection and microcloning of the long arm of human chromosome 7. Mol. Biol. Rep. 12:3.

Kao, F.-T. 1987. Chromosome microdissection and microcloning in human molecular genetics. Somatic Cell Mol. Genet. 13:375.

Kao, F.-T. 1993. Microdissection and microcloning of human chromosome regions in genome and genetic disease analysis. BioEssays 15:141.

Lamont, S. J. 1996. Chicken genome mapping strategies: Expressed sequence tags, heterologous genes, and genetic regulatory elements. In: R. H. Miller, V. G. Pursel, and H. D. Norman (Ed.) Beltsville Symposia in Agricultural Research XX: Biotechnology's Role in the Genetic Improvement of Farm Animals. pp 75–80. American Society of Animal Science, Savoy, IL.

Langford, C. F., H. Telenius, N.G.A. Miller, P. D. Thomsen, and E. M. Tucker. 1993. Preparation of chromosome-specific paints and complete assignment of chromosomes in the pig flow karyotype. Anim. Genet. 24:261.

Larsen, F., G. Gunderson, R. Lopez, and H. Prydz. 1992. CpG islands as gene markers in the human genome. Genomics 13:1095.

Leeb, T., S. Schmolzl, G. Brem, and B. Brenig. 1993. Genomic organization of the porcine skeletal muscle ryanodine receptor (ryr1) gene coding region 4624 to 7929. Genomics 18:349.

Lewin, H. A., and M. L. Mirsky. 1996. Mapping genes for disease resistance: Lessons from the bovine leukemia virus. In: R. H. Miller, V. G. Pursel, and H. D. Norman (Ed.) Beltsville Symposia in Agricultural Research XX: Biotechnology's Role in the Genetic Improvement of Farm Animals. pp 103–113. American Society of Animal Science, Savoy, IL.

Li, Y., J. M. Robl, and F. A. Ponce de Leon. 1992. A novel approach for the construction of chromosome-specific DNA libraries. Proc. 10th Eur. Colloquium of Domest. Anim. p 11. Utrecht Univ., The Netherlands.

Lichter, P., C. C. Tang, K. Call, G. Hermanson, G. A. Evans, D. Housman, and D. C. Ward. 1990. High-resolution mapping of human chromosome 11 by in situ hybridization with cosmid clones. Science (Wash DC) 247:64.

Lincoln, S. E., M. J. Daly, and E. S. Lander. 1991. PRIMER: A Computer Program for Automatically Selecting PCR Primers. MIT Center for Genome Research and Whitehead Institute for Biomedical Research, Cambridge, MA.

Lisitsyn, N. A., J. A. Segre, K. Kusumi, N. M. Lisitsyn, J. H. Nadeau, W. N. Frankel, M. H. Wigler, and E. S. Lander. 1994. Direct isolation of polymorphic markers linked to a trait by genetically directed representational difference analysis. Nature Genet. 6:57.

Ludecke, H.-J., G. Senger, U. Claussen, and B. Horsthemke. 1989. Cloning defined regions of the human genome by microdissection of banded chromosome and enzymatic amplification. Nature (Lond.) 338:348.

MacLennan, D. H., C. Duff, F. Zorzato, J. Fujii, M. Phillips, R. G. Korneluk, W. Frodis, B. A. Britt, and R. G. Worton. 1990. Ryanodine receptor gene is a candidate for predisposition to malignant hyperthermia. Nature (Lond.) 343:559.

MacLennan, D. H., and M. S. Phillips. 1992. Malignant hyperthermia. Science (Wash DC) 256:789.

Mickelson, J. R., C. M. Knudson, C. F. Kennedy, K. P. Campbell, and C. F. Louis. 1992. Structural and functional correlates of a mutation in the malignant hyperthermia-susceptible pig ryanodine receptor. FEBS Lett. 301:49.

Miller, J. R., S. C. Dixon, N.G.A. Miller, E. M. Tucker, J. Hindkjaer, and P. D. Thomsen. 1992. A chromosome 1-specific DNA library from the domestic pig (*Sus scrofa domestica*). Cytogenet. Cell Genet. 61:128.

Murray, J. C., K. K. Buetow, J. L. Weber, S. Ludwigsen, T. Scherpbier-Heddema, F. Manion, J.

Quillen, V. C. Sheffield, S. Sunden, G. M. Duyk, J. Weissenbach, G. Gyapay, C. Dib, J. Morrissette, G. M. Lathrop, A. Vignal, R. White, N. Matsunami, S. Gerken, R. Melis, H. Albertsen, R. Plaetke, S. Odelberg, D. Ward, J. Dausset, D. Cohen, and H. Cann. 1994. A comprehensive human linkage map with centimorgan density. Science (Wash DC) 265:2049.

Nurminsky, D. I., and D. L. Hartl. 1993. Amplification of the ends of DNA fragments cloned in bacteriophage P1. Biotechniques 15:201.

Olsen, A. S., J. Combs, E. Garcia, J. Elliott, C. Amemiya, P. de Jong, and G. Threadgill. 1993. Automated production of high density cosmid and YAC colony filters using a robotic workstation. Biotechniques 14:116.

Otsu, K., V. K. Khanna, A. L. Archibald, and D. H. MacLennan. 1991. Cosegregation of porcine malignant hyperthermia and a probable causal mutation in the skeletal muscle ryanodine receptor gene in backcross families. Genomics 11:744.

Paszek, A. A., L. B. Schook, C. F. Louis, J. R. Mickelson, G. H. Flickinger, M. A. Murtaugh, J. Mendiola, C. W. Beattie, G. A. Rohrer, L. J. Alexander, L. Andersson, H. Ellegren, M. Johansson, P. Mariani, L. Marklund, B. Hoyheim, W. Davies, M. Fredholm, A. L. Archibald, and C. S. Haley. 1995. First Int. Workshop on Porcine Chromosome 6. Anim. Genet. 26:377.

Pierce, J. C., and N. L. Sternberg. 1992. Using bacteriophage P1 system to clone high molecular weight genomic DNA. Methods Enzymol. 216:549.

Pirrotta, V., H. Jackle, and J. E. Edstrom. 1983. Microcloning of microdissected chromosome fragments. In: A. Hollaender and J. K. Setlow (Ed.) Genetic Engineering. Principles and Methods. Vol. 5. pp 1–17. Plenum Press, New York.

Ponce de Leon, F. A., S. Ambady, G. A. Hawkins, S. M. Kappes, M. D. Bishop, J. M. Robl, and C. W. Beattie. 1996. Development of a bovine X chromosome linkage group and painting probes to assess cattle, sheep, and goat X chromosome segment homologies. Proc. Natl. Acad. Sci. USA 93:3450.

Rempel, W. E., L. Ming-Yu, J. R. Mickelson, and C. F. Louis. 1995. The effect of skeletal muscle ryanodine receptor genotype on swine performance and carcass quality traits. Anim. Sci. 60:249.

Richa, J., and C. W. Lo. 1989. Introduction of human DNA into mouse eggs by injection of dissected chromosome fragments. Science (Wash DC) 245:175.

Rohrer, G. A., L. J. Alexander, J. W. Keele, T. P. Smith, and C. W. Beattie. 1994. A microsatellite linkage map of the porcine genome. Genetics 136:231.

Rothschild, M. F., C. Jacobson, D. A. Vaske, C. K. Tuggle, T. Wang, T. H. Short, G. R. Eckardt, S. Sasaki, A. Vincent, D. G. McLaren, O. Southwood, H. van der Steen, A. Mileham, and G. Plastow. 1995. The estrogen receptor locus is associated with a major gene influencing litter size in pigs. Proc. Natl. Acad. Sci. USA 93:201.

Rouquier, S., M. A. Batzer, and D. Giorgi. 1994. Application of bacterial artificial chromosomes to the generation of contiguous physical maps: A pilot study of human ryanodine receptor gene (RYR1) region. Anal. Biochem. 217:205.

Rouquier, S., D. Giorgi, B. Trask, A. Bergmann, M. S. Phillips, D. H. MacLennan, and P. deJong. 1993. A cosmid and yeast artificial chromosome contig containing the complete ryanodine receptor (RYR1) gene. Genomics 17:330.

Saunders, R.D.C., D. M. Glover, M. Ashburner, I. Sider-Kiamos, C. Louis, M. Monastirioti, C. Savakis, and F. Kafatos. 1989. PCR amplification of DNA microdissected from a single polytene chromosome band: A comparison with conventional microcloning. Nucleic Acids Res. 17:9027.

Scalenghe, F., E. Turco, J. E. Edstrom, V. Pirrotta, and M. Melli. 1981. Microdissection and cloning of DNA from a specific region of Drosophila melanogaster polytene chromosomes. Chromosoma 82:205.

Schmitz, A., B. Chaput, P. Fouchet, M. N. Guilly, G. Frelat, and M. Vaiman. 1992a. Swine chromosomal DNA quantification by bivariate flow karyotyping and karyotype interpretation. Cytometry 13:703.

Schmitz, A., P. Chardon, I. Gainche, B. Chaput, M. N. Guilly, G. Frelat, and M. Vaiman. (1992b). Pig standard bivariate flow karyotype and peak assignment for chromosomes X, Y, 3, and 7. Genomics 14:357.

Shizuya, H., B. Birren, U. Kim, V. Mancino, T. Slepak, T. Tachiri, and M. Simon. 1992. Cloning and stable maintenance of 300-kilobase-pair fragments of human DNA in Escherichia coli using an F-factor-based vector. Proc. Natl. Acad. Sci. USA 89:8794.

Soller, M. 1991. Mapping quantitative trait loci affecting traits of economic importance in animal populations using molecular markers. In: L. B. Schook, H. A. Lewin, and D. G.

McLaren (Ed.) Gene Mapping Techniques and Applications. pp 21–49. Marcel Dekker, Inc., New York.

Soller, M. 1994. Marker assisted selection—An Overview. Anim. Biotechnol. 5:193.

Telenius, H., N. P. Carter, C. E. Bebb, R. Nordenskjold, B.A.J. Ponder, and A. Tunnacliffe. 1992. Degenerate oligonucleotide-primed PCR (DOP-PCR): General amplification of target DNA by a single degenerate primer. Genomics 13:718.

VanDevanter, D. R., N. M. Choongkittaworn, K. A. Dyer, J. Aten, P. Otto, C. Behler, E. M. Bryant, and P. S. Rabinovitch. 1994. Pure chromosome-specific PCR libraries from single sorted chromosomes. Proc. Natl. Acad. Sci. USA 91:5858.

Van Dilla, M. A., L. L. Deaven, K. L. Albright, N. A. Allen, M. R. Aubuchon, M. F. Bartholdi, N. C. Brown, E. W. Campbell, A. V. Carrano, L. M. Clark, L. S. Cram, B. D. Crawford, J. C. Fuscoe, J. W. Gray, C. E. Hildebrand, P. J. Jackson, J. H. Jett, J. L. Longmire, C. R. Lozes, M. L. Luedemann, J. C. Martin, J. S. McNinch, L. J. Meincke, M. L. Mendelsohn, J. Meyne, R. K. Moyzis, A. C. Munk, J. Perlman, D. C. Peters, A. J. Silva, and B. J. Trask. 1986. Human chromosome-specific DNA libraries: Construction and availability. Bio/Technology 4:537.

Vecchionacce, H. 1981. Porcine stress syndrome and pork quality. Ph.D. Dissertation. Iowa State Univ., Ames.

Webb, A. J., A. E. Carden, C. Smith, and P. Imlah. 1982. Porcine stress syndrome in pig breeding. Proc. 2nd World Congr. Genet. Livest. Prod. 5:588.

Webb, A. J., S. P. Simpson, and O. I. Southwood. 1985. Porcine stress syndrome: Recent advances. Royal Agric. Soc. England Animal Health and Productivity Symposium, London. pp 143–161.

Yang, H., H.-R. Jung, S. Solinas-Toldo, E. Lang, R. Bolt, R. Fries, and G. Stranzinger. 1992. A reciprocal whole-arm translocation, rcp(1;6)(1p6p;1q6q) in a boar, localization of the breakpoints, and reassignment of the genes for glucose phosphate isomerase (GPI) and calcium release channel (CRC). Cytogenet. Cell Genet. 61:67.

Yerle, M., Y. Lahbib-Mansais, C. Mellink, A. Goureau, P. Pinton, G. Echard, J. Gellin, C. Zilstra, N. De Haan, A. A. Bosma, B. Chowdhary, F. Gu, I. Gustavsson, P. D. Thomsen, K. Christensen, G. Rettenberger, H. Hameister, A. Schmitz, B. Chaput, and G. Frelat. 1995. The PiGMaP consortium cytogenetic map of the domestic pig (*Sus scrofa domestica*). Mamm. Genome 6:176.

Yerle, M., A. Schmitz, D. Milan, B. Chaput, L. Monteagudo, M. Vaiman, G. Frelat, and J. Gellin. 1993. Accurate characterization of porcine bivariate flow karyotype by PCR and fluorescence *in situ* hybridisation. Genomics 16:97.

Zhang, W., D. L. Kuhlers, and W. E. Rempel. 1992. Halothane gene and swine performance. J. Anim. Sci. 70:1307.

Zilstra, C., A. A. Bosma, and N. A. de Haan. 1994. Comparative study of pig-rodent somatic cell hybrids. Anim. Genet. 25:319.

Chicken Genome Mapping Strategies: Expressed Sequence Tags, Heterologous Genes, and Genetic Regulatory Elements[1]

S. J. Lamont

Department of Animal Science, 201 Kildee Hall, Iowa State University, Ames, IA 50011-3150, U.S.A.

Abstract Mapping the chicken genome is rapidly progressing by using a variety of mapping technologies. These technologies include RFLP analysis with candidate gene probes, expressed sequence tags (EST) and random cloned DNA, blood group typing, and PCR analysis with RAPD markers, CR1 repeat-elements, and microsatellite markers. The emphasis of our laboratory on avian immunogenetics has led us to focus our recent mapping efforts on use of three strategies anticipated to contribute to knowledge of genetic control of immune response and disease resistance. The first is RFLP-mapping of EST from a mitogen-stimulated splenocyte-derived library in a collaborative project aimed at contributing anchor loci to build a consensus map from the two international chicken genome maps and mapping loci of potential involvement in immune function. Loci detected by cDNA probes from the Compton laboratory were mapped in the East Lansing mapping reference population. The second strategy is the use of gene probes from heterologous species to identify homologous sequences in the chicken. The human G9a clone, which maps in the human major histocompatibility complex (MHC), hybridized to chicken DNA fragments that mapped to three distinct loci, including one locus linked to the MHC. The third strategy is fine-mapping the regulatory elements of the MHC class II genes by sequence homology comparisons and subsequent functional characterization. Transfection studies provided functional confirmation of both positive and negative regulatory elements in the 5′-flanking region of the chicken MHC class II genes and the ability to modulate these regulatory elements by exogenous factors. The use of highly polymorphic marker systems for loci that are widely dispersed throughout the genome, along with strategies that focus mapping efforts on sequences of putative biological impact, will yield rapid progress in developing a useful genetic map of the chicken.

Key Words: Chicken, Gene Mapping, EST, Candidate Gene, Regulatory Element

Introduction

The research efforts directed toward mapping the chicken genome are concentrated in an informal, international consortium of laboratories with active interests in chicken molecular genetics. The researchers have agreed upon use of two backcross populations originating in Compton, U.K., and East Lansing, Michigan, USA (Bumstead and Palyga, 1992; Crittenden et al., 1993) as the international mapping reference panels, and more than a dozen laboratories worldwide are using these panels. Mapping data from participating laboratories are collated at Compton and East Lansing. The public has Internet access to a summary of chicken gene mapping data

---

[1]Journal Paper No. J-16544 of the Iowa Agric. and Home Econ. Exp. Sta., Ames, IA. Projects No. 2576, 3073.

from a database, ChickGBASE, established at the Roslin Institute, Edinburgh, U.K. (Burt et al., 1995) via the World Wide Web (http://www.ri.bbsrc.ac.uk/chickmap/ChickMapHomePage.html or http://poultry.mph.msu.edu).

A recent review of chicken genome mapping indicated 461 different markers mapped on either or both of the East Lansing and Compton reference panels (Burt et al., 1995). These markers represented many different types: "classical" markers identifiable by phenotypic expression; CR1 repeats; expressed sequence tags (EST); endogenous viral loci; cloned genes; random clones; random amplified polymorphic DNAs; minisatellites; and microsatellites (Bumstead and Palyga, 1992; Crittenden et al., 1993; Levin et al., 1993, 1994a,b; Bumstead et al., 1994; Crooijmans et al., 1995). Recently, the availability of automated genotyping and new panels of chicken microsatellites and minisatellites (Khatib et al., 1993; Bruford et al., 1994; Cheng and Crittenden, 1994; Crooijmans et al., 1994; Groen et al., 1994; Levin et al., 1995) has led to an explosion mapping these loci, with up to 500 microsatellite loci expected to be placed on the map within the next year. Use of marker types, such as repetitive sequences, that are widely dispersed in the chicken genome (Bruford and Burke, 1994) are an effective strategy for generation of a saturated map that can be used to locate quantitative trait loci (QTL) for use in marker-assisted selection (Beattie, 1994; Soller, 1994).

The use of "saturation" strategies in genome mapping can be complemented by use of "focused" mapping strategies. The latter strategies focus on mapping and analysis of sequences about which there is pre-existing information on function, location, or source, which suggests a biological effect of the sequences. Because a major aim of chicken genome mapping efforts is the identification of markers for biological performance (Cheng, 1994), the focus of investigations on sequences of putative biological importance will aid in locating such markers. Several other papers at this symposium discussed saturation mapping strategies. This paper gives examples of focused mapping strategies by describing the Iowa State University avian immunogenetics laboratory's use of heterologous probes, expressed sequence tags, and fine-mapping of functional genetic regulatory elements.

Heterologous Probes

Two major categories of genetic loci have been described (O'Brien et al., 1993). Type I loci are coding gene loci. Type II markers are DNA sequences that are highly polymorphic in the species under investigation but may not be conserved enough to recognize locus homology in distant species. The predominant Type II marker example is that of microsatellite loci. Mapping of Type I markers, which detect specific genes, can be of value in providing information on syntenic groups of homologous loci conserved throughout evolution and can also serve as candidate genes for studies of quantitative trait loci (QTL). Our laboratory has had a long-term interest in the molecular structure and allelic variation of the major histocompatibility complex (MHC) genes (Pitcovski et al., 1989; Warner et al., 1989; Xu et al., 1989; Lamont et al., 1990; Chen and Lamont, 1992; Sung et al., 1993). The MHC genes have a major effect on immune response and resistance to disease. The structural and functional equivalents of mammalian class I and class II MHC antigens have been identified (reviewed in Kroemer et al., 1990; Lamont and Dietert, 1990; Lamont, 1993). An additional locus, *Rfp-Y*, containing both class I and class II genes, has been found to be unlinked to the classical MHC that bears the B blood group (Miller et al., 1994). The locations of genes homologous to mammalian class III region genes, however, have remained elusive. Some protein polymorphisms associated with the mammalian MHC class III region (complement component factor B and glyoxalase I) were demonstrated to be unlinked to the serological EaB locus in chickens (Rubinstein et al., 1981; Koch, 1986). To map chicken homologs of mammalian MHC class III genes and to expand the chicken MHC syntenic group, we screened several mammalian MHC clones for efficiency of hybridization to chicken DNA and potential for use as a Type I marker. One of these clones, pG9a-4C7 (Milner and Campbell, 1993), gave repeatable RFLP in both reference mapping populations. Three different loci were mapped (Spike and Lamont, 1995) and one locus was linked to the chicken MHC locus in the Compton reference population, as determined by the serologically detected *Ea-B*. This use of a Type I marker to provide the first evidence of a gene of the class III region mapping to the chicken MHC will help expand knowledge of the evolutionary conservation of synteny in the MHC and facilitate understanding of the complete functions of this gene complex (Trowsdale, 1995).

Expressed Sequence Tags

An expressed sequence tag (EST) is a segment of a sequence from a cDNA clone that corresponds to an mRNA. The use of ESTs is a rapid approach to cDNA characterization that speeds the tagging of genes and provides genetic markers (Adams et al., 1991). Choice of a cDNA library from a tissue of specific function and developmental stage optimizes the identification of ESTs for expressed genes related to traits of interest. We mapped ESTs, consisting of clones derived from a mitogen-stimulated, T-cell enriched cDNA library (Bumstead et al., 1994) in the East Lansing reference population (Crittenden et al., 1993). The choice of EST from an immune cell-derived library was intended to optimize mapping of loci that are potentially related to immune function. Of 250 clones, however, only 21% had identifiable mammalian homologs and another 10% matched previously cloned chicken genes, with many involved in transcription (Bumstead et al., 1995). The 25 clones used in the current study had at least one sequence tag with homology to known genes, and some had already been placed on the Compton map (Bumstead et al., 1994, 1995). Forty percent of the 25 cDNA clones screened gave polymorphic RFLP in the East Lansing reference population with at least one of seven restriction enzyme-digests. Ten loci were mapped (Spike et al., 1996). Each locus was linked to previously published or recently mapped loci, indicating a high degree of marker saturation in the East Lansing map. Two of the cDNAs co-segregated with genes of interest in immunogenetics: *ev1* (endogenous viral gene 1; Crittenden, 1991) and *EaR* (erythrocyte antigen R, which segregates with susceptibility to avian leukosis virus subgroup B and E; Bacon et al., 1988). Another marker, *c372*, had an expressed sequence tag almost identical to chicken macrophage migration inhibitory factor (Wistow et al., 1993), which is a T-cell lymphokine that initiates macrophage inflammatory response. This EST mapping approach contributed anchor loci for the building of a consensus map of the two reference populations and provided markers that may prove of value in QTL analysis of immune response. The compilation of a comprehensive consensus map is currently under way (L. B. Crittenden, personal communication, 1995).

Functional Fine Mapping of Genetic Regulatory Elements

Most mapping strategies have the goal of placing additional marker loci on the map. Variation in chromosomal structure (DNA sequence variation) is viewed as the potential application point of the mapping information in enhancement of animal or population performance. For a few well-characterized gene systems (e.g., the MHC), sufficient structural information is available to extend investigations to a second level of mapping; that is, functional characterization of genetic regulatory elements. Fine-mapping the function of regulatory elements that control expression of structural genes opens the possibility of enhancing biological traits by modulating gene expression levels. We have analyzed regulatory elements of the chicken MHC class IIβ gene. The 5′ flanking region of the gene was cloned into a reporter vector (chloramphenicol acetyltransferase, CAT) and transfected into a chicken macrophage cell line (Chen et al., 1995). Positive orientation-dependent promoter activity of the chicken DNA occurred if the vector contained an SV40 enhancer (Chen et al., 1993). Deletion analysis revealed positive and strongly negative regulatory elements in this promoter fragment. The presence of elements responsive to interferon and to glucocorticoids were identified by sequence analysis. The functions of these elements were confirmed by addition of exogenous substances to transfected cell cultures and the subsequent observation of interferon enhancement and glucocorticoid suppression of reporter gene expression (Chen, 1995). This first functional characterization of a chicken MHC class IIβ gene promoter provides information needed for modulation of gene expression levels to enhance disease resistance.

Implications

The rapid progress in mapping of the chicken genome suggests that application of marker-assisted selection to alter allelic frequencies in populations may soon become a reality. International cooperation and the use of multiple strategies for mapping with both saturation and focused approaches are yielding markers for traits of biological importance. The map generated for the two reference families will be expanded by analysis of additional resource populations that possess phenotypic and allelic variation for agriculturally important traits. And functional analyses of genetic regulatory elements will enable enhancement of biological performance by modulating gene expression as a complement to selecting populations for beneficial genotypes. Chickens are one of the few livestock species in which highly inbred and congenic lines are available. These genetic resources may enable the first practical application in a livestock species of genetically directed representational difference analysis (Lisitsyn et al., 1994) to directly isolate polymorphic markers linked to a trait.

Acknowledgments

The research collaborations of N. Bumstead, T. A. Burke, R. D. Campbell, S. Carpenter, Y. Chen, H. Cheng, L. B. Crittenden, J. Dodgson, C.-H. Hsu, H. S. Lillehoj, M.

Qureshi, H. Sillibourne, C. Spike, and Y. Wannemuehler, which contributed to the studies presented in this paper, are gratefully acknowledged.

Literature Cited

Adams, M. D., J. M. Kelley, J. D. Gocayne, M. Dubnick, M. H. Polymeropoulos, H. Xiao, C. R. Merril, A. Wu, B. Olde, R. F. Moreno, A. R. Kerlavage, W. R. McCombie, and J. C. Venter. 1991. Complementary DNA sequencing: Expressed sequence tags and human genome project. Science (Wash DC) 252:1651.

Bacon, L. D., E. J. Smith, and L. B. Crittenden. 1988. Detection of subgroup E endogenous avian leukosis virus (ALV) susceptibility in chickens by hemagglutination using alloantisera. Poult. Sci. 67(Suppl. 1):50 (Abstr.).

Beattie, C. W. 1994. Livestock genome maps. Trends Genet. 10(9):334.

Bruford, M. W., and T. Burke. 1994. Minisatellite DNA markers in the chicken genome. I. Distribution and abundance of minisatellites in multilocus DNA fingerprints. Anim. Genet. 25:381.

Bruford, M. W., O. Hanotte, and T. Burke. 1994. Minisatellite DNA markers in the chicken genome. II. Isolation and characterization of minisatellite loci. Anim. Genet. 25:391.

Bumstead, N., and J. Palyga. 1992. A preliminary linkage map of the chicken genome. Genomics 13:690.

Bumstead, N., H. Wain, N. Salmon, and J. Sillibourne. 1995. Genomic mapping of immunological genes. In: J. F. Davison, N. Bumstead, and P. Kaiser (Ed.) Advances in Avian Immunological Research. pp 75-85. Carfax Publ., Abingdon, U.K.

Bumstead, N., J. R. Young, C. Tregaskes, J. Palyga, and P.P.J. Dunn. 1994. Linkage mapping and partial sequencing of 10 cDNA loci in the chicken. Anim. Genet. 25:337.

Burt, D. W., N. Bumstead, J. J. Bitgood, F. A. Ponce de Leon, and L. B. Crittenden. 1995. Chicken genome mapping: A new era in avian genetics. Trends Genet. 11(5):190.

Chen, Y. 1995. Identification and characterization of a chicken major histocompatibility complex class IIβ gene promoter. Ph.D. Dissertation, Iowa State Univ., Ames.

Chen, Y., S. L. Carpenter, and S. J. Lamont. 1995. Transfection of a putative chicken MHC class II gene promotor region into chicken macrophages. In: T. F. Davison, N. Bumstead, and P. Kaiser (Ed.) Advances in Avian Immunological Research. pp 129–133. Carfax Publ., Abingdon, U.K.

Chen Y., S. Carpenter, and S. J. Lamont. 1993. Identification of a chicken MHC class II gene promoter. FASEB J. (Suppl.):286A.

Chen, Y., and S. J. Lamont. 1992. Major histocompatibility complex class I restriction fragment length polymorphism analysis in highly inbred chicken lines and lines selected for major histocompatibility complex and immunoglobulin production. Poult. Sci. 71:999.

Cheng, H. H. 1994. The chicken genetic map: A tool for the future. Poult. Dig. (June):24–28.

Cheng, H. H., and L. B. Crittenden. 1994. Microsatellite markers for genetic mapping in the chicken. Poult. Sci. 73:539.

Crittenden, L. B. 1991. Retroviral elements in the genome of the chicken: Implications for poultry genetics and breeding. Crit. Rev. Poult. Biol. 3:73.

Crittenden, L. B., L. Provencher, L. Santangelo, I. Levin, H. Abplanalp, R. W. Briles, W. E. Briles, and J. B. Dodgson. 1993. Characterization of a red jungle fowl by white leghorn backcross reference population for molecular mapping of the chicken genome. Poult. Sci. 72:334.

Crooijmans, R.P.M.A., A.J.A. van Kampen, J. J. van der Poel, and M.A.M. Groenen. 1994. New microsatellite markers on the linkage map of the chicken genome. J. Hered. 85:410.

Crooijmans, R.P.M.A., J. J. van der Poel, and M.A.M. Groenen. 1995. Functional genes mapped in the chicken genome. Anim. Genet. 26:73.

Groen, A. F., R.P.M.A. Crooijmans, A.J.A. van Kampen, S. van der Beek, J. J. van der Poel, and M.A.M. Groenen. 1994. Microsatellite polymorphism in commercial broiler and layer lines. Proc. 5th World Congr. Genet. Appl. Livest. Prod. 21:94.

Khatib, H., E. Genislav, L. B. Crittenden, N. Bumstead, and M. Soller. 1993. Sequence-tagged microsatellite sites as markers in chicken reference and resource populations. Anim. Genet. 24:355.

Koch, C. 1986. A genetic polymorphism of the complement factor B in chickens not linked to the major histocompatibility complex (MHC). Immunogenetics 23:364.

Kroemer G., A. Bernot, G. Behar, A.-M. Chausse, L.-N. Gastine, F. Guillemot, I. Park, P. Thoraval, R. Zoorob, and C. Auffray. 1990. Molecular genetics of the chicken MHC: Current status and evolutionary aspects. Immunol. Rev. 113:116.

Lamont, S. J. 1993. The major histocompatibility complex in chickens: In: R. J. Etches and A. M. Verrinder Gibbins (Ed.) Manipulation of the Avian Genome. p 185. CRC Press, Boca Raton, FL.

Lamont, S. J., and R. R. Dietert. 1990. Immunogenetics. Chapter 22. In: R. Crawford (Ed.) Poultry Breeding and Genetics. Elsevier Publ., Amsterdam, The Netherlands.

Lamont, S. J., B. M. Gerndt, C. M. Warner, and L. D. Bacon. 1990. Analysis of restriction fragment length polymorphisms of the major histocompatibility complex of $15I_5$-*B*-congenic chicken lines. Poult. Sci. 69:1195.

Levin, I., H. H. Cheng, C. Baxter-Jones, and J. Hillel. 1995. Turkey microsatellite DNA loci amplified by chicken-specific primers. Anim. Genet. 26:107.

Levin, I., L. B. Crittenden, and J. B. Dodgson. 1993. Genetic map of the chicken Z chromosome using random amplified polymorphic DNA (RAPD) markers. Genomics 16:224.

Levin, I., L. B. Crittenden, and J. B. Dodgson. 1994a. Mapping DNA polymorphisms using PCR primers derived from the sequence of an avian CR1 element. J. Hered. 85:73.

Levin, I., L. Santangelo, H. Cheng, L. B. Crittenden, and J. B. Dodgson. 1994b. An autosomal genetic linkage map of the chicken. J. Hered. 85:79.

Lisitsyn, N. A., J. A. Segre, K. Kusumi, N. M. Lisitsyn, J. H. Nadeau, W. N. Frankel, M. H. Wigler, and E. S. Lander. 1994. Direct isolation of polymorphic markers linked to a trait by genetically directed representational difference analysis. Nature Genet. 6:57.

Miller, M. M., R. Goto, A. Bernot, R. Zoorob, C. Auffray, N. Bumstead, and W. E. Briles. 1994. Two *Mhc* class I and two *Mhc* class II genes map to the chicken Rfp-Y system outside the *B* complex. Proc. Natl. Acad. Sci. USA 91:4397.

Milner, C. M., and R. D. Campbell. 1993. The G9a gene in the human major histocompatibility complex encodes a novel protein containing ankyrin-like repeats. Biochem. J. 290:811.

O'Brien, S. J., J. E. Womack, L. A. Lyons, K. J. Moore, N. A. Jenkins, and N. G. Copeland. 1993. Anchored reference loci for comparative genome mapping in mammals. Nature Genet. 3:103.

Pitcovski, J., S. J. Lamont, A. W. Nordskog, and C. M. Warner. 1989. Analysis of *B-G* and immune response genes in the Iowa State University S1 chicken line by hybridization of sperm deoxyribonucleic acid with a major histocompatibility complex class II probe. Poult. Sci. 68:94.

Rubinstein, P., L. de Haas, I. Y. Pevzner, and A. W. Nordskog. 1981. Glyoxalase 1 (GLO) in the chicken: Genetic variation and lack of linkage to the MHC. Immunogenetics 13:493.

Soller, M. 1994. Marker assisted selection—an overview. Anim. Biotechnol. 5(2):193.

Spike, C. A., N. Bumstead, L. B. Crittenden, and S. J. Lamont. 1996. RFLP mapping of expressed sequence tags in the chicken. J. Hered. (In press).

Spike, C. A., and S. J. Lamont. 1995. Genetic analysis of 3 loci homologous to human G9a: Evidence for linkage of a class III gene with the chicken MHC. Anim. Genet. 26:185.

Sung, A. M., A. W. Nordskog, S. J. Lamont, and C. M. Warner. 1993. Isolation and characterization of cDNA clones for chicken major histocompatibility complex class II molecules. Anim. Genet. 24:227.

Trowsdale, J. 1995. Both man and birds and beast: Comparative organization of MHC genes. Immunogenetics 41:1.

Warner, C. M., B. Gerndt, Y. Xu, Y. Bourlet, C. Auffray, S. Lamont, and A. W. Nordskog. 1989. Restriction fragment length polymorphism (RFLP) analysis of MHC class II genes from inbred chicken lines. Anim. Genet. 20:225.

Wistow, G. J., M. P. Shaughnessey, D. C. Lee, J. Hodin, and P. S. Zelenka. 1993. A macrophage migration inhibitory factor is expressed in the differentiating cells of the eye lens. Proc. Natl. Acad. Sci. USA 90:1272.

Xu, Y., J. Pitcovski, L. Peterson, C. Auffray, Y. Bourlet, B. Gerndt, A. W. Nordskog, S. J. Lamont, and C. M. Warner. 1989. Isolation and characterization of three class II major histocompatibility complex genomic clones from the chicken. J. Immunol. 142:2122.

Development and Use of the Sheep Genome Map

N. E. Cockett*[,1,2] and J. F. Medrano†

**Department of Animal, Dairy and Veterinary Sciences, Utah State University, Logan, UT 84322-4700, U.S.A. and †Department of Animal Science, University of California, Davis, CA, 95616, U.S.A.*

Abstract Genome scanning is a method in which many genetic markers, spread across the genome, are tested for genetic linkage with traits of interest. Many researchers are using genome scanning to identify economic trait loci (ETL) in livestock. An advantage of this method is that no prior knowledge of biological pathways or similar traits is needed. By including enough genetic markers into the analysis, all chromosomes can be systematically examined for association. Approximately 200 equally spaced markers will be necessary to completely scan the genome of livestock species. Using the genome scanning approach, markers have been identified that are linked to two ETL in sheep; these traits include the Booroola fecundity gene *FecB* responsible for high ovulation rate, and the callipyge gene (*CLPG*), associated with muscle hypertrophy in the hind quarters. In both studies, once the first marker was linked to the trait, additional markers previously mapped to the corresponding regions in sheep, cattle, and humans were tested for linkage. In this way, the chromosomal areas around the ETL were finely mapped. It should be noted that while genetic markers associated with the traits have been identified in these studies, the causative genes are not known at this time and methods other than genome scanning will be needed to isolate the trait loci.

Key Words: Sheep, Genome Map, Genetic Markers, Karyotype

Introduction

Traditional plant and animal breeding practices have made great progress in improving the original wild-type species, resulting in the domesticated varieties, lines, and breeds we have today. These improvements were accomplished by selecting plants and animals possessing desirable phenotypes and eliminating those that were undesirable. In recent years, genetic analysis of organisms has been revolutionized by recombinant DNA technology, and particularly by the recognition that DNA polymorphisms can be used as genetic markers. This new technology allows the direct identification and manipulation of DNA sequences, and the application of marker-assisted selection (MAS) to complement traditional selection methods.

A persistent problem in this new direction of genetic research has been the difficulty in identifying important genes that are suitable for manipulation. Not only are relatively few of this genes actually isolated, but in many biological pathways the important genes have not been identified. However, in the past 5 years, significant amounts of information have been reported about genes and genetic markers in

---

[1]Author to whom correspondence should be addressed.

[2]This research was supported by the Utah Agric. Exp. Stat., Utah State Univ. Approved as journal paper no. 4837.

livestock species. Central to obtaining this information has been the development of genome maps as tools for studying the genetic control of economically important traits. These maps contain assignments of markers and genes to specific regions along the chromosomes and are useful for organizing systematic searches for chromosomal regions containing genes of economic importance. Once these regions have been identified, additional markers in the area can be used to locate the locus of interest with better resolution. It has been hypothesized that approximately 150 equally-spaced markers will be necessary to completely scan the genome of livestock species (Georges and Massey, 1991). Because of marker clustering, many more markers will be needed to achieve an adequate coverage for a basic 10 to 20 centimorgan (cM) genetic map. Additional markers will be essential to increase map resolution in order to proceed with the isolation of economically important genes and to apply marker-assisted selection strategies across families. Thus, further development of the genome maps for livestock species is warranted.

Sheep Genome Map

In summarizing the status of the ovine genome map, approximately 477 loci have been mapped to all 26 autosomal chromosomes and both sex chromosomes (P. D. Pearce and T. E. Broad, personal communication; Crawford et al., 1995). Of these assignments, 64 loci (13% of all assignments) were done through *in situ* hybridization, 106 (22%) through somatic cell hybrids, and 307 (65%) through family linkage analyses. A major contributor to the sheep genome map was the AgResearch group at Invermay, New Zealand. In two publications (Crawford et al., 1994, 1995), they described the mapping of 33 candidate genes and 213 microsatellites. Although the Invermay group identified a number of microsatellites (86) that were derived from sheep sequences (Crawford et al., 1994, 1995), a number of investigators successfully used bovine microsatellites for mapping sheep (e.g., Cockett et al., 1994; Crawford et al., 1995). In these studies, approximately 75% of the bovine primer pairs amplified genomic DNA of sheep and approximately 50% of these amplified microsatellites were polymorphic.

A relatively new approach for identifying genetic markers is the random amplified polymorphic DNA (RAPD) assay (Williams et al., 1990). This assay is based on the polymerase chain reaction (PCR) and has been optimized to identify markers for the sheep genome (Cushwa and Medrano, 1994). Screening 131 RAPD primers across eight of the New Zealand International Mapping Families (IMF) an average of 85 polymorphisms per family were identified. During the screening process, RAPD markers that were polymorphic in the sires and dams were then tested in the progeny. On average 49% of RAPD polymorphisms were polymorphic in three or more families, and a total of 47 markers were mapped by linkage analysis (unpublished results). In addition, three candidate Y chromosome-linked RAPD markers were identified. These Y-chromosome RAPD fragments were present in all tested IMF males (n=62), absent in all tested IMF females (n=64), and conserved in all sheep breeds examined (Finn, Hampshire, Rambouillet, Javanese, and Suffolk). The markers were then converted into sequence characterized amplified regions (SCARs) (Paran and Michelmore, 1993). One of the SCAR markers was physically mapped to sheep and goat Y chromosomes by *in situ* hybridization (W. T. Cushwa, K. Beer, A. Ponce De Leon, J. F. Medrano, unpublished results).

To aid in mapping the sheep genome, a sheep reference population was established at the USDA, ARS U.S. Meat Animal Research Center (MARC) in Clay Center, Nebraska (C. Beattie, personal communication). Four large half-sib families have produced 247 offspring and include two Romanov (3/4 Romanov; 1/4 Rambouillet) × Rambouillet rams mated to 21 Romanov ewes, producing 135 lambs, and two Romanov (3/4 Romanov; 1/4 Suffolk) × Suffolk rams mated to 22 Romanov ewes, producing 112 offspring. Other sheep reference families include the AgResearch IMF pedigrees of New Zealand and two full-sib families from INRA in France; both of these reference populations have incorporated the Booroola Merino breed into their mating schemes.

Sheep Karyotype Central to the development of a genome map is a well-characterized species karyotype. Sheep have 54 chromosomes, with 26 pairs of autosomes and two sex chromosomes. Most of the chromosomes are telocentric with the exception of three characteristic pairs of large metacentric chromosomes (1, 2, and 3) and a very small Y metacentric chromosome. In the past, there has been some confusion in ovine chromosomal assignments because of contradictions in nomenclature for sheep chromosomes. Although three standardized sheep karyotypes have been presented, each of the publications has accumulated errors, which has perpetuated the confusion. The first standard GTG karyotype was published after the Reading, England conference in 1976 (Ford et al., 1980). However, text descriptions and photographs were switched for sheep chromosomes 3p and 3q, 8 and 9, and 19 and 20. The second standard karyotype, a G-banded ideogram, was published in 1985 (Long, 1985). The 3p and 3q error was rectified but the other chromosomes followed the Reading publication, thereby continuing the previous errors. In addition, there were errors in numbering the smaller chromosomes (21, 23, 24, and 25). The 1989 ISCNDA report (Di Berardino et al., 1990) presented G- and R-banded ideograms for cattle. Unfortunately, chromosomes 4 and 6, and 24 and 26 were interchanged. Because the sheep R-banded standard karyotype was correlated by inference to the cattle karyotype, subsequent problems in ovine nomenclature arose from these switches. For example, the β-casein locus has been reported on both sheep chromosomes 4 and 6 depending on the standard that was used. Ansari et al. (1992) using the Reading standard mapped β-casein to chromosome 6 and Hayes et al. (1993) using the ISCNDA standard mapped the locus to chromosome 4. To resolve this confusion, the International Sheep Standardization Committee was established in 1993. Sequential Q- to G- and Q- to R-stained chromosomes have been prepared by the committee, and in the resulting publication representative Q-, G- and R-banded karyotypes and ideograms are presented (Ansori et al., 1996).

A classical cytogenetic map gives visual reality to genetic and physical maps and to the chromosomes themselves (Patterson, 1993). The traditional weakness of cytogenetic maps has been their low resolution; for example, it is impossible to tell from a cytogenetic map if a translocation involves a deletion. However, new technologies of *in situ* hybridization and heterologous chromosome painting can overcome this weakness and allow the integration of linkage and cytogenetic maps. These techniques were recently used to evaluate syntenic relationships between humans and pigs (Johansson et al., 1995; Rettenberger et al., 1995), providing clear examples of high-level resolution that we can expect to see in the future for all domestic species.

Other Sheep Mapping Resources Two somatic cell hybrid panels containing sheep chromosomes exist. One panel is composed of hamster × sheep cell hybrids comprising chromosomes 1, 3, and C and Robertsonian translocations of (6:24) and (9:10) in a predominantly monochromosomal state (Burkin et al., 1993). The other panel contains 25 hamster × sheep hybrid cell lines (Saidi-Mehtar et al., 1989).

A large insert yeast artificial chromosome (YAC) library has been constructed from sheep DNA (Broom and Hill, 1994). This library contains 13,400 clones with an average insert size of 600 to 700 kb. An estimated 15% of the clones are chimeric. This library will be important for facilitating the transition from linkage to physical mapping, particularly important when identifying causal genes that control economic trait loci.

An informational database that includes mapped loci in sheep is now available. SheepBase contains an up-to-date compilation of published data from sheep genome mapping projects and provides both physical and linkage maps of the sheep genome, together with information on individual loci and associated references. The current version of SheepBase contains about 430 loci gathered from 105 publications. There are about 170 Type I loci and 260 Type II loci currently in the database. Information is presented using the WWW interface and can be accessed through a number of nodes (Table 1). SheepBase was designed and implemented by Alan Hillyard, Jackson Laboratory, Bar Harbor, USA and data entry and editing has been performed by Jude Sise, AgResearch, New Zealand. An international editorial committee has been established (Table 1) and will make policy decisions on the reference and mapping data held in SheepBase, compile the integrated genetic map, and review new submitted data.

Candidate Genes The primary goal of many animal scientists is to identify genes affecting economic trait loci (ETL). One method of identifying these ETLs is through the use of "candidate genes." The candidate gene approach uses selected genes of known function that are hypothesized to influence the physiology or pathology of specific traits. Genetic polymorphisms are first identified in the candidate gene, often

Table 1. SheepBase, an informational database containing mapped loci in sheep

Access nodes
AgResearch (New Zealand):
http://dirk.invermay.cri.nz
Jackson Laboratory (U.S.):
http://anita.jax.org/sheepgbase/manager.html
National Agricultural Library (U.S.):
http://tetra.gig.usda.gov:8400/
Editorial committee
Grant Montgomery, AgResearch, New Zealand
(montgomeryg@inveramy.cri.nz)
Tom Broad, AgResearch, New Zealand
(broadte@agresearch.cri.nz)
Frank Nickolas, University of Sydney, Australia
(frankn@doolittle.vetsci.su.oz.au).
Noelle Cockett, Utah State University, U.S.
(fanoelle@cc.usu.edu)

using restriction fragment length polymorphism (RFLP) analysis. Animals in resource families characterized for the trait(s) are then tested for the polymorphism. If suitable families segregating for the trait(s) are available, linkage relationships among loci (RFLP alleles and the ETL) can be estimated using the lod score method (Morton, 1955). This approach was used by a number of researchers to identify ETL in sheep as well as the other livestock species. Associations include adipocyte P2 and lipoprotein lipase with carcass traits in sheep (Muggli-Cockett et al., 1993), the estrogen receptor polymorphism associated with prolificacy in Meishan pigs (Rothschild et al., 1994), the growth hormone and parathyroid hormone association with growth traits in beef cattle (Rocha et al., 1992), and prolactin (Cowan et al., 1990), α_{s1}, β and κ-caseins (Ng-Kwai-Hang et al., 1983; Cowan et al., 1992; Sorenson et al., 1994) and α-lactalbumin (Bleck and Bremel, 1993) affecting milk production traits in dairy and beef cattle. Additionally, the examination of candidate genes at the level of DNA sequencing has identified mutations causing bovine leukocyte adhesion deficiency (Kehrli et al., 1992) and uridine monophosphate synthase deficiency (Schwenger et al., 1993) in cattle, periodic hyperkalemic paralysis in horses (Rudolph et al., 1992), and porcine stress syndrome (Fuji et al., 1991).

The critical decision in the candidate gene approach is selection of the candidate gene. Sometimes similar traits are present in other species and associated genes may have been already isolated. However, the search often requires identification of rate-limiting steps in the biological pathways that regulate the trait. Genes controlling these steps must then be isolated and characterized. Polymorphisms within these functional genes are often difficult to identify and may require complete DNA sequencing of the gene from several animals.

Genome Scanning A second method for identifying ETL is "genome scanning." In this method many genetic markers, spread throughout the genome, are tested for genetic linkage with the traits of interest. In contrast to the candidate gene approach, no prior knowledge of biological pathways or similar traits is needed for genome scanning. By including enough genetic markers into the analysis, all chromosomes can be systematically examined for association with the trait of interest. It is estimated that approximately 200 equally spaced markers will be necessary to completely scan the genome of livestock species. Using the genome scanning approach, markers have been identified that are linked to the sheep Booroola fecundity gene (Montgomery et al., 1993) and the *callipyge* gene (Cockett et al., 1994) in sheep (Table 2), as well as traits in other livestock species such as the horned gene (Georges et al., 1993b), Weaver's disease (Georges et al., 1993a), and genes controlling milk production (Georges et al., 1995) in cattle, and genes affecting growth and carcass traits in swine (Andersson et al., 1994) and beef cattle (Beever et al., 1990). Once a marker is linked to the trait, additional markers previously mapped to the region are tested for association. In this way, the chromosomal area is finely mapped. While genetic markers associated with the traits have been identified in these studies, the causative genes are not known at this time.

Bulked Segregant Analysis This technology represents a unique and powerful approach for the detection of genetic markers closely linked to specific genomic regions (Giovannoni et al., 1991; Michelmore et al., 1991). In this approach, two pools of DNAs are constructed from selected individuals of a segregating population. The two pools differ in allelic composition within the genomic regions of interest, while they are genetically similar outside these regions. Differences in the

Table 2. Economic trait loci in sheep identified by genome scanning

Trait	Chromosome	Reference
Booroola fecundity (*FecB*)	OAR 6	Montgomery et al. (1993)
Callipyge (*CLPG*)	OAR 18	Cockett et al. (1994)

electrophoretic patterns of RAPD-PCR products between the pool identify polymorphisms potentially linked to the targeted region. This approach was used to screen for genetic markers closely linked to the *high growth* locus (*hg*) in mice (Horvat and Medrano, 1994), which controls weight gain and mature body size. A previous mapping study positioned *hg* in an interval on mouse chromosome 10 (Horvat and Medrano, 1995). Eight RAPD primers out of 888 screened primers yielded polymorphisms linked to an 8 cM interval harboring the *hg* locus, demonstrating that the bulked segregant analysis efficiently identified markers linked to a targeted locus. In domestic animals, this approach may be useful in identifying markers linked to sheep ETL (e.g., Booroola fecundity gene or callipyge), or in searching for markers in gaps from existing genetic maps. Different pooling strategies may be used depending on the available segregating populations, either based on markers bordering the locus of interest (Giovannoni et al., 1991) or based on phenotypic data (Michelmore et al., 1991). In either case, the application of bulk segregant analysis with RAPD-PCR screening seems to be an efficient technique to target specific genomic regions and increase map resolution.

Mapped Traits in Sheep Recently, a mutation was identified that causes muscle hypertrophy in sheep. Animals expressing the mutation have dramatically increased musculature in the hind quarters and loin area. The name *callipyge* (from Greek *calli*, beautiful and *pugē*, buttocks) and symbol *CLPG* were proposed for the gene (Cockett et al., 1994). Flanking markers for the *callipyge* locus were recently identified (Cockett et al., 1994). In this study, offspring of two paternal half-sib pedigrees were first genotyped for four variable number of tandem repeat (VNTR) markers. While linkage was excluded for three of the markers (lod<3.0), the fourth VNTR marker, GMBT16, gave a combined lod score of 7.9 at 20% recombination. GMBT16 had previously been assigned by *in situ* hybridization to ovine chromosome 18 and to bovine chromosome 21, the evolutionary homolog of ovine chromosome 18 (Georges et al., 1991). Markers mapped to bovine chromosome 21 in other studies were then tested in the families and two microsatellite markers, CSSM18 and TGLA122, were shown to be linked to the *CLPG* locus. Final position for *CLPG* was within the interval of CSSM18 and TGLA122 at 3 and 17.5 cM, respectively, with a maximum lod score of 26.2.

Several studies have demonstrated that the *callipyge* gene segregates in paternal half-sib pedigrees as a dominant trait (Jackson and Green, 1993; Snowder et al., 1994), with indications that the trait is under the control of a single autosomal gene. Further crossing experiments revealed a non-Mendelian segregation pattern of the callipyge phenotype (S. Jackson, personal communication). When heterozygous callipyge females (*CLPG/clpg*) were mated to normal rams (*clpg/clpg*), none of the 49 resulting offspring exhibited the callipyge phenotype. All offspring were genotyped for the DNA markers flanking the *callipyge* locus, and a normal 50%:50% segregation ratio of the corresponding chromosomal segment was observed. Thus, the *CLPG* allele was not expressed when the mutation was inherited from the female parent. In crosses between heterozygous parents (*CLPG/clpg*), only 35% (23/66) callipyge offspring were produced. This proportion deviates significantly from the 75% callipyge offspring expected under Mendelian inheritance of a dominant mutation in an F_2 generation. These offspring were also genotyped with the flanking markers; as before, lambs that inherited the maternal *CLPG* and paternal *clpg* alleles were normal in appearance. In addition, animals homozygous for the callipyge mutation (*CLPG/CLPG*) did not express the trait. Characterization of this novel form of genetic inheritance was possible because of the availability of flanking genetic markers for determining genotypes.

Genetic markers have been linked to another ETL in sheep, the Booroola fecundity gene, *FecB* (Montgomery et al., 1993). The *FecB* mutation is codominant for ovulation rate and partially dominant for litter size, with one copy of *FecB* increasing litter size by approximately one extra lamb. Twelve pedigrees were used in the linkage study; each pedigree included a heterozygous sire (*FecB*/+) carrier mated to homozygous wild-type (+/+) dams. Ovulation rate of female offspring was determined by laparoscopy, allowing classification of the females as either *FecB*/+ or +/+. After screening 144 markers through the pedigrees, linkage to *FecB* was first detected with the microsatellite OarAE101, with a maximum lod score of 17.33 at a distance of 13 cM. Using linkage information for OarAE101 from the ovine genome map, a second marker, OarHH55, was tested and found linked to *FecB* at 20 cM with a maximum lod score of 9.38. These markers, and by inference the *FecB* locus, were later assigned to sheep chromosome 6 (Crawford et al., 1995).

Summary

The sheep genome map is growing at an exponential rate; in 1989 only 41 assignments to specific chromosomes were reported, in 1993 there were 107 assignments, in 1994 there were 235 mapped loci, and now in 1995 there are over 475 assignments. The map should continue to grow at this rapid pace because the tools necessary for mapping are now in place. Genomic DNA from sheep reference families are available for distribution, both RAPD and microsatellite markers are being developed, somatic cell libraries have been constructed, and *in situ* hybridization techniques are undergoing optimization. The advancement of the ovine genome map has been instrumental in the identification of genetic markers associated with two economically important loci in sheep, the *callipyge* and Booroola fecundity genes. The mapping of additional loci controlling wool, growth, and carcass traits is expected in the next few years.

Literature Cited

Andersson, L., C. S. Haley, H. Ellegren, S. A. Knott, M. Johansson, K. Andersson, L. Andersson-Eklund, I. Edfors-Lija, M. Fredholm, I. Hansson, J. Hakansson, and K. Lundstrom. 1994. Genetic mapping of quantitative trait loci for growth and fatness in pigs. Science (Wash DC) 263:1771.

Ansari, H. A., D. W. Maher, P. D. Pearce, and T. E. Broad. 1996. Resolving ambiguities in the karyotype of domestic sheep (*Ovis aries*). II. G-, Q-, and R-banded ideograms, and chromosome-specific molecular markers. Chromosoma (In press).

Ansari, H. A., P. D. Pearce, D. W. Maher, A. A. Malcolm, N. J. Wood, S. H. Phua, and T. E. Broad. 1992. Resolving ambiguities in the sheep karyotype: a molecular cytogenetic approach. Proc. 10th European Colloquium on Cytogenetics of Domestic Animals 100:21.

Beever, J. E., P. D. George, R. L. Fernando, C. J. Stormont, and H. A. Lewin. 1990. Associations between genetic markers and growth and carcass traits in a paternal half-sib family of Angus cattle. J. Anim. Sci. 68:337.

Bleck, G. T., and R. D. Bremel. 1993. Correlation of the α-lactoalbumin (+15) polymorphism to milk production and milk composition of Holsteins. J. Dairy Sci. 76:2292.

Broom, M. F., and D. F. Hill. 1994. Construction of a large-insert yeast artificial chromosome library from sheep DNA. Mamm. Genome 5:817.

Burkin, D. J., H. G. Moarse, T. E. Broad, P. D. Pearce, H. A. Ansari, P. E. Lewis, and C. Jones. 1993. Mapping the sheep genome: Production of characterized sheep × hamster cell hybrids. Genomics 16:466.

Cockett, N. E., S. P. Jackson, T. L. Shay, D. Nielson, S. S. Moore, M. R. Steel, W. Barendse, R. D. Green, and M. Georges. 1994. Chromosomal localization of the callipyge gene in sheep (*Ovis aries*) using bovine DNA markers. Proc. Natl. Acad. Sci. USA 91:3019.

Cowan, C. M., M. R. Dentine, R. L. Ax, and L. A. Schuler. 1990. Structural variation around the prolactin gene linked to quantitative traits in an elite Holstein sire family. Theor. Appl. Genet. 79:577.

Cowan, C. M., M. R. Dentine, and T. Coyle. 1992. Chromosome substitution effects associated with κ-casein and β-lactoglobulin in Holstein cattle. J. Dairy. Sci. 75:1097.

Crawford, A. M., K. G. Dodds, A. J. Ede, C. A. Pierson, G. W. Montgomery, H. G. Garmonsway, A. E. Beattie, K. Davies, J. F. Maddox, S. W. Kappes, R. T. Stone, T. C. Nguyen, J. M. Penty, E. A. Lord, J. E. Broom, J. Buitkamp, W. Schwaiger, J. T. Epplen, P. Matthew, M. E. Matthews, D. J. Hulme, K. J. Beh, R. A. McGraw, and C. W. Beattie. 1995. An autosomal genetic linkage map of the sheep genome. Genetics 140:703.

Crawford, A. M., G. W. Montgomery, C. A. Pierson, T. Brown, K. G. Dodds, S.L.F. Sunden, H. M. Henry, A. J. Ede, P. A. Swarbrick, T. Berryman, J. M. Penty, and D. F. Hill. 1994. Sheep linkage mapping: Nineteen linkage groups derived from the analysis of paternal half-sib families. Genetics 137:373.

Cushwa, W. T., and J. F. Medrano. 1994. Identification of RAPD genetic markers in sheep. Proc. 5th World Congr. Genet. Appl. Livest. Prod. XXI:133.

Di Berardino, D., H. Hayes, R. Fries, and S. Long. 1990. (ISCNDA) The Second International Conference on Standardization of Domestic Animal Karyotypes. Cytogenet. Cell Genet. 53:65.

Ford, C. E., D. L. Pollock, and I. Gustavsson. 1980. Reading Conference (1976): Proceedings of the First International Conference for the Standardisation of Banded Karyotypes of Domestic Animals. Hereditas 92:145.

Fujii, J., K. Otsu, F. Zorzato, S. de Leon, V. K. Khanna, J. E. Weiler, P. J. O'Brien, and D. H. MacLennan. 1991. Identification of a mutation in porcine ryanodine receptor associated with malignant hyperthermia. Science (Wash DC) 253:448.

Georges, M., A. Gunawardana, D. Threadgill, M. Lathrop, I. Olsaker, A. Mishara, L. Sargeant, A. Schoeberlein, M. Steele, C. Terry, D. Threadgill, X. Zhao, T. Holm, R. Fries, and J. Womack. 1991. Characterization of a set of variable number of tandem repeat markers conserved in Bovidae. Genomics 11:24.

Georges, M., A. B. Dietz, A. Mishra, D. Nielsen, L. S. Sargeant, A. Sorensen, M. R. Steele, X. Zhao, H. Leipold, J. E. Womack, and M. Lathrop. 1993a. Microsatellite mapping of a gene causing weaver disease in cattle will allow the study of an associated quantitative trait locus. Proc. Natl. Acad. Sci. USA 90:1058.

Georges, M., R. Drinkwater, T. King, A. Mishra, S. S. Moore, D. Nielsen, L. S. Sargeant, A. Sorensen, M. R. Steele, X. Zhao, H. Leipold, J. E. Womack, and J. Hetzel. 1993b. Microsatellite mapping of a gene affecting horn development in *Bos taurus*. Nature Genet. 4:206.

Georges, M., and J. M. Massey. 1991. Velogenetics, or the synergistic use of marker assisted selection and germ-line manipulation. Theriogenology 35:151.

Georges, M., D. Nielson, M. Mackinnon, A. Mishra, R. Okimoto, A. T. Pasquino, L. Sargeant, A. Sorensen, M. Steele, X. Zhao, J. E. Womack, and I. Hoeschele. 1995. Mapping quantita-

tive trait loci controlling milk production in dairy cattle by exploiting progeny testing. Genetics 139:907.
Giovannoni, J. J., R. A. Wing, M. W. Ganal, and S. D. Tanksley. 1991. Isolation of molecular markers from specific chromosomal intervals using DNA pools from existing mapping populations. NAR 19:6553.
Hayes, H., E. Petit, C. Bouniol, and P. Popescu. 1993. Localisation of the alpha S2-casein gene (CASAS2) to the homologous cattle, sheep and goat chromosomes 4 by in situ hybridization. Cytogenet. Cell Genet. 64:28.
Horvat, S., and J. F. Medrano. 1994. Targeting a specific genomic interval to identify RAPD markers linked to the high growth (*hg*) locus in mice. Proc. 5th World Congr. Genet. Appl. Livest. Prod. XXI:71.
Horvat, S., and J. F. Medrano. 1995. Interval mapping of the *high growth* (*hg*), a major locus that increases weight gain in mice. Genetics 139:1737.
Jackson, S. P., and R. D. Green. 1993. Muscle trait inheritance, growth performance and feed efficiency of sheep exhibiting a muscle hypertrophy phenotype. Proc. West. Sect. Am. Soc. Anim. Sci. 44:364.
Johansson, M., H. Ellegren, and L. Andersson. 1995. Comparative mapping reveals extensive linkage conservation—but with gene order rearrangements—between the pig and the human genomes. Genomics 25:682.
Kehrli, M. E., Jr., D. E. Schuster, A. B. Dietz, M. Ackermann, R. Gilbert, and J. Pohlenz. 1992. Bovine leukocyte adhesion deficiency (BLAD) among Holstein cattle: Molecular biology, immunology and clinical phenotypes. In: Proc. XXIII Int. Conf. Anim. Genet. p 44. Int. Soc. Anim. Genet., Interlaken, Switzerland.
Long, S. 1985. Standard nomenclature for the G-band karyotype of the domestic sheep (*Ovis aries*). Hereditas 103:165.
Michelmore, R. W., I. Paran, and R. V. Kesseli. 1991. Identification of markers linked to disease-resistance genes by bulked segregant analysis: A rapid method to detect markers in specific genomic regions by using segregating populations. Proc. Natl. Acad. Sci. USA 88:9828.
Montgomery, G. W., A. M. Crawford, J. M. Penty, K. G. Dodds, A. J. Ede, H. M. Henry, C. A. Pierson, E. A. Lord, S. M. Galloway, A. E. Schmack, J. A. Sise, P. A. Swarbrick, V. Hanrahan, F. C. Buchanan, and D. F. Hill. 1993. The ovine Booroola fecundity gene (*FecB*) is linked to markers from a region of human chromosome 4q. Nature Genet. 4:410.
Morton, N. E. 1955. Sequential tests for the detection of linkage. Am. J. Hum. Genet. 7:277.
Muggli-Cockett, N. E., G. D. Snowder, and T. L. Shay. 1993. Investigations of adipocyte P2 and lipoprotein lipase gene expression in ovine adipose tissue. Proc. VII World Conf. Anim. Prod. 2:8.
Ng-Kwai-Hang, K. F., J. F. Hayes, J. E. Moxley, and H. G. Monardes. 1983. Association of genetic variants of casein and milk serum proteins with milk, fat and protein production in dairy cattle. J. Dairy Sci. 67:835.
Patterson, D. 1993. Mapping the way ahead. Nature Genet. 4:323.
Rettenberger, G., C. Klett, U. Zechner, J. Kunz, W. Vogel, and H. Hameister. 1995. Visualization of the conservation of synteny between humans and pigs by heterologous chromosomal painting. Genomics 26:372.
Rocha, J. L., J. F. Baker, J. E. Womack, J. O. Sanders, and J. F. Taylor. 1992. Statistical associations between restriction fragment length polymorphism and quantitative traits in beef cattle. J. Anim. Sci. 70:3360.
Rothschild, M. F., C. Jacobson, D. A. Vaske, C. K. Tuggle, T. H. Short, S. Sasaki, G. R. Eckardt, and D. G. McLaren. 1994. A major gene for litter size in pigs. Proc. 5th World Congr. Genet. Appl. Anim. Prod. XXI:225.
Rudolph, J. A., S. H. Spier, G. Byrns, C. V. Rojas, D. Bernoco, and E. P. Hoffman. 1992. Periodic paralysis in Quarter horses: A sodium channel mutation disseminated by selective breeding. Nature Genet. 2:1444.
Saidi-Mehtar, N., M. Imam Ghali, S. Heuertz, and M. C. Hors-Cayla. 1989. Sheep gene mapping by somatic cell hybridization. Cytogenet. Cell Genet. 51:1072.
Schewenger, B., S. Schober, and D. Simon. 1993. DUMPS cattle carry a point mutation in the uridine monophosphate synthase gene. Genomics 16:241.
Snowder, G. D., J. R. Busboom, N. E. Cockett, F. Hendricks, and V. T. Mendenhall. 1994. Effect of the *callipyge* gene on lamb growth and carcass characteristics. Proc. 5th World Congr. Genet. Appl. Livest. Prod. XVIII:51.

Sorenson, K. N., N. E. Cockett, D. D. Kress, D. C. Anderson, D. D. Doornbos, and D. Bailey. 1994. Allele frequencies of the αs1-casein, κ-casein and β-lactoglobulin genes in beef cattle and estimation of associations with milk fat percentage, milk protein percentage, and milk yield. Proc. West. Sect. Am. Soc. Anim. Sci. 45:3.

Integrating the Syntenic, Physical, and Linkage Maps

James E. Womack

Department of Veterinary Pathobiology, Texas A&M University, College Station, TX 77843-4467, U.S.A.

Abstract Multiple mapping methods have been applied to animal genomes, each resulting in a unique map with its own intrinsic value. Integration of maps is important because it enhances the value of the sum total of genomic information and also provides an internal test for accuracy and reliability of this information. Integration of maps simply requires mapping the same locus by multiple methods. In practice, however, this is not always a simple task. Type I markers, because of their value in comparative mapping, are most often used for synteny and cytogenetic mapping. Type II markers, by definition highly polymorphic, are more likely to comprise the backbone of linkage maps. Integration of maps has been accomplished by incorporating Type I markers into linkage maps. This is an extremely important exercise for comparative mapping and the eventual extrapolation of data from the map-rich genomes of mouse and man to the map-poor genomes of livestock species. New techniques to identify polymorphisms in Type I markers and the use of interspecific backcrosses can be used to enhance allelic segregation. Linkage maps have been integrated with cytogenetic maps through the *in situ* hybridization of cosmids containing microsatellites. Recent success with *in situ* PCR suggests that integration can also be accomplished by direct cytogenetic mapping of the microsatellites that constitute the linkage map.

Key Words: Genome Mapping, Integration, Synteny, Linkage, Cytogenetic, Comparative Mapping

Introduction

Gene mapping in animals has been an integral component of genetic research for most of this century. Sturtevant (1913) introduced animal biologists to the underlying principles of genomic research with his recombination map of the Drosophila X-chromosome. Hutt (1936) published a chicken linkage map almost 60 years ago, the first for a farm animal. Only a few genes were mapped in other animal species over the next several decades, however, except for the comprehensive genetic map produced for the laboratory mouse. Both biological and technical constraints retarded the proliferation of maps of other mammalian genomes. Linkage mapping requires the analysis of co-segregation of variant loci in numbers of meiotic products. Large collections of mutants and the application of immunological and biochemical techniques to gene products provided the necessary allelic variation for the mouse linkage map. Hutt (1936) also capitalized on availability of mutant stocks and morphological variants for his chicken map. Crossing, followed by backcrossing or intercrossing of inbred or partially inbred strains, provided the meiotic products for both the chicken and mouse maps. The recent genomics revolution, most apparent in modern human medical genetics but also presently emerging as a major force in livestock breeding, was made possible by several technical advances. First, analysis of hybrid somatic cells circumvented both the requirements for meiotic segregation and intraspecies allelic variation. Popularized in the 1970s, somatic cell techniques still account for more assignments of human genes to chromosomes than any other method. Second, a variety of molecular techniques revealed DNA level polymorphism that far exceeds the resolution possible in the electrophoretic or immunological analysis of gene products. Human geneticists had the wisdom to pool their segregation analyses in a common set of meiotic products, the widely used CEPH families. Mouse geneticists expanded their resources to include recombinant-inbred strains and more recently, interspecific backcrosses. Following the lead of human and mouse geneticists, animal scientists now appreciate the potential value of genome maps in livestock improvement (Soller and Beckmann, 1982; Smith and Simpson, 1986; Beckmann and Soller, 1988) and have embarked on a mission to map genomes of domestic animals. Genome maps of farm animals, like those of mouse and human, are being constructed by different methods and for different reasons. Each map has its own purpose and its own value as a unique entity. Nonetheless, integration by including common unique anchor sequences in the various maps enhances the sum total of genomic information. Integration also provides an important check for accuracy of assignments.

Why Map Genomes?

To Study Chromosomal Evolution Most mammalian genome maps are constructed with the initial goal of advancing the fundamental biological knowledge of the species. Biologists are curious about the organization of animal genomes, how this organization came to be, and how it influences gene expression. It is becoming increasingly apparent that the well-funded and well-organized human genome initiative will ultimately provide the foundation of our basic knowledge of mammalian genomes. The laboratory mouse will be a major contributor to this knowledge base as well. Other mammals, including the livestock species, have important roles to play, however, in unraveling the pathways of mammalian genome evolution. Mapping homologous genes in multiple species provides more information about chromosomal evolution than we can retrieve with today's best cytogenetic technologies.

To Permit Marker-Assisted Selection The current focus of animal genome efforts in many laboratories is toward the more practical goal of marker-assisted selection (MAS). DNA markers of positive alleles for economic trait loci (ETL), many of which may be quantitative trait loci (QTL), have the potential to rapidly accelerate the efficiency of selective breeding (Soller and Beckmann, 1982; Smith and Simpson, 1986; Weller et al., 1990). The ideal marker for selecting a phenotypic trait is the actual gene responsible for the trait. Thus, the "candidate gene" approach seeks allelic variation in genes directly responsible for economically important phenotypic variation. This approach requires a fundamental knowledge of the physiological and developmental basis of the trait followed by extensive screening for variation in the candidate genes related to the known or suspected physiological mechanisms. Sequence variation related to a positive or negative phenotype must then be demonstrated. Ideally, this sequence variation can be incorporated into the marker assay. Shuster et al. (1992), for example, identified the genetic defect responsible for leukocyte adhesion deficiency (LAD) in Holstein cattle as a missense mutation coding amino acid 128 in CD18. The polymerase chain reaction (PCR) was employed to distinguish the mutant and normal alleles, providing the ultimate genetic marker of this economic trait locus. By detecting heterozygous carriers and avoiding their use as breeding stock, the mutant allele can be eliminated from a herd in a single generation.

Unfortunately, the physiological basis and consequently the genes for most economic traits are unknown. Loci responsible for variation in such traits, even quantitative trait loci, can be genetically mapped, however (Lander and Botstein, 1989; Paterson et al., 1989; Georges et al., 1993a,b, 1995; Andersson et al., 1994). Markers mapped in close proximity to ETL can be used for selection if the recombination frequency (map distance) is sufficiently small and the "phase" of marker and ETL alleles is known. The frequency of meiotic recombination between linked loci defines the unit of a genetic map, 1% recombination represented as one centimorgan (cM). Thus, a marker 10 cM from an ETL can be used to select the ETL with 90% probability of success since 10% of the gametes produced by a parental animal will be recombinants of the marker and ETL alleles. Efficiency can be dramatically increased by identifying markers flanking the ETL since recombination in the region spanned by two markers can be detected. One goal of animal gene mapping is therefore a map of highly polymorphic markers spaced at intervals of 20 cM or less across every cattle chromosome.

Linked markers can be used in selection only with knowledge of phase of the marker allele and the ETL allele. This is not a problem in pedigrees where the phase is initially known and the probability of recombination is small. Linked markers cannot be used at the population level (i.e., no pedigree information), however, unless the marker and trait are in linkage disequilibrium. This reemphasizes the value of a marker in the gene responsible for the trait of interest.

To Facilitate Positional Cloning A third and perhaps economically the most important goal of gene mapping is to identify and clone genes responsible for ETL. The terms "positional cloning" and "map-based cloning" have replaced "reverse genetics" to denote the application of mapping information to clone a gene responsible for a trait in the absence of information about the biochemical basis of the trait. The dense human linkage map coupled with availability of large insert libraries and immensely valuable deleted chromosomes have led to the cloning of several disease genes. The successful search for the cystic fibrosis gene (Rommens et al., 1989) is an excellent example. The task of cloning genes for ETL in livestock is a formidable one, however, for a number of reasons. Our maps are not so dense. Large insert libraries for livestock are only beginning to be developed and used. Chromosomal deletions of economically important genes have not been identified and propagated. Moreover, funding for enhancing these resources does not seem to be imminent. Our task is further complicated by the quantitative nature of most of our traits of interest. Alternative strategies to conventional positional cloning must be developed.

Methods of Gene Mapping

Synteny Mapping

Synteny simply means on the same strand, or, in a genetic sense, on the same chromosome. A synteny map is nothing more than a list of genes that are demonstrated to be located on the same chromosome. Since synteny mapping in most species has preceded complete assignment of genes to specific chromosomes, designations U1, U2, and so on were given to "unassigned" syntenic groups. These should be considered temporary place holders that eventually give way to specific chromosomal assignments. Unfortunately, a blatant misuse of the word "synteny" has crept into recent genetic literature. Such statements as "... there was a greater degree of synteny between cattle and human than between cattle and mouse," or "... only a limited degree of synteny has been demonstrated between human, mouse and swine" are genetically meaningless and are apparent misuses of the term synteny as an abbreviation for Nadeau's (1989) useful definition of synteny conservation.

The common approach to synteny mapping is through somatic cell genetics. Hybrid somatic cells can be constructed in such a configuration that the chromosomes of one species are preferentially lost. Each clone of hybrid cells will have a partial genome of that species along with the complete genome of the other, usually a rodent cell line. Since chromosome loss is largely random, each clone will have a different subset of chromosomes retained from the species being mapped. Analysis of pairs of genes across such a panel of hybrid cell lines will reveal concordance or discordance of retention, suggesting their location on the same or different chromosomes, respectively. Gene products or DNA sequences may be mapped by synteny analysis, the only criterion being the ability to detect the presence or absence of the gene or gene product of the targeted species against the rodent background. Enzyme electrophoresis, Southern blotting, and PCR have all been effective analytical tools for synteny mapping.

Somatic cell genetics does not ordinarily assign markers to specific chromosomal regions. Consequently, genes in a synteny map are usually not ordered. Somatic cell methods employing rearranged chromosomes are an exception to this generalization and have been used very effectively in the human map. Unfortunately, panels of hybrid cell lines with rearranged chromosomes have not yet been systematically employed in mapping non-human genomes.

In Situ Hybridization

The physical localization of genes or other cloned DNA segments to specific sites on chromosomes is effectively achieved by *in situ* hybridization. This technique employs the labeling of a probe with a visually detectable marker, followed by hybridization of the probe to denatured DNA of the otherwise intact chromosome. The specificity of the reaction is dictated by the uniqueness of the probe used. Whole genomes, highly repetitive elements, and unique sequences have all been successfully localized by *in situ* hybridization. While radioactive probes predominated in the early application of this technology, fluorescent probes are now generally used. Trask (1991) reviewed the advantages of fluorescence *in situ* hybridization (FISH) over isotopic labeling. It provides superior spatial resolution and generally requires the visualization of fewer labeled chromosomes. It is faster and the probe is more stable. The sensitivities are similar, each requiring a few kilobase (kb) pairs of uninterrupted sequence as the hybridizing probe. Schemes have been developed that allow multiple probes with different color signals to be used on the same chromosomes.

Cosmids, cloning vectors that accommodate large DNA sequences (up to 45 kb), are effective probes for FISH. These large inserts often contain repetitive DNA and, consequently, the target DNA must first be prehybridized with unlabeled total genomic DNA. This method has been effectively used to anchor the growing bovine linkage map to chromosomes, as described below.

The above technologies result in physical maps that describe the localization of genes in physical relationships to other genes and to the chromosomes on which they reside. Higher-resolution physical maps that place markers in contiguous clones (contig maps) are forthcoming in livestock but will most likely span small regions of special interest rather than whole chromosomes, as is targeted in the human genome initiative and is a prerequisite to total genome sequencing.

Linkage Mapping

Linkage mapping defines biological relationships of markers and is evaluated in terms of meiotic recombination. Linkage is therefore measurable only in gametic products. This requires detection of variant markers that distinguish maternal and paternal chromosomes in heterozygous individuals. A linkage map is made by calculating the percentage of recombinants in the parental arrangements of alleles on a chromosome. This requires animals that are heterozygous for multiple loci and sufficient numbers of offspring (or gametes) for a statistically meaningful analysis. The segregation of three or more loci permits ordering of genes on the map on the assumption that double recombinants are rare relative to single region recombinants.

O'Brien (1992) categorized markers for mapping purposes as class I or II. Class I markers are expressed sequences (genes). They are usually conserved from one species to another and are thus the material for comparative gene mapping. Unfortunately they are often not polymorphic and therefore are difficult to incorporate into linkage maps. Class II markers are more widely used for linkage mapping. These are markers whose value lies in their high level of polymorphism. As such, they are not necessarily expressed sequences but often anonymous stretches of DNA.

Two types of polymorphic DNA markers dominate the linkage maps of most plant and animal species. Restriction fragment length polymorphisms (RFLPs) identify loci that differ in the presence or absence of specific restriction sites. Electrophoresis of enzyme digested DNA, blotting onto a solid membrane, and hybridization to a labeled probe (Southern, 1975) often reveals polymorphism in DNA fragment sizes. The major advantage of RFLPs as linkage markers is that they often define polymorphisms in or around Type I markers. Thus, they are useful for comparative mapping and are also a potential source of variation in candidate genes for economic traits. Blotting and probe development are laborious and time-consuming, however.

The discovery of microsatellites (Weber and May, 1989; Fries et al., 1990) revolutionized linkage mapping in most animal species. These islands of di-, tri-, or tetranucleotide repeats are apparently ubiquitous throughout animal genomes. The number of repeats is often polymorphic, with from two to a dozen or more alleles present in a breeding population. They are defined by unique flanking sequences that provide primers for amplification by PCR. Microsatellites have become the predominant Type II marker in the growing linkage maps of livestock species.

Other types of markers such as enzyme polymorphisms, blood group antigens, or phenotypic traits such as coat color or horn development also appear on linkage maps. These all add interest and biological significance to the map but are not generally sufficiently polymorphic in reference families to provide the skeletal frame work.

Linkage mapping also requires observable meiotic products, usually in the form of offspring from individuals segregating the markers discussed above. Large full sib-ships or half sib-ships are ideal for linkage analysis. Since a comprehensive map requires large numbers of markers scored on a common set of meiotic products, sharing of family material among laboratories is the most effective arrangement for generating a linkage map. Sets of shared reference families are available for most species, including cattle (Barendse et al., 1994). Offspring are not essential for analyzing linkage, however, since multiple DNA loci can now be analyzed for allelic variation directly on meiotic products. Van Eijk et al. (1993) have ordered three loci on bovine Chr 23 by PCR after primer extension preamplification (PEP) of individual sperm.

Integrating Maps

Each mapping experiment results in a unique map with its own limitations and experimental error. Since individual genomic variation prohibits the construction of a single definitive map of the genome of a species, maps are at best approximations of some mythical biological reality. Individual genomic variation within a species is limited, however, and consequently such an approximation as a map provides a very useful tool. Individuals that constitute a species generally have the same number of morphologically indistinguishable chromosomes with the exception of sexual dimorphism in one pair. The arrangement of repetitive and unique DNA sequences, including genes, is highly conserved within these chromosomes. The conservation of chromosomal morphology and genetic content, including gene order, permits the extrapolation of genomic information from individual to individual. Meiotic recombination rates vary between sexes and probably from individual to individual within sexes. Nonetheless, with rare exceptions linkage maps should reflect a common order of loci on a particular chromosome, even though recombination rates between loci may vary. Furthermore, this order should be consistent with cytogenetic maps of the same chromosome and the total content of both the linkage and cytogenetics maps of a chromosome should be consistent with the synteny map.

Maps are integrated when they share common unique loci often referred to as anchor loci simply because they anchor one map to the other. There are several reasons to integrate maps, not the least of which is to check accuracy. All maps are subject to experimental error and a comparison of anchor locus position in maps produced from different experiments is important to the mapping process. Partially anchored maps also aid in the extrapolation of non-anchored data from one map to another. Consider the hypothetical linkage maps in Figure 1, which are integrated by the common anchor loci, A, B, and C. Markers x, y, z, p, q, and r are not integrated, yet the location of x and y can be extrapolated to the region above B in map II and r can be extrapolated to the region below C in map I. Neither the linear order of x, y, and z nor the order of p and q can generally be extrapolated from such maps since these markers were not all analyzed in a common set of meiotic products. Order is a most likely estimate even in the individual experiments.

Integration of cytogenetic and linkage maps provides an evaluation of chromosomal coverage of the linkage map. As illustrated in Figure 2, cytogenetic mapping of the most telomeric marker on the bovine chr 1 linkage map suggests nearly complete coverage of the chromosome in that direction. Complete chromosome coverage of the linkage map in the centromeric direction is likely, although not directly demonstrated. Recombination fractions cannot be unequivocally correlated with chromosomal length or base pairs of DNA.

Linkage maps are the principal tools for locating ETL. The eventual identification and isolation of the genes, however, will likely require chromosome-based biological resources. Libraries constructed from dissected chromosome regions or cell lines with informative deletions will require extrapolation of linkage information to the cytogenetic map.

Cytogenetic and linkage maps are integrated by mapping a common anchor locus by both methods. Cosmids or phage containing polymorphic linkage markers have been mapped by FISH (Solinas Toldo et al., 1993; Vaimen et al., 1993) to effectively anchor bovine linkage and cytogenetic maps. Troyer et al. (1994) have demonstrated PCR amplification directly on chromosome spreads and subsequently applied the method to microsatellites to anchor pig chromosome 10 to its respective linkage map (Xie et al., 1995). Periodic anchoring of linkage maps to chromosomes by these or other methods is an important and necessary step in the development of integrated genome maps.

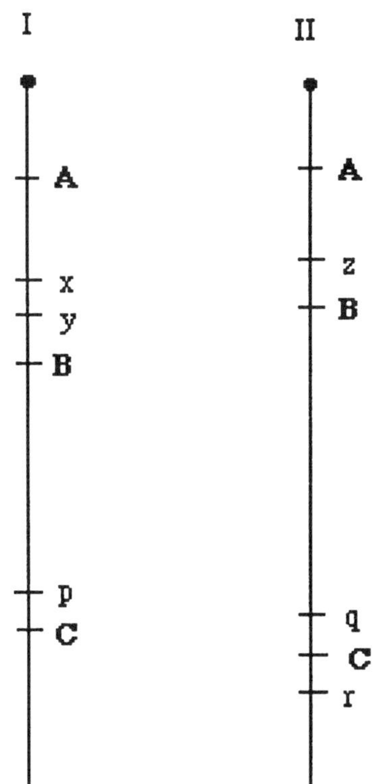

Figure 1. Two hypothetical linkage maps of the same animal chromosome illustrating integration by the anchor loci A, B, and C. Positions of non-anchored loci x, y, z, p, q, and r can be extrapolated from map to map but only with serious limitations.

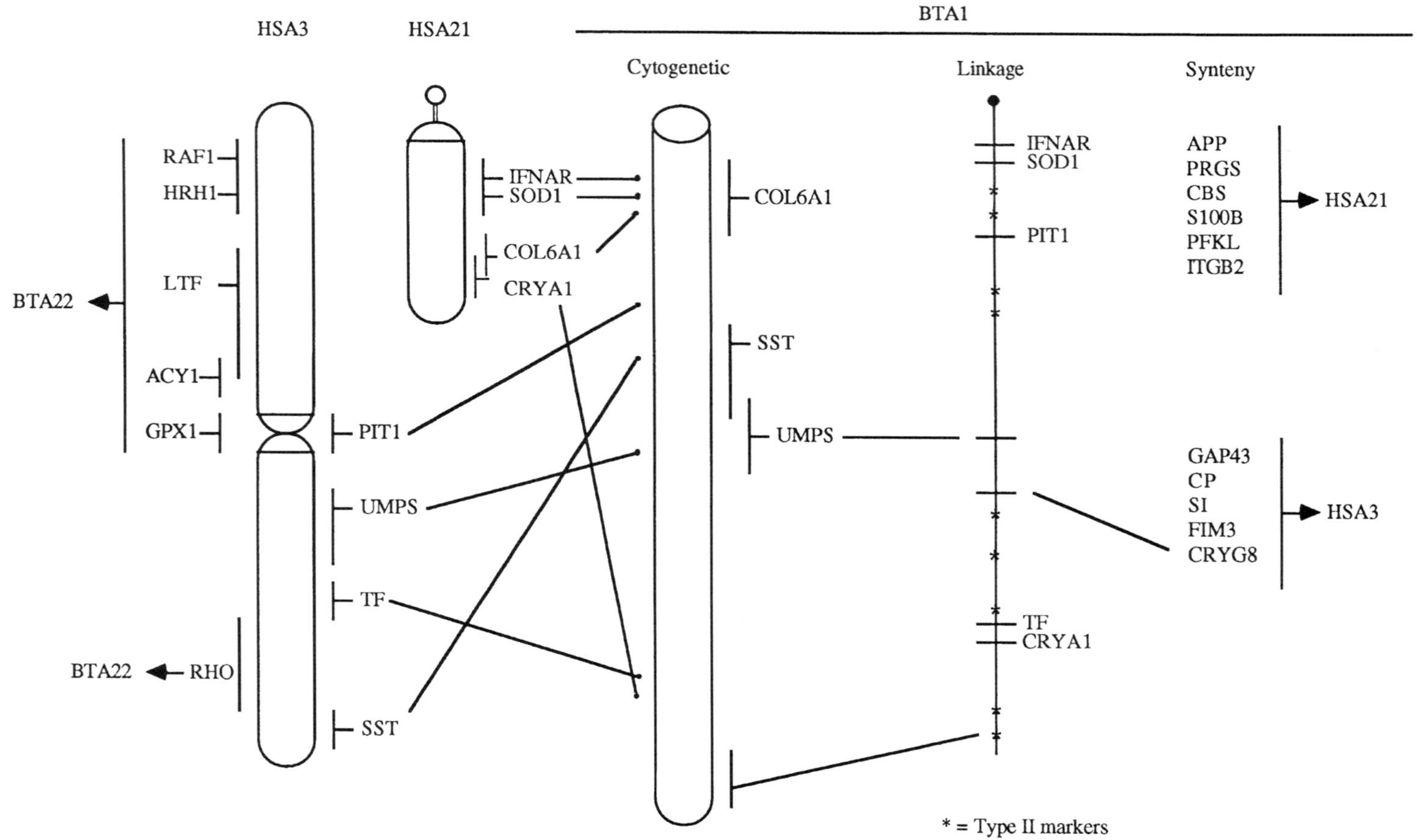

Figure 2. Current state of integration of bovine (BTA) chromosome 1 map including comparative relationship with human (HSA) chromosomes 3 and 21. The bovine cytogenetic, synteny and linkage maps all provide important comparative information. Conservation of synteny does not imply conservation of linkage.

Comparative Mapping

The ultimate integration of maps is in the interspecies comparisons we have come to call comparative gene mapping. Comparative maps have revealed that mammalian chromosome evolution has been highly conservative. Humans and mice are destined to be the "map-rich" species among mammals, and much is to be learned about cattle, sheep, and pigs by identifying conserved map regions relative to these species. Sheep and cattle genomes are closely related (Echard et al., 1994) and mappers of each should benefit from the comparative analysis of one genome with respect to the other. The same probably holds for poultry genomes (chicken, turkey, quail), although it is still unknown how much these will benefit from mammalian gene maps.

While extensive conservation of synteny has been observed between cattle and humans (Womack and Moll, 1986; Threadgill et al., 1991), conservation of linkage (conservation of gene order as defined by Nadeau, 1989) may not be so prevalent. Barendse et al. (1994) incorporated a sufficient number of Type I loci into the linkage map to demonstrate the presence of several rearrangements of gene order within conserved syntenies. This reinforces the need for comparative maps to include comparative gene order as well as comparative synteny to fully benefit from the human and mouse maps. As illustrated in Figure 2, cattle chromosome 1 has extensive synteny conservation with HSA 21 and 3. Integrated cytogenetic and linkage maps are required, however, to show the rearrangement of order of HSA21 and HSA3 genes on BTA1. O'Brien et al. (1993) addressed the value and also some of the problems of comparative gene mapping and proposed a list of 321 comparative anchor loci. These loci are spaced at an average of 10 cM intervals on the human map and may provide a common set of markers for integrating maps of different mammalian species in an efficient manner.

Comparative Positional Cloning

Sufficient numbers of mapped markers now exist in most farm animals for extensive genome coverage of families segregating ETL. The polled locus (Georges et al., 1993b), weaver (Georges et al., 1993a) and milk QTL (Georges et al., 1995) in cattle, the Booroola fecundity gene in sheep (Montgomery et al., 1993) and growth and fatness QTL in pigs (Andersson et al., 1994) are examples of mapped ETL awaiting the identification of the genes responsible for the traits. The mapping of ETL to 10-20 cM regions of chromosomes will become increasingly common over the next few years. The next step, finding the genes, remains a major obstacle in farm animals.

The recent increased emphasis on identifying and mapping all the expressed genes of mice and humans provides some optimism, however, and places an even greater premium on comparative mapping in livestock. The mapping of an ETL to a 10 cM region narrows the list of candidate genes to 200 if we assume 20 genes per centimorgan and equivalency of centimorgans across species. This also assumes the availability of comparative mapping information, including gene order, at the 10 cM level. The pool of candidate genes or expressed sequence tags (ESTs) will come from the rapidly expanding human and mouse maps (Figure 3). Candidate tissues for ETL expression can further reduce this number. The rate of development of EST maps in humans and mice is difficult to predict, but nonetheless the trend is

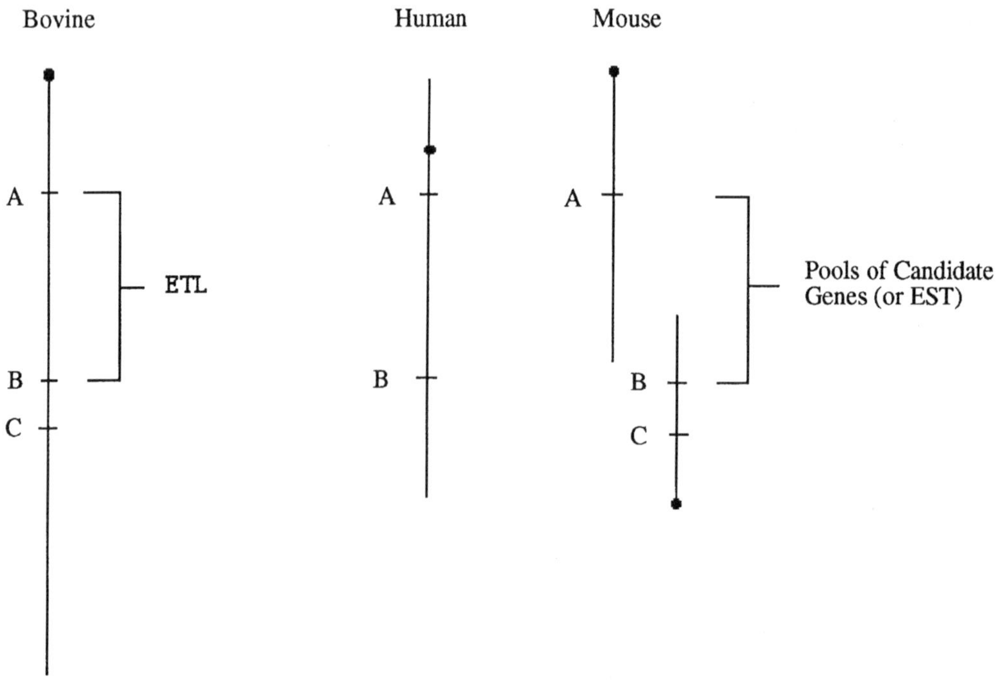

Figure 3. Hypothetical illustration of basic strategy for comparative positional cloning. Mapping of comparative anchor reference loci A, B, and C is essential to extrapolating candidate genes for animal ETL from the human and mouse maps.

encouraging. New methods for trapping and identifying transcribed sequences are emerging. Comparative maps may very well provide our best hope for map-based cloning and cannot be neglected in our meager but determined initiatives to map genomes of farm animals.

Literature Cited

Andersson, L., C. S. Haley, H. Ellegren, S. A. Knott, M. Johansson, K. Andersson, L. Andersson-Eklund, I. Edfors-Lilja, M. Fredholm, I. Hansson, J. Håkansson, and K. Lundström. 1994. Genetic mapping of quantitative trait loci for growth and fatness in pigs. Science (Wash DC) 236:1771.

Barendse, W., S. M. Armitage, L. M. Kossarek, A. Shalom, B. W. Kirkpatrick, A. M. Ryan, D. Clayton, L. Li, H. L. Neibergs, N. Zhang, W. M. Grosse, J. Weiss, P. Creighton, F. McCarthy, M. Ron, A. J. Teale, R. Fries, R. A. McGraw, S. S. Moore, M. Georges, M. Soller, J. E. Womack, and D.J.S. Hetzel. 1994. A genetic linkage map of the bovine genome. Nature Genet. 6:227.

Beckmann, J. S., and M. Soller. 1988. Detection of linkage between marker loci and loci affecting quantitative traits in crosses between segregating populations. Theor. Appl. Genet. 76:228.

Echard, G., T. E. Broad, D. Hill, and P. Pearce. 1994. Present status of the ovine gene map (*Ovis aries*); comparison with the bovine map (*Bos taurus*). Mamm. Genome 5:324.

Fries, R., A. Eggen, and G. Stranzinger. 1990. The bovine genome contains polymorphic microsatellites. Genomics 8:403.

Georges, M., A. B. Dietz, A. Mishra, D. Nielsen, L. Sargeant, A. Sorensen, M. R. Steele, X. Zhao, H. Leipold, and J. E. Womack. 1993a. Microsatellite mapping of the gene causing weaver disease in cattle will allow the study of an associated quantitative trait locus. Proc. Natl. Acad. Sci. USA 90:1058.

Georges, M., R. Drinkwater, T. King, A. Mishra, S. S. Moore, D. Nielsen, L. S. Sargeant, A. Sorensen, M. R. Steele, X. Zhao, J. E. Womack, and J. Hetzel. 1993b. Microsatellite mapping of a gene affecting horn development in *Bos taurus*. Nature Genet. 4:206.

Georges, M., D. Nielsen, M. Mackinnon, A. Mishra, R. Okimoto, A. T. Pasquino, L. S. Sargeant, A. Sorensen, M. R. Steele, X. Zhao, J. E. Womack, and I. Hoeschele. 1995. Mapping quantitative trait loci controlling milk production in dairy cattle by exploiting progeny testing. Genetics 139:907.

Hutt, F. B. 1936. Genetics of the fowl. VI. A tentative chromosome map. Neue Forschungen in Tierzucht und Abstammungslehre (Duerst Festschrif). pp 105–112. Bern, Verbandsdruckerei.

Lander, E. S., and D. Botstein. 1989. Mapping Mendelian factors underlying quantitative traits using RFLP linkage maps. Genetics 121:185.

Montgomery, G. W., A. M. Crawford, J. M. Penty, K. G. Dodds, A. J. Ede, H. M. Henry, C. A. Pierson, E. A. Lords, S. M. Galloway, A. E. Schmack, J. A. Sise, P. A. Swarbrick, V. Hanrahan, F. C. Buchanan, and D. F. Hill. 1993. The ovine Booroola fecundity gene (*FecB*) is linked to markers from a region of human chromosome 4q. Nature Genet. 4:410.

Nadeau, J. H. 1989. Maps of linkage and synteny homologies between mouse and man. Trends Genet. 5:82.

O'Brien, S. J. 1992. Mammalian genome mapping; lessons and prospects. Curr. Opin. Genet. Dev. 1:105.

O'Brien, S. J., J. E. Womack, L. A. Lyons, K. J. Moore, N. A. Jenkins, and N. G. Copeland. 1993. Anchored reference loci for comparative genome mapping in mammals. Nature Genet. 3:103.

Paterson, A. H., E. S. Lander, J. D. Hewitt, S. Peterson, S. E. Lincoln, and S. D. Tanksley. 1989. Resolution of quantitative traits into Mendelian factors by using complete linkage map of restriction fragment length polymorphisms. Nature (Lond.) 335:721.

Rommens, J. M., M. C. Iannuzzi, B.-S. Kerem, M. L. Drumm, G. Melmer, M. Dean, R. Rozmahel, J. L. Cole, D. Kennedy, N. Hidaka, M. Zsiga, M. Buchwald, J. R. Riordan, L.-C. Tsui, and F. S. Collins. 1989. Identification of the cystic fibrosis gene: Chromosome walking and jumping. Science (Wash DC) 245:1059.

Shuster, D. E., M. E. Kehrli, Jr., M. R. Ackermann, and R. O. Gilbert. 1992. Identification and prevalence of a genetic defect that causes leukocyte adhesion deficiency in Holstein cattle. Proc. Natl. Acad. Sci. USA 89:9225.

Smith, C., and S. P Simpson. 1986. The use of genetic polymorphism in livestock improvement. J. Anim. Breed Genet. 103:205.

Solinas Toldo, R. Fries, P. Steffen, H. L. Neibergs, W. Barendse, J. E. Womack, D.J.S. Hetzel, and G. Stranzinger. 1993. Physically mapped, cosmid-derived microsatellite markers as anchor loci on bovine chromosomes. Mamm. Genome 4:720.

Soller, M., and J. S. Beckmann. 1982. Restriction fragment length polymorphisms and genetic improvement. Proc. 2nd World Congr. Genet. Appl. Livest. Prod. 6:396.

Southern, E. 1975. Detection of specific sequences among DNA fragments separated by gel electrophoresis. J. Mol. Biol. 98:503.

Sturtevant, A. H. 1913. The linear arrangement of six sex-linked factors in *Drosophila*, as shown by their mode of association. J. Exp. Zool. 14:43.

Threadgill, D. S., J. P. Krause, S. A. Krawetz, and J. E. Womack. 1991. Evidence for the evolutionary origin of human chromosome 21 from comparative gene mapping in the cow and mouse. Proc. Natl. Acad. Sci. USA 88:154.

Trask, B. J. 1991. Fluorescence in situ hybridization. Trends Genet. 7:150.

Troyer, D. L., H. Xie, D. W. Goad, and D. Z. Skinner. 1994. Use of a new technique to map the porcine α interferon gene to chromosome 1. Mamm. Genome 5:112.

Vaimen, D., I. Bahri-Darwich, D. Mercier, M. Yerle, A. Eggen, H. Leveziel, G. Guerin, J. Gellim, and E. P. Cribu. 1993. Mapping of new bovine microsatellites on cattle chromosome 15 with somatic cell hybrids, linkage analysis, and fluorescence in situ hybridization. Mamm. Genome. 4:676.

Van Eijk, M.J.T., I. Russ, and H. A. Lewin. 1993. Order of bovine *DRB3*, *DYA*, and *PRL* determined by sperm typing. Mamm. Genome 4:113.

Weber, J. L., and P. E. May. 1989 Abundant class of human DNA polymorphisms which can be typed using the polymerase chain reaction. Am. J. Hum. Genet. 44:388.

Weller, J. L., Y. Kashi, and M. Soller. 1990. Power of daughter and granddaughter designs for determining linkage between marker loci and quantitative trait loci in dairy cattle. J. Dairy Sci. 73:2525.

Womack, J. E., and Y. D. Moll. 1986. A gene map of the cow: Conservation of linkage with mouse and man. J. Hered. 77:2.

Xie, H., L. J. Alexander, G. A. Rohrer, C. W. Beattie, and D. L. Troyer. 1995. Use of DISC-PCR and FISH to assign a linkage group to pig chromosome 10. Mamm. Genome 6:139.

Mapping Genes for Disease Resistance: Lessons from the Bovine Leukemia Virus

Harris A. Lewin and Michael L. Mirsky

Laboratory of Immunogenetics, Department of Animal Sciences, University of Illinois at Urbana-Champaign, 1201 W. Gregory Drive, Urbana, IL 61801, U.S.A.

Abstract Despite dramatic improvements in management practices, infectious diseases are still a major liability to producers of food animals. The development of low-resolution linkage maps of farm animal genomes allows for new opportunities to map genes that affect disease resistance and susceptibility. A proven strategy for mapping loci that affect host resistance to infectious diseases (IDL) is classical linkage analysis. For mapping IDL by linkage analysis, it is necessary to have a suitable population of animals scored for a disease phenotype(s). Disease traits can be categorical or have a continuous distribution. Mapping IDL can be achieved by genome scanning or the candidate gene approach, which may use comparative mapping information. New technologies with the potential for cloning of segments that are tightly linked to IDL include genome mismatch scanning and representational difference analysis. These new methods could eliminate the need for scoring large numbers of markers, thus conserving time and effort. These strategies are, at least in theory, applicable to outbred populations and family structures that are presently being used to map economic trait loci in dairy and beef cattle.

Key Words: Genetic Resistance, Infectious Diseases, Bovine Leukemia Virus

Historical Overview

Despite the dramatic improvements in sanitation, nutrition, diagnostic tests, and veterinary care, infectious diseases are still a significant threat to the sustainability of animal agriculture. Available data suggest that infectious diseases cause at least a $5 to 6 billion loss per annum to the nation's food animal producers (Drummond et al., 1981). Furthermore, the persistence of zoonotic diseases such as salmonellosis and brucellosis present a continuing hazard to human health. The huge monetary losses due to infectious diseases and the implications for human health provide a framework for the recent emphasis on identifying genes that influence resistance and susceptibility to infectious diseases (IDL) in farm animals.

Historically, the identification of disease resistance in farm animals has its roots in animal husbandry. In F. B. Hutt's classic book *Genetic Resistance to Disease in Domestic Animals* (1958), carefully documented evidence for the genetic control of resistance to infectious diseases in different species is cited, ranging from American foulbrood in bees to tuberculosis in Zebu cattle. In 1964, when most of the present generation of animal geneticists were still in grade school, Hutt wrote:

> The reason why genetic resistance to disease has been so little utilized in domestic animals is not any lack of such resistance, but, rather, inability or reluctance on the part of most official controllers of disease to recognize its potentialities. Disinfectants, vaccination, sulpha drugs and antibiotics are orthodox; genes are not (Hutt,1964).

Thus, Professor Hutt introduced the concept that diseases could be controlled by genetic selection. Thirty-one years later, with a new era ushered in by molecular biology, we are on the threshold of putting F. B. Hutt's ideas into practice.

Strategies for Mapping and Cloning Infectious Disease Loci

The approaches for identification of IDL continue to evolve rapidly in parallel with advances in molecular genetics. Strategies differ, depending on whether one is working with inbred or non-inbred species (reviewed by Lewin et al., 1991 and by Malo and Skamene, 1994). The scheme presented will be limited to non-inbred species that may be under selection (Figure 1). First, it is desirable to have accurate diagnostic tests and a resistance phenotype that is clearly defined. Second, a suitable study population segregating the disease must be identified. Carefully designed population studies often are a first step in defining a disease association with a candidate gene. Subsequently, or in lieu of a population study, family studies are conducted to confirm the association between genetic markers and a disease. Families exhibiting variation in resistance and susceptibility to a disease provide a sensitive means of determining whether resistance has a genetic component. Planned matings may be implemented for direct challenge with pathogens and subsequent matings performed with animals resisting the challenge. Backcross animals directly challenged with the pathogen may then be used for classical linkage studies. The latter approach is quite feasible with smaller animals, such as chickens, that have a short generation interval. Although direct challenge experiments are possible with larger animals, the expense and the time required to produce answers generally dictates that studies employ existing families or herds segregating resistance to a disease under natural conditions of exposure. After a suitable population is identified and the disease trait scored or measured, the mode of inheritance can be determined. Appropriate statistical procedures are then used to determine whether it is a monogenic or polygenic trait that is inherited in a dominant, recessive, or additive fashion. When it has been determined that the disease trait is under genetic control, then mapping of the gene(s) can proceed.

Genetic mapping of IDL in non-inbred species most commonly employs linkage analysis in families. Linkage mapping can be accomplished by whole genome scanning if a map consisting of an adequate number of polymorphic markers is available for the species of interest. When a putative orthologous gene has been mapped in another species, comparative information can be used to facilitate mapping (discussed below). If the genomic region of interest contains a cloned candidate gene, the clone or a tightly linked microsatellite may be used as a probe to test directly the hypothesis of linkage.

Two new techniques, genetically directed representational difference analysis (GDRDA; Lisitsyn et al., 1994) and genome mismatch scanning (Nelson, 1995), permit the identification of markers closely linked to an unknown gene. In non-inbred species, both methods seek to resolve small chromosome segments that are identical by descent and that also contain a putative disease-associated locus. In theory GDRDA should be applicable to all traits that are segregating in large half-sib families.

If a marker is closely linked to an IDL the polymorphism provides a point of entry into the genome for cloning of the gene responsible for the trait. Two approaches for isolating a gene when map position is known are "positional cloning" and selection of "candidate genes" (reviewed by Collins, 1995). In the former, one identifies large genomic inserts having polymorphic markers that flank the unknown gene. To conduct positional cloning, one would like to begin with markers that are separated by no more than 1 to 2 cM (approximately 1 to 2 million base pairs). Following the

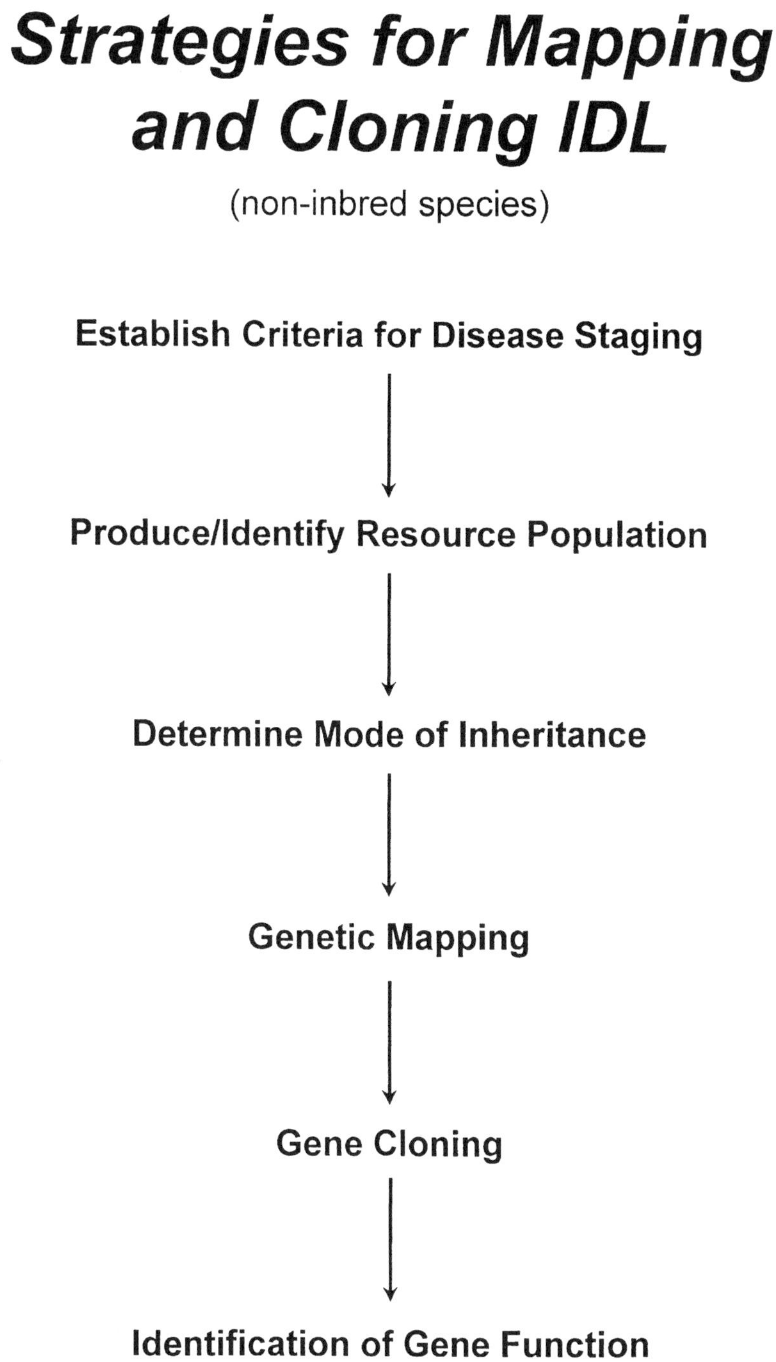

Figure 1. Strategies for mapping and cloning infectious disease loci (IDL).

isolation of large inserts containing flanking markers and subcloning, regions containing coding sequences could be isolated by exon trapping (Buckler et al., 1991). For positional cloning one must also have an assay system that indicates whether the gene isolated has the potential to be the unknown gene of interest. At present, this is the most serious limitation in the positional cloning of quantitative trait loci, because the tissues expressing the gene(s) controlling the trait may not be known.

The candidate gene approach permits short-cutting the laborious process of positional cloning. This strategy uses information on known genes mapped to a given chromosomal interval. The candidate gene approach is particularly powerful when combined with comparative mapping information. An excellent example of the combined approach was the identification of the mutation in the calcium receptor channel gene as the cause of halothane susceptibility in pigs (Fujii et al., 1991).

Use of Comparative Gene Maps for Mapping Infectious Disease Loci

Genes influencing resistance and susceptibility to infectious diseases have been mapped and cloned in a number of species (reviewed by Lewin et al., 1991 and by Malo and Skamene, 1994). Comparative mapping has been used to great advantage for this purpose. In general, it is more difficult to demonstrate conserved function than conserved linkage, yet there are a number of noteworthy examples (Table 1). Two examples will be discussed here for brevity. The *mx1* gene of the mouse has been shown dramatically to influence susceptibility of inbred mouse strains to influenza virus, an orthomyxovirus (Lindenmann, 1964). A deletion of exons 9-11 in the *mx1* gene has been found in most susceptible strains (Staeheli et al., 1988). The orthologue of *mx1* has been mapped in humans (*MXA* on HSA21) and in several farm animal species (Malo and Skamene, 1994). Pathogenic orthomyxoviruses that cause significant diseases exist in many of these species (i.e., fowl plague, equine influenza, and swine influenza). Thus, the *mx1* gene may have commercial applications (Müller et al., 1992).

Table 1. Comparative mapping of cloned infectious disease loci

	Chromosome[a]			
Gene[b]	Human	Mouse	Cattle	Pig
MHC	6	17	23	7
MXA	21	16	1	13
XLA/BTK	X	X	X?	X?
NRAMP1	2	1	2?	15?
CD18	21	10	1	13?
FGR	1	4	2	6?

[a]Question mark after a chromosome indicates assumed chromosomal location from existing comparative mapping information.
[b]Human nomenclature.

Another example of comparative mapping facilitating the identification of orthologous IDL was the discovery of the mutation responsible for bovine leukocyte adhesion deficiency (Shuster et al., 1992a,b). Bovine leukocyte adhesion deficiency is an autosomal recessive disease that results in multiple functional abnormalities of neutrophils. The *CD18* gene encodes the β subunit of the heterodimeric β_2 integrins that are required for neutrophils to adhere to endothelial surfaces that serve as a portal to sites of inflammation. Cattle homozygous for the allele of *CD18* that encodes an Asp→Gly substitution at residue 128 usually die of bacterial infections within a few months of age. A similar disease was first identified in humans and clearly associated with a defect in the expression of *CD18* (Arnaout et al., 1990). Other IDL that have potential for being important in multiple species include *XLA/BTK* (X-linked immunodeficiency; Rawlings et al., 1993; Thomas et al., 1993) and *NRAMP1* (susceptibility to *Salmonella typhimurium, Leishmania donovani,* and *Mycobacterium* spp.; Malo et al., 1994).

The Major Histocompatibility Complex and Disease Resistance

One of the first genetic systems to be studied by comparative molecular genetics was the major histocompatibility complex (MHC). Because of the high degree of conservation of MHC genes and gene products among the vertebrates, the MHC is ideal for comparative studies. Literally hundreds of MHC-disease associations have been described (Tiwari and Terasaki, 1985; Klein, 1986; Lewin et al., 1991). In humans, associations have been principally with autoimmune diseases, whereas in farm animals, associations are primarily with infectious diseases (Tiwari and Terasaki, 1985; Lewin et al., 1991). Advances in our understanding of the biochemistry of peptide-MHC interactions (Rothbard and Gefter, 1991) and the elucidation of the three-dimensional structure of MHC class I (Bjorkman et al., 1987) and class II (Stern et al., 1994) molecules have revolutionized thinking about MHC-disease associations. We now know that MHC class I and class II molecules are receptors for peptides that are derived from intracellular and extracellular proteins (reviewed by Germain, 1994). Processed peptides are bound by MHC in the cytoplasm and transported to the cell surface where they are recognized as a single structure by specific receptors on T lymphocytes. T cells that express CD4 (mostly the T-helper subset) recognize peptides that are complexed with class II molecules, whereas T cells that express CD8 (mostly the cytotoxic T-cell subset) recognize peptides complexed with class I molecules. Genetic differences between individual responses to a given peptide are related to their MHC alleles' affinities for that peptide and the presence of T cells that are capable of recognizing the peptide-MHC complex. Individuals without alleles capable of binding a peptide may not respond immunologically to the peptide. Alternatively, the allele(s) may bind the peptide but T cells that can recognize that MHC-peptide combination have been deleted or functionally inactivated during the process of T-cell education in the thymus. In experimental systems, it has been demonstrated that resistance to a lethal challenge with an infectious agent may depend on the host's ability to recognize and respond to a single immunodominant peptide possessed by that pathogen (Braciale et al., 1989; Del Val et al., 1991).

Given the importance of MHC genes in immune responsiveness and the high degree of conservation of MHC genes between species it is understandable that the MHCs of farm animals have been under intensive investigation. Perhaps the best characterized MHC of any non-primate, non-rodent species is the bovine MHC, or *BoLA*, which is located on bovine chromosome 23 (BTA23; reviewed by Lewin, 1996). There are more than 50 structural genes mapped to BTA23, most of them contained within the bovine MHC. On a comparative basis, BTA23 has synteny conservation with HSA6p and MMU17. BTA23 and HSA6 are more highly conserved; the gene for prolactin (*PRL*) is located on a different chromosome (MMU13) than the MHC of the mouse. However, there are differences in the organization of MHC genes in cattle and humans. Cattle have two class II regions separated by at least 15 cM (van Eijk et al., 1993). We recently showed by sperm typing that recombination rate varies significantly in the interval between *DYA* and *PRL* (Park et al., 1995). In two bulls tested, significantly different recombination rates of 16.9% and 31.0% were detected.

Lessons from the Bovine Leukemia Virus

In several species of farm animals, clear-cut associations with infectious diseases have been documented (reviewed by Lewin et al., 1991). Most notable is the association between the B^{21} haplotype and resistance to Marek's Disease in chickens (Briles et al., 1977). It was the convincing disease associations with retroviruses in the mouse and the aforementioned association with Marek's Disease in chickens that sparked our initial interest in the possible role of the bovine MHC in resistance to the bovine leukemia virus (BLV).

The bovine leukemia virus (BLV) is an exogenous C-type retrovirus of cattle and is the etiologic agent of enzootic bovine leukosis (EBL; reviewed by Burny et al., 1980). Economic losses associated with BLV infection have been estimated to be in excess of $80 million annually to the North American dairy industry (Da et al., 1993). Efforts to develop a vaccine for BLV have been unsuccessful; control and eradication of the virus have been dependent on the identification and elimination of cattle with antibodies to the virus or clinical signs of disease.

Bovine leukemia virus infection consists of at least three distinct stages of disease. The majority of BLV-infected cattle are free of clinical signs although serum antibodies to the major viral envelope glycoprotein (gp51) persist throughout life. About 30% of cattle infected with BLV progress to persistent lymphocytosis (PL), characterized by 2- to 10-fold expansion of circulating B-cells. A small percentage (< 5%) of BLV-infected cattle eventually develop clinical disease, manifested as either B-cell leukemia or lymphosarcoma.

Epidemiologic studies demonstrating a familial basis for the development of PL and lymphoma in BLV-infected herds provided the earliest evidence that host genes influence the outcome of BLV infection (Croshaw et al., 1963; Bendixen, 1965). Associations between *BoLA* haplotype and PL were found for class I antigens by several groups studying different herds and breeds of cattle (Lewin and Bernoco, 1986; Palmer et al., 1987; Lewin, 1988; Stear et al., 1988). The weak relative risk associated with *BoLA-A* alleles led us to propose the involvement of *BoLA* class II genes in mediating resistance and susceptibility to PL (Lewin, 1988).

To test the hypothesis that class II genes were more intimately involved in susceptibility to PL, RFLP studies of class II genes were conducted using a case-control design (van Eijk et al., 1992). When animals were grouped by *BoLA-A/DRB2/DQB* haplotype, resistance and susceptibility to PL were more strongly associated with *DRB2* alleles than with *BoLA-A* alleles. Following the conclusion of the class II RFLP study, it was shown that *DRB2* is not expressed, or is expressed at very low levels, *in vivo*, whereas the *DRB3* locus is expressed at a high level (Burke et al., 1991). Because *DRA* is not very polymorphic we thought that any effect of the *DR* locus on disease progression must be mediated through alleles of the *DRB3* gene. To further explore this relationship, exon 2 of *DRB3* was sequenced from haplotypes previously associated with resistance and susceptibility to PL (Xu et al., 1993). Exon 2 of the class II β-chain contains polymorphic and non-polymorphic amino acids relevant to peptide binding and T-cell receptor interactions (Stern et al., 1994). Among the resistance-associated *DRB3.2* alleles, the only common motif was the presence of glutamic acid (E) and arginine (R) residues at positions 70 and 71, respectively. Among the susceptibility-associated alleles, all contained valine-aspartic acid-threonine (VDT) at positions 75-77; position 78 was occupied by valine in the resistance-associated *DRB3.2*11* allele and by tyrosine in *DRB3.2*16* and *DRB3.2*8*. The relationships between these motifs and disease phenotype were confirmed by allele-specific PCR and PCR-RFLP analysis in a population study (Xu et al., 1993). The ER motif was associated with dominant resistance to PL in this study. The presence of the VDTY motif at positions 75-78 tended to be predictive of susceptibility (11/19 were PL$^+$).

Because molecular mimicry of host MHC molecules by pathogens has been implicated in evasion of the immune response, the sequenced *DRB3.2* alleles were examined for homologies with BLV proteins (Xu et al., 1993). Of three amino acid identities of three or more residues, one was the four-amino-acid peptide VDTY, which occupies positions 643-646 of the BLV reverse transcriptase (*pol*). Only 357 sequence matches with the tetrapeptide VDTY were obtained in a search of the GenBank Peptide Sequence Databases (3,700 expected), and 53% of these involved class II or retroviral reverse transcriptase genes. These findings, in combination with the apparent association between VDTY at positions 75-78 and susceptibility to PL, suggested that molecular mimicry could play a role in *DRB3*-mediated susceptibility.

Model for MHC-Mediated Resistance to Persistent Lymphocytosis

The use of x-ray diffraction to determine the crystalline structure of *HLA-DR1* molecules complexed with peptide (Stern et al., 1994) has shown that PL-associated positions 70, 71, and 78 of the *DRB3* molecule could contribute directly to peptide binding. Furthermore, residues 70 and 71 occupy an apical position on the alpha helix of the β-chain, and recognition of the peptide-MHC complex by the T-cell receptor can be disrupted by altering residues at these positions. The association of the ER motif at positions 70-71 with resistance to PL suggests that these residues are involved in the selection of an immunodominant epitope necessary for limiting disease progression in BLV-infected cattle.

The strong associations between *DRB3* alleles and development of PL led us to propose a model in which the class II molecule functions to limit the spread of BLV, thereby minimizing the opportunity for polyclonal B-cell expansion leading to PL (Figure 2). According to this model, animals having *DRB3* alleles associated with resistance to PL process and present viral antigen, resulting in an immune response that rapidly eliminates infected cells. Alternatively, the presentation of viral antigens by *DRB3* alleles associated with susceptibility to PL would fail to elicit this response, resulting in greater dissemination of BLV. This model predicts that cattle genetically resistant to the development of PL will have fewer detectable BLV-infected cells than PL-susceptible cows.

To test this model we developed a highly sensitive PCR-based method that allowed us to detect and quantify BLV-infected peripheral blood mononuclear cells at different stages of infection (Mirsky et al., 1993). Seven pairs of naturally infected adult Holstein cows were selected for study. The cows were paired by age and date of seroconversion, without regard to lymphocyte count or PL status. One of the pair was selected for the resistance-associated *A14/DRB3.2*11* haplotype, whereas the other member of the pair was selected for one of the PL-susceptibility-associated haplotypes. Using the new method to sort and test B-cells for the presence of the virus (Mirsky et al., 1993), cattle with the PL-resistant haplotype were found to have significantly fewer circulating BLV-infected $CD5^+$ and $CD5^-$ B-cells than their age- and seroconversion-matched PL-susceptible counterparts (M. L. Mirsky, C. Olmstead, Y. Da, and H. A. Lewin, unpublished data). Cows with susceptibility-associated genotypes had an average 10-fold increase in the number of infected $CD5^+$ B-cells and 75-fold increase in the number of infected $CD5^-$ B-cells compared with cows with resistance-associated genotypes. These results clearly suggest that resistant animals are able to limit the spread of BLV *in vivo*, in support of the proposed model.

The mechanism(s) by which *DRB3* or a closely linked gene limits the spread of BLV has not yet been determined. Both resistance and susceptibility to PL could be accounted for by determinant selection. In this model, animals with *DRB3* alleles containing the ER amino acid motif at positions 70 and 71 are able to process and present an immunologically relevant BLV peptide(s) to T-helper (T_H) cells. Animals

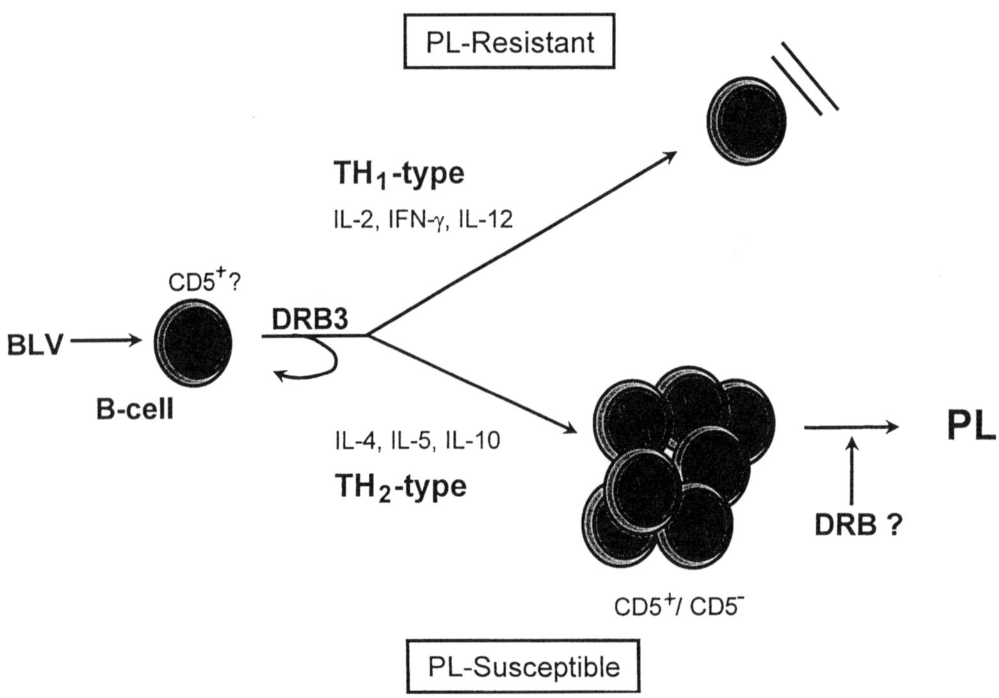

Figure 2. Model for *BoLA-DRB3*-associated resistance and susceptibility to persistent lymphocytosis in cattle infected with BLV. See text for description of the model.

lacking the ER motif may not be able to respond as efficiently to BLV infection, thus resulting in susceptibility to PL. Alternatively, the class II alleles expressed by an individual may determine whether a type 1 (predominantly cell-mediated) or type 2 (predominantly antibody-mediated) T_H response is elicited to BLV (Swain et al., 1995). A similar mechanism of T_H1/T_H2 interplay has been proposed to explain differences in disease progression in humans infected with HIV-1 (Clerici and Shearer, 1993). There is evidence in the mouse that the dominant T-cell cytokine profile is dependent on the class II allele presenting an antigen (Murray et al., 1993). The model we are currently testing is one in which early T_H-mediated cellular immune responses are influenced by *DRB3* alleles and act to limit spread of BLV within the B-cell compartment.

Economic Considerations

The frequency of *BoLA* alleles associated with resistance to PL increased significantly over a 7-year time period in our herd, in which the prevalence of infection is 60 to 70% (Da et al., 1993). When cattle reach approximately 6 years of age, at a time when we have shown that PL has its maximum negative effects on milk production, we observed a sharp rise in the phenotype frequency of the *A14*-bearing resistance-associated *BoLA* haplotype (Da et al., 1993). In 8-year-old cohorts the frequency of this allele doubled, from about 30% to 60%. Thus, in a herd with a high prevalence of BLV infection, *A14* was also associated with longevity. Longevity is, at least in part, a function of resistance to disease, milk production, and reproductive success. These results suggest that *BoLA-DRB3* is an excellent candidate locus for marker-assisted selection.

Summary and Conclusions

Several IDL have been mapped and cloned but there is still great potential for discovering new genes that could dramatically influence resistance to infectious diseases of farm animals. The technologies are now in hand to achieve this goal. To use these technologies responsibly the possible negative effects of selection for disease resistance on productivity must be considered. For specific diseases, such as EBL, altering MHC allele frequencies can, perhaps, produce a desirable outcome. The potential for lowered health care costs to producers adequately justifies continued research on the identification and utilization of IDL.

Acknowledgments

This work was supported in part by grants from the United States Department of Agriculture, National Research Initiative (AG 92-37204-7956) and from the National Cancer Institute, National Institutes of Health (RO1CA59148).

Literature Cited

Arnaout, M. A., N. Dana, S. K. Gupta, D. G. Tenen, and D. M. Fathallah. 1990. Point mutations impairing cell surface expression of the common β subunit (CD18) in a patient with leukocyte adhesion molecule (Leu-CAM) deficiency. J. Clin. Invest. 85:977.

Bendixen, H. J. 1965. Bovine enzootic leukosis. Adv. Vet. Sci. Comp. Med. 10:129.

Bjorkman, P. J., M. A. Saper, B. Samraoui, W. S. Bennett, J. L. Strominger, and D. C. Wiley. 1987. Structure of the human class I histocompatibility antigen, HLA-A2. Nature (Lond.) 329:506.

Braciale, T. J., M. T. Sweetser, L. A. Morrison, D. J. Kittlesen, and V. L. Braciale. 1989. Class I major histocompatibility complex-restricted cytolytic T lymphocytes recognize a limited number of sites on the influenza hemagglutinin. Proc. Natl. Acad. Sci. USA 86:277.

Briles, W. E., H. A. Stone, and R. K. Cole. 1977. Marek's disease: Effects of B histocompatibility alloalleles in resistant and susceptible chicken lines. Science (Wash DC) 195:193.

Buckler, A. J., D. D. Chang, S. L. Graw, J. D. Brook, D. A. Haber, P. A. Sharp, and D. E. Housman. 1991. Exon amplification: A strategy to isolate mammalian genes based on RNA splicing. Proc. Natl. Acad. Sci. USA 88:4005.

Burke M. G., R. T. Stone, and N. E. Muggli-Cockett. 1991. Nucleotide sequences and Northern analysis of a bovine major histocompatibility class II DRβ-like cDNA. Anim. Genet. 22:343.

Burny, A., C. Bruck, H. Chantrenne, Y. Cleuter, D. Dekegel, J. Ghysdael, R. Kettmann, M. Leclercq, J. Leunen, M. Mammerickx, and D. Portetelle. 1980. Bovine leukemia virus: Molecular biology and epidemiology. In: G. Klein (Ed.) Viral Oncology. p 231. Raven Press, New York.

Clerici, M., and G. M. Shearer. 1993. A $T_H1 \rightarrow T_H2$ switch is a critical step in the etiology of HIV infection. Immunol. Today 14:107.

Collins, F. S. 1995. Positional cloning moves from perditional to traditional. Nature Genet. 9:347.

Croshaw, J. E., Jr., D. A. Abt, R. R. Marshak, W.C.D. Hare, J. Switzer, J. Ipsen, and R. M. Dutcher. 1963. Pedigree studies in bovine lymphosarcoma. Ann. N. Y. Acad. Sci. 108:1193.

Da, Y., R. D. Shanks, J. A. Stewart, and H. A. Lewin. 1993. Milk and fat yields decline in bovine leukemia virus-infected Holstein cattle with persistent lymphocytosis. Proc. Natl. Acad. Sci. USA 90:6538.

Del Val, M., H. J. Schlicht, H. Volkmer, M. Messerle, M. J. Reddehase, and U. H. Koszinowski. 1991. Protection against lethal cytomegalovirus infection by a recombinant vaccine containing a single nonameric T-cell epitope. J. Virol. 65:3641.

Drummond, R. O., G. Lambert, H. E. Smalley, Jr., and C. E. Terrill. 1981. Estimated losses of livestock to pests. In: D. Pimentel (Ed.) CRC Handbook of Pest Management in Agriculture. p 111. CRC Press, Boca Raton, FL.

Fujii, J., K. Otsu, F. Zorzato, S. De Leon, V. K. Khanna, J. E. Weiler, P. J. O'Brien, and D. H. MacLennan. 1991. Identification of a mutation in porcine ryanodine receptor associated with malignant hyperthermia. Science (Wash DC) 253:448.

Germain, R. N. 1994. MHC-dependent antigen processing and peptide presentation: Providing ligands for T lymphocyte activation. Cell 76:287.

Hutt, F. B. 1958. Genetic Resistance to Disease in Domestic Animals. Comstock Publishing Associates, Ithaca, NY.

Hutt, F. B. 1964. Animal Genetics. The Ronald Press Company, New York.

Klein, J. 1986. The population. In: Natural History of the Major Histocompatibility Complex. p 609. John Wiley & Sons, New York.

Lewin, H. A. 1988. Association between BoLA-A locus alleles and subclinical progression of bovine leukemia virus infection. Anim. Genet. Suppl. 1:55.

Lewin, H. A. 1996. Genetic organization, polymorphism and function of the bovine major histocompatibility complex. In: L. B. Schook and S. J. Lamont (Ed.) The Major Histocompatibility Complex Region of Domestic Animal Species. CRC Press, Ann Arbor, MI. (In press).

Lewin, H. A. and D. Bernoco. 1986. Evidence for BoLA-linked resistance and susceptibility to subclinical progression of bovine leukaemia virus infection. Anim. Genet. 17:187.

Lewin, H. A., P. A. Clamp, J. E. Beever, and L. B. Schook. 1991. Mapping genes for resistance to infectious diseases. In: L. B. Schook, H. A. Lewin, and D. G. McLaren (Ed.) Gene Mapping: Techniques and Applications. p 283. Marcel Dekker, New York.

Lindenmann, J. 1964. Inheritance of resistance to influenza virus in mice. Proc. Soc. Exp. Biol. Med. 116:506.

Lisitsyn, N. A., J. A. Segre, K. Kusumi, N. M. Lisitsyn, J. H. Nadeau, W. N. Frankel, M. H. Wigler, and E. S. Lander. 1994. Direct isolation of polymorphic markers linked to a trait by genetically directed representational difference analysis. Nature Genet. 6:57.

Malo, D., and E. Skamene. 1994. Genetic control of host resistance to infection. Trends Genet. 10:365.

Malo, D., K. Vogan, S. Vidal, J. Hu, M. Cellier, E. Schurr, A. Fuks, N. Bumstead, K. Morgan, and P. Gros. 1994. Haplotype mapping and sequence analysis of the mouse *Nramp* gene predict susceptibility to infection with intracellular parasites. Genomics 23:51.

Mirsky, M. L., Y. Da, and H. A. Lewin. 1993. Detection of bovine leukemia virus proviral DNA in individual cells. PCR Methods Applications 2:333.

Müller, M., B. Brenig, E.-L. Winnacker, and G. Brem. 1992. Transgenic pigs carrying cDNA copies encoding the murine Mx1 protein which confers resistance to influenza virus infection. Gene 121:263.

Murray, J. S., J. Madri, T. Pasqualini, and K. Bottomly. 1993. Functional CD4 T cell subset interplay in an intact immune system. J. Immunol. 150:4270.

Nelson, S. F. 1995. Genomic mismatch scanning: Current progress and potential applications. Electrophoresis 16:279.

Palmer, C., M. Thurmond, J. Picanso, A. W. Brewer, and D. Bernoco. 1987. Susceptibility of cattle to bovine leukemia virus infection associated with BoLA type. In: Proc. 91st Annu. Mtg. of U. S. Anim. Health Assoc. p 218.

Park, C., I. Russ, Y. Da, and H. A. Lewin. 1995. Genetic mapping of *F13A* to BTA23 by sperm typing: Difference in recombination rate between bulls in the *DYA-PRL* interval. Genomics 27:113.

Rawlings, D. J., D. C. Saffran, S. Tsukada, D. A. Largaespada, J. C. Grimaldi, L. Cohen, R. N. Mohr, J. F. Bazan, M. Howard, N. G. Copeland, N. A. Jenkins, and O. N. Witte. 1993. Mutation of unique region of Bruton's tyrosine kinase in immunodeficient XID mice. Science (Wash., DC) 261:358.

Rothbard, J. B., and M. L. Gefter. 1991. Interactions between immunogenic peptides and MHC proteins. Annu. Rev. Immunol. 9:527.

Shuster, D. E., B. T. Bosworth, and M. E. Kehrli, Jr. 1992a. Sequence of the bovine CD18-encoding cDNA: Comparison with the human and murine glycoproteins. Gene 114:267.

Shuster, D. E., M. E. Kehrli, Jr., M. R. Ackermann, and R. O. Gilbert. 1992b. Identification and prevalence of a genetic defect that causes leukocyte adhesion deficiency in Holstein cattle. Proc. Natl. Acad. Sci. USA 89:9225.

Staeheli, P., R. Grob, E. Meier, J. G. Sutcliffe, and O. Haller. 1988. Influenza virus-susceptible mice carry *Mx* genes with a large deletion or a nonsense mutation. Mol. Cell. Biol. 8:4518.

Stear, M. J., C. K. Dimmock, M. J. Newman, and F. W. Nicholas. 1988. BoLA antigens are associated with increased frequency of persistent lymphocytosis in bovine leukaemia virus infected cattle and with increased incidence of antibodies to bovine leukemia virus. Anim. Genet. 19:151.

Stern, L. J., J. H. Brown, T. S. Jardetzky, J. C. Gorga, R. G. Urban, J. L. Strominger, and D. C. Wiley. 1994. Crystal structure of the human class II MHC protein HLA-DR1 complexed with an influenza virus peptide. Nature (Lond.) 368:215.

Swain, S. L., L. M. Bradley, M. Croft, S. Tonkonogy, G. Atkins, A. D. Weinberg, D. D. Duncan, S. M. Hedrick, R. W. Dutton, and G. Huston. 1995. Helper T-Cell subsets: Phenotype, function and the role of lymphokines in regulating their development. Immunol. Rev. 123:115.

Thomas, J. D., P. Sideras, C.I.E. Smith, I. Vorechovsky, V. Chapman, and W. E. Paul. 1993. Colocalization of X-linked agammaglobulinemia and X-linked immunodeficiency genes. Science (Wash DC) 261:355.

Tiwari, J. L., and P. I. Terasaki. 1985. HLA and Disease Associations. Springer-Verlag, New York.

van Eijk, M.J.T., I. Russ, and H. A. Lewin. 1993. Order of bovine *DRB3*, *DYA*, and *PRL* determined by sperm typing. Mamm. Genome 4:113.

van Eijk, M.J.T., J. A. Stewart-Haynes, J. E. Beever, R. L. Fernando, and H. A. Lewin. 1992. Development of persistent lymphocytosis in cattle is closely associated with *DRB2*. Immunogenetics 37:64.

Xu, A., M.J.T. van Eijk, C. Park, and H. A. Lewin. 1993. Polymorphism in *BoLA-DRB3* exon 2 correlates with resistance to persistent lymphocytosis caused by bovine leukemia virus. J. Immunol. 151:6977.

Comparative Genomics: The Next Generation[1]

Leslie A. Lyons, Marilyn M. Raymond, and Stephen J. O'Brien

Laboratory of Viral Carcinogenesis, National Cancer Institute, Frederick Cancer Research and Development Center, Frederick, MD 21702, U.S.A.

Abstract Recent advances in human gene mapping technologies have led to increased emphasis in developing representative genetic maps for several species, particularly domestic plants and animals. These maps are being compiled with two distinct goals: to provide a resource for genetic analysis and to help dissect the evolution of genome organization by comparing linkage relationships of homologous genes. We recently proposed a list of reference anchor loci suitable for comparative gene mapping in mammals and other vertebrate classes. The comparative anchors, which are Type I (coding gene) loci, provide a comparative framework, but the addition of Type II (microsatellite) loci is also necessary for mapping phenotypes in animal gene maps. Strategies for combining the advantages of Type I and Type II locus markers are presented with emphasis on application of these concepts to the construction of a gene map of the domestic cat.

Key Words: Comparative Genomes, Gene Mapping, Interspecific, Cat

Introduction

Nearly 2,000 years ago Norwegian Vikings set sail to explore uncharted lands and to compile geographic maps of a new world. Their efforts and those of subsequent explorers changed both the cultures they encountered and their own societies. There exist some intriguing parallels between these early pioneers and today's cartographers of our genomes. Both drew maps that exert unmeasured influence and both share in the excitement of new discoveries. The difference lies in the object of their toil. Modern gene mappers concentrate on the source code of life, the 100,000 different genes that prescribe the development of a living species and contribute to the biological diversity on our planet. But species genomes are themselves the product of many millions of years of evolutionary processing that have been shaped through chance, natural selection, and genetic drift. The field of comparative gene mapping examines the similarities and differences in gene linkage organizations between species related by descent from an ancient ancestor. Much as comparative anatomy formed the framework for interpreting functional morphology and physiology, comparative genetics offers the prospect of functional interpretation of genome organization by evaluating evolutionary divergence. As the techniques of gene mapping improve, the prospect of building informative gene maps in virtually any species becomes realistic (O'Brien, 1991, 1993b; O'Brien et al., 1993, 1994).

[1]This article originally appeared in *Animal Biotechnology* (Anim. Biotechnol. 5:103) and is reprinted with permission.

The Comparative Genome World

Comparative gene mappers are not only explorers of new frontiers but time travelers, too. During the Paleocene, approximately 60 to 70 million years ago, an explosive mammalian radiation occurred, leading to the present 19 orders of mammals, representing nearly 5,000 extant species (Nowak, 1991). Gene maps are currently being developed for many species, covering such diverse orders as Primate, Rodentia, Carnivora, Artiodactyla, Perissodactyla, Lagomorpha, Metatheria, and Prototheria (O'Brien, 1993a; O'Brien et al., 1994). Besides the map for *Homo sapiens*, which is fueled by the human genome initiative, the mouse is the most investigated genome. Over 2,000 coding genes have been mapped to the 20 chromosome pairs of the estimated 1,600 cM mouse genome (Copeland, et al., 1993). An order of magnitude less developed, the cow and the cat maps represent Artiodactyla and Carnivora, respectively, while the horse, rabbit, marsupial, and monotreme maps are gaining momentum in the comparative mapping arena. Other mammalian orders, Insectivora, Chiroptera, Edentata, Cetacea, Pinnipedia, and Sirenia, have yet to see maps develop for their representatives. Gene maps for non-mammalian vertebrates (fish, amphibia, and aves) have allowed comparisons across the vertebrata subphylum (Morizot, 1990). In all there are over 100 substantial gene maps already begun and their comparative distinctiveness provides valuable insight into the history and functional organization of modern genomes (O'Brien, 1991, 1993a,b; O'Brien et al., 1988).

Syntenic Comparative Maps

The development of comparative maps for several higher vertebrates started with somatic cell hybrid lines. In this procedure, cells of the mapped species are fused with competent rodent cells, usually mouse or Chinese hamster (O'Brien and Nash, 1982; Womack and Moll, 1986; O'Brien et al., 1988). The hybrid cells maintain their entire rodent chromosomal complement but segregate the map-species' chromosomes in different combinations. Hybrid panels can be identified that represent all of the donor's genetic complement in different combinations. These clones can then be karyotyped to identify the portion of donor complement present. Some species are difficult to karyotype since their genetic complement consists of high numbers of acrocentric chromosomes, such as the cow and dog (Womack and Moll, 1986; Ostrander et al., 1993). Similar banding patterns and the lack of a size reference for the acrocentrics can make confident identification in hybrids extremely difficult. Hybrid clone typing is facilitated by the detection of isozyme markers in hybrid cell lines. Genes for isozymes are thereby mapped by correlating concordant occurrence of the gene marker and a specific chromosome in a hybrid panel.

In situ hybridization of radiolabeled DNA segments to metaphase spreads of chromosomes allows the direct physical localization of genes in either somatic cell hybrid lines or directly in the species' cell lines (Greig and Willard, 1992; Lopez et al., 1994). Analysis of interphase chromosomes has allowed even more precise localizations. With the development of different fluorescent dye tags, mapping of several genes simultaneously allows the establishment of gene order and a crude measure of distance by these same *in situ* techniques (Trask, 1991). The fluorescent labeling of YACs, cosmids, and entire cDNA libraries from single chromosomes or from chromosomal segments has led to a cytogenetic technique termed chromosome painting. *In situ* hybridization of a mixture of molecular clones derived from a single chromosome illuminates the homologous chromosomes of the test species, showing blocks of chromosomal conservation (Jauch et al., 1992; O'Brien, 1993b). This method can produce a direct visualization of the gross chromosomal rearrangements between any two species. In cases in which the chromosomal complements are highly rearranged, such as the gibbon or canid karyotypes (O'Brien, et al., 1988, 1994; Ostrander, et al., 1993), chromosome painting can greatly simplify analysis and interpretation, raising the prospect of tracking genomic rearrangements between species without actual map construction.

Comparative Linkage Maps

The fundamental approach to establishing a genetic linkage map is basically the same strategy developed for the first gene map of *Drosophila* nearly a century ago: find variation, establish informative matings, and assess linkage by reduced recombination frequency. Several recent advances have enormously facilitated these basic approaches. Interspecific backcrosses make use of evolutionary time to facilitate the detection of polymorphism for virtually any conserved gene segment (Copeland et al., 1993). A parental cross is established between closely related different species that will produce fertile, mostly heterozygous, F_1 progeny. These crosses are made either naturally or by assisted reproduction techniques, including artificial insemination and *in vitro* fertilization. The recombination during F_1 gametogenesis is detected by typing offspring of an $F_1 \times$ parental backcross. Thus, pedigree size need not be large because all matings are informative and phase is known. Interspecific backcrosses are being employed in the construction of maps for cows (*Bos taurus* × *B. gaur*), pigs (wild boar × large white), cats (*Felis bengalensis* × *F. catus*), mice (*Mus domesticus* × *M. spretus*), fish (*Xiphophorus maculatus* × *X. helleri*), and other species.

Type I (coding gene) markers are nearly all typed with DNA methods, notably Southern blotting or polymerase chain reaction (PCR)-formatted DNA typing. Ideally, most markers should utilize PCR, primarily because of economy of genomic DNA required. Conserved genes can be typed by employing PCR-primers that span introns and less conserved regions of the genes, such as 3' untranslated regions of cDNA clone sequences. The full realm of polymorphism detection systems can then be applied: restriction fragment length polymorphisms (RFLPs), frequent cutting (4 bp) restriction enzymes, single-stranded conformation polymorphism (SSCP), and oligo-nucleotide typing. Polymerase chain reaction has opened the doors to the direct analysis of single gametes (Zhang et al., 1992). Single sperm typing can facilitate the examination of recombination at very short distances within a region of interest and oocyte typing can help orient linkage maps by gene-centromere mapping (Cui et al., 1992; Zhang et al., 1992).

Comparable Comparatives

A comparison of syntenic or linkage organization between multiple species requires a common language or standards to be agreed upon by the comparative gene mapping community. The same landmark loci should be mapped in each species so that regions of homology across genomes can be recognized. The best landmarkers would be Type I markers, evolutionary conserved, coding genes. Because they are under selective constraints, these loci are often not polymorphic within a species, although the interspecific backcrosses nearly always yield DNA variation. We recently proposed a series of 321 reference anchor loci for comparative gene mapping (O'Brien et al., 1993). These were selected from Type I loci mapped in four mammal species gene maps (human, mouse, cat, and cow) with the idea that the designated anchor loci would be included in developing mammal gene maps. The selection criteria of the anchor loci were: 1) previously mapped in more than one species (mouse, human, cow, or cat); 2) evenly spaced 5 to 10 cM apart (mouse set the standards in this aspect); 3) genes with physiological significance (reproduction, neurology, disease, carcinogenesis, etc.); 4) molecular clones available to the public domain; 5) represented murine or human reference markers; and 6) they flanked the Smallest Conserved Evolutionarily Unit Segments (SCEUSs) so far identifiable across humans, mice, cows, and cats. For mapping phenotypes of specific interest in various species, we also require a map of high-resolution polymorphic (Type II) loci as well.

Type II loci are selected for their high polymorphic content and are essential for the future application of the map to positional cloning of adjacent coding loci, which generally prove to be low in polymorphic content. Three classes of Type II loci are currently utilized: 1) minisatellite loci (a tandem repeat of a 16 to 60 bp fragment); 2) microsatellite loci (a short tandem repeat of a 1 to 6 bp motif); and 3) SSCP of PCR amplified segments of the genome. The ideal Type II locus would be abundant, uniformly distributed in the genome, highly polymorphic, genotyped easily, and could be dispersed with ease in the scientific community. Microsatellite loci best fulfill these criteria and have rapidly become the polymorphic marker of choice in gene mapping. Microsatellite maps are under development in human, mouse, cow, pig, sheep, cat, dog, and several other species.

The Cat Map: A Representative Carnivore

Our laboratory has applied the above approaches to building a syntenic and linkage map of the cat genome including both Type I and Type II markers. Over 100 Type I loci have been assigned to feline chromosomes by hybrid panel analysis (Figure 1) (O'Brien, 1993c) and we are in the process of building a linkage map using an interspecies backcross between domestic cat (*F. catus*) and Asian leopard cat (*Prionailures bengalensis*) or Geoffroy's cat (*Oncifelis geoffroyi*). The crosses are accomplished both by natural matings and by laparoscopic artificial insemination (Howard et al., 1992a,b). Approximately 30% of the attempted AI procedures (over 100 attempted) produced kittens using methods developed for domestic cats. The F_1 males of each cross adhere to Haldane's rule and are sterile, but the F_1 females have been productive and are crossed back to domestic cats. To date over 30 backcross cats have been produced. Based upon the computation illustrated in Figure 2 this many individuals would be sufficient to construct a 10 cM resolution linkage map.

Presently, six domestic × Geoffroy's F_1 females (Safari cats) and eight mature and three immature domestic × Leopard cat (Bengal cats) F_1 females are being used to produce backcross offspring. Analysis of the comparative mapping anchor loci for

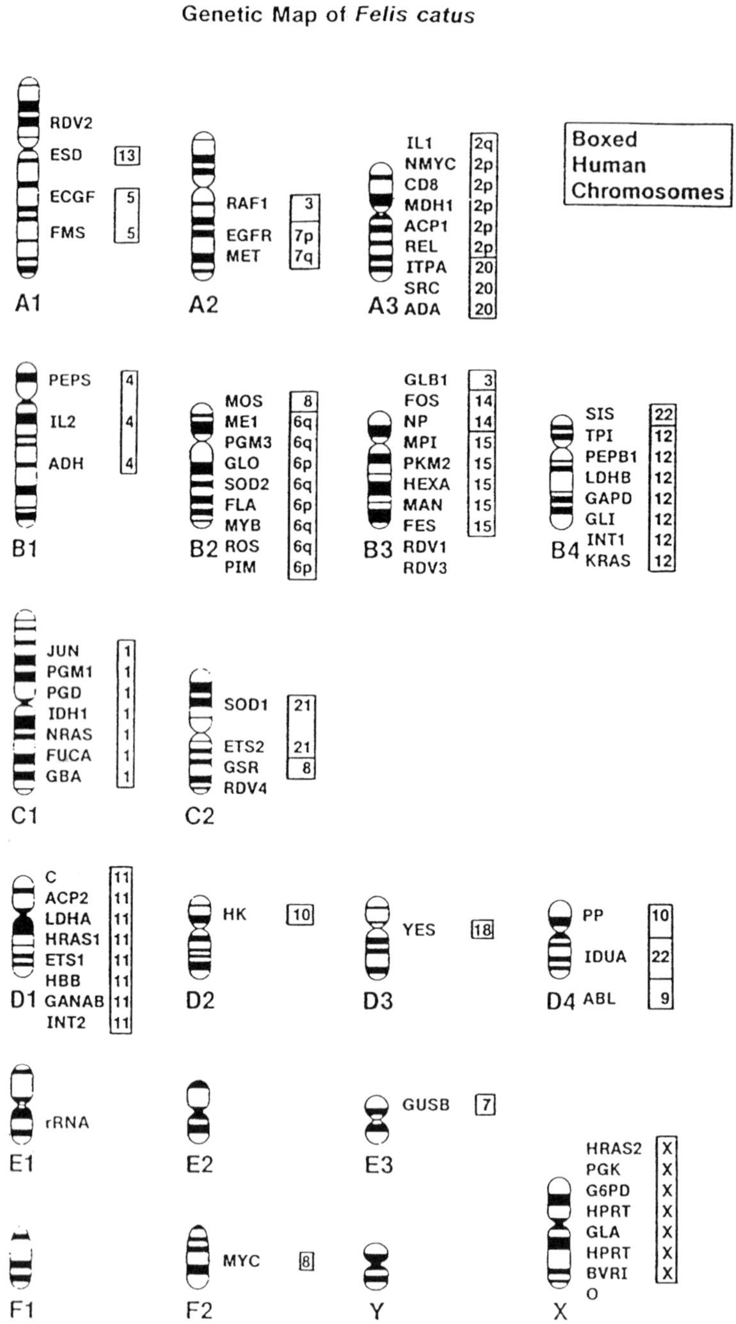

Figure 1. Syntenic map of domestic cat based on analysis of rodent × cat hybrid panels plus *in situ* hybridization (O'Brien, 1993c). Gene symbols follow human nomenclature (McAlpine et al., 1993). The chromosomal position of the homologous locus in the human gene map is listed next to each locus and groups of syntenic segments are enclosed in boxes.

human chromosome 12 and X are proceeding. Over 80% of the tested Type I markers show strong hybridization signals by Southern blot analysis in the cat. These species are sufficiently divergent to show RFLP polymorphism in 80 to 90% of the restriction enzymes tested for each marker. *HOX3@* and *GLI* had been previously mapped to cat chromosome B4 using the somatic cell hybrid panels (Figure 1). Linkage analysis in the interspecific backcross pedigrees have established a three-point linkage between *HOX3@-Gli-Col2A1*, making *Col2A1* the first Type I marker mapped using these methods.

Type II markers are also being investigated in the cat. In order to isolate microsatellites, a small insert (250 to 500 bp) genomic library of *Felis catus* (female) was screened with radiolabeled $(dC\text{-}dA)_{15}/(dG\text{-}dT)_{15}$ oligonucleotides at high

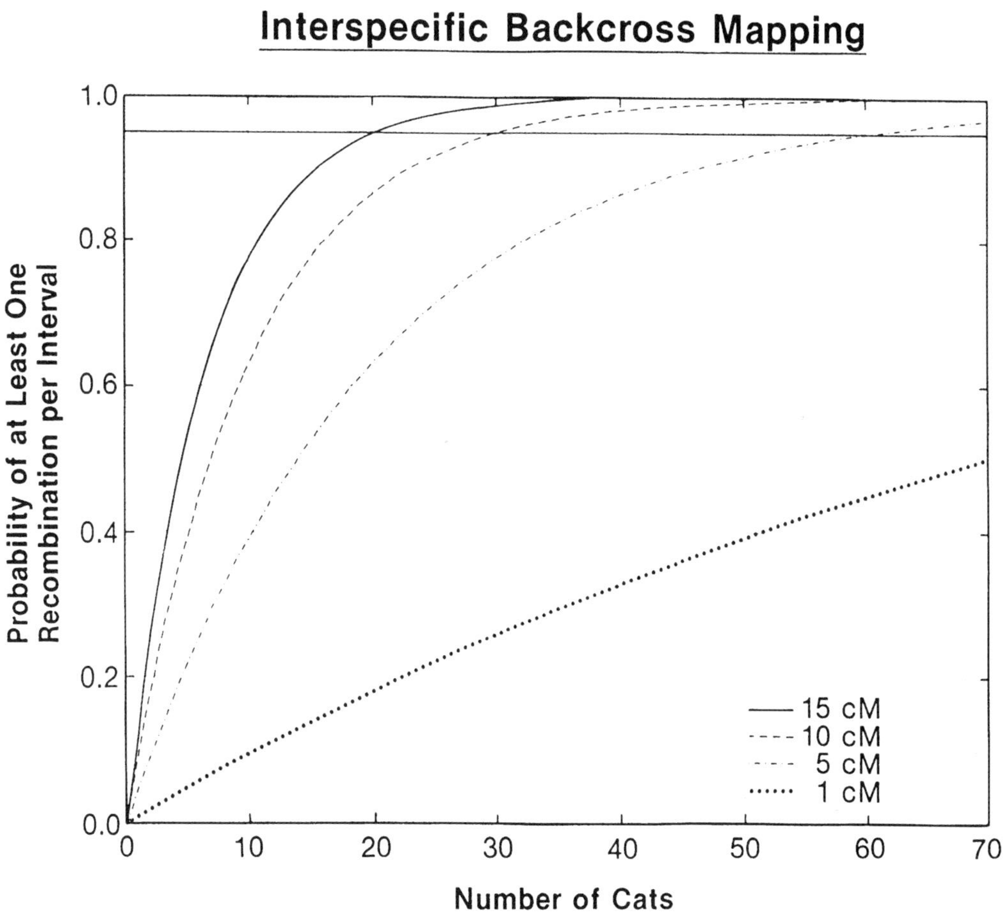

Figure 2. Relationship between recombination interval between markers, θ, ($0 \leq \theta \leq .5$), and the number of backcross cats (n) required to expect at least one recombination with 95% confidence. Curves are derived by solving the following equation: $P(n, \theta) = \int_0^{\theta} ne^{-nx}dx = 1 - e^{-n\theta}$. By example, for θ = .10 (10 cM), n = 31 backcross cats is sufficient to be 95% confident that a recombination event will be observed. Assumptions include evenly spaced markers within the interval, no recombination hotspots or suppressions across the interval, and fully informative crosses.

stringency. Positive recombinants were sequenced using an Applied Biosystems DNA Sequencer and primer pairs were designed for PCR amplification of microsatellite loci. Two hundred primer pairs have been designated (Menotti-Raymond and O'Brien, 1993). Initial characterization of PCR products of 60 loci amplified in unrelated domestic cats, Geoffroy's cat and Asian Leopard cat, and run in 4% agarose gels demonstrated that 73% of the primer pairs generated as product of anticipated size in the Asian Leopard cat and 75% in the Geoffroy's cat. Mendelian inheritance has been observed in the initial genotyping of microsatellite loci in the three-generation pedigree of backcrossed individuals that are being used in the generation of the recombination map. The frequency of the dinucleotide $(dC\text{-}cA)_n$–$(dG\text{-}dT)_n$ in the cat genome seems to be approximately one repeat in 40,000 bp, which is the approximate frequency observed in humans.

The methods and approaches that have developed extensive gene maps of mouse and human are now being applied to several mammal and vertebrate species as well. The integration of syntenic and linkage maps of coding genes with highly polymorphic microsatellite maps have made the construction of gene maps achievable in virtually any mammalian species. Because genetic information is cumulative, it is inevitable that our insight into the evolution of genome organization will blossom in the near future. Further, comparative phenotypic inference about diseases, phenotypes, and virtually any genetic trait offers the prospect of expanding the one-dimensional human or mouse maps into two added dimensions; the first to include comparison with gene maps for other living species and the second back in evolutionary times to the coalescence of genomes of common ancestors of modern species.

Literature Cited

Copeland, N. G., N. A. Jenkins, D. J. Gilbert, J. T. Eppig, L. J. Maltais, J. C. Miller, W. F. Dietrich, A. Weaver, S. E. Lincoln, and R. G. Steen. 1993. A genetic linkage map of the mouse: Current applications and future prospects. Science (Wash DC) 262:57.

Cui, X., J. Gerwin, W. Navidi, H. Li, M. Kuehn, and N. Arnheim. 1992. Gene-centromere linkage mapping by PCR analysis of individual oocytes. Genomics 13:713.

Greig, G. M., and H. F. Willard. 1992. Beta satellite DNA: Characterization and localization of two subfamilies from the distal and proximal short arms of the human acrocentric chromosomes. Genomics 12:573.

Howard, J. G., M. A. Barone, A. M. Donoghue, and D. E. Wildt. 1992a. The effect of preovulatory anesthesia on ovulation in laparoscopically inseminated domestic cat. J. Reprod. Fertil. 96:175.

Howard, J. G., A. M. Donoghue, M. A. Barone, K. L. Goodrowe, E. S. Blumer, K. Snodgrass, D. Starnes, M. Tucker, M. Bush, and D. E. Wildt. 1992b. Successful induction of ovarian activity and laparoscopic intrauterine artificial insemination in the cheetah (*Acinonyx jubatus*). J. Zoo Wild. Med. 23:288.

Jauch, A., J. Weinberg, R. Stanyon, N. Arnold, S. Tofanelli, T. Ishida, and T. Cremer. 1992. Reconstruction of genomic rearrangements in great apes and gibbons by chromosome painting. Proc. Natl. Acad. Sci. USA 89:8611.

Lopez, J. V., N. Yuhki, R. Masuda, W. S. Modi, and S. J. O'Brien. 1994. *Numt,* a recent transfer and tandem amplification of mitochondrial DNA to the nuclear genome of the domestic cat. J. Mol. Evol. 39:174.

McAlpine, P. J., T. B. Shows, C. Boucheix, M. A. Pericak-Vance, and W. A. Anderson. 1993. The 1992 catalog of mapped genes and report of the nomenclature committee. Genome Priority Rep. 1:11.

Menotti-Raymond, M., and S. J. O'Brien. 1993. Dating the genetic bottleneck of the African cheetah. Proc. Natl. Acad. Sci. USA 90:3172.

Morizot, D. C. 1990. Use of fish gene maps to predict ancestral vertebrate genome organization. In: Isozymes: Structure, Function and Use in Biology and Medicine. pp 207–234. Wiley-Liss, New York.

Nowak, R. M. 1991. Mammals of the World. 5th Ed. Vol. 1. p 1628. Johns Hopkins University Press, Baltimore, MD.

O'Brien, S. J. 1991. Mammalian genome mapping: Lessons and prospects. Curr. Opin. Genet. Dev. 1:105.

O'Brien, S. J. 1993a. Genetic Maps: Locus Maps of Complex Genomes. 5th Ed. Cold Spring Harbor Laboratory Press, Cold Spring Harbor, NY.

O'Brien, S. J. 1993b. The genomics generation. Curr. Biol. 3:395.

O'Brien, S. J. 1993c. Genetic map of *Felis catus* (domestic cat). In: S. J. O'Brien (Ed.) Genetic Maps: Locus Maps of Complex Genomes, Nonhuman Vertebrates. 6th Ed. pp 4.250–4.253. Cold Spring Harbor Laboratory Press, Cold Spring Harbor, NY.

O'Brien, S. J., and W. G. Nash. 1982. Genetic mapping in mammals: Chromosome map of the domestic cat. Science (Wash DC) 216:257.

O'Brien, S. J., J. Peters, A. G. Searle, J. E. Womack, P. A. Johnson, and J.A.M. Graves. 1994. Comparative gene mapping committee report. In: A. J. Cuticchia, and P. L. Pearson (Ed.) Human Gene Mapping, 1993: A Compendium. Johns Hopkins University Press, Baltimore, MD.

O'Brien, S. J., H. N. Seuanez, and J. E. Womack. 1988. Mammalian genome organization: An evolutionary view. In: A. Campbell (Ed.) Annual Review of Genetics. Vol. 22. pp 323–351. Annual Reviews Inc., Palo Alto, CA.

O'Brien, S. J., J. E. Womack, L. A. Lyons, K. J. Moore, N. A. Jenkins, and N. G. Copeland. 1993. Anchored reference loci for comparative genome mapping in mammals. Nature Genet. 3:103.

Ostrander, E. A., G. F. Sprague, Jr., and J. Rine. 1993. Identification and characterization of dinucleotide repeat (CA)n markers for genetic mapping in dog. Genomics 16:207.

Trask, B. J. 1991. Fluorescence *in situ* hybridization: Applications in cytogenetics and gene mapping. Trends Genet. 77:149.

Womack, J. E., and Y. D. Moll. 1986. Gene map of the cow: Conservation of linkage with mouse and man. J. Hered. 77:2.

Zhang, L., X. Cui, K. Schmitt, R. Hubert, W. Navidi, and N. Arnheim. 1992. Whole genome amplification from a single cell: Implications for genetic analysis. Proc. Natl. Acad. Sci. USA 89:5847.

Session III

Embryo and Gamete Technology

In Vitro Maturation and Fertilization of Pig Oocytes

Billy N. Day and Hiroaki Funahashi

Animal Sciences Unit, University of Missouri-Columbia, Columbia, MO 65211, U.S.A.

Abstract Research progress in the *in vitro* maturation (IVM) and *in vitro* fertilization (IVF) of porcine oocytes is being made in several laboratories. The use of recently refined technology has resulted in an acceptable rate of blastocyst formation and the birth of live piglets. Detrimental effects of elevated NaCl concentration in the maturation medium as well as the importance of sufficient intracellular glutathione content for cytoplasmic maturation of oocytes leading to subsequent pronuclear formation have been demonstrated. In addition, it has been determined that pretreatment of spermatozoa and enhancement of culture conditions during IVM will increase the incidence of monospermic penetration. However, a relatively high incidence of polyspermy and a large variation in the rate of embryonic development to day 30 of pregnancy occur even when oocytes are matured and fertilized by use of the most recent technological advancements. Traditional systems for IVM are considerably different from the *in vivo* ovarian follicular environment prior to the LH surge. Culture conditions of oocyte-cumulus complexes prior to exposure to gonadotropins may have important effects on developmental ability following IVM and IVF. Additional research is needed to ascertain the mechanisms involved in the establishment of developmental competence in oocytes during the final phases of folliculogenesis and to mimic these changes *in vitro* to achieve a higher incidence of monospermic penetration following IVM/IVF.

Key Words: Pig, Oocyte, Preimplantation Embryo, Maturation, Fertilization, *In Vitro*

Introduction

In vitro production of porcine embryos has the potential to provide an effective system for mass production of embryos, not only for *in vitro* manipulation (such as production of transgenic pigs) but also for use in research studies of embryological development and early pregnancy loss. The efficient production of porcine embryos *in vitro* would reduce the cost for procurement of embryos for these purposes. With this increased quantity of embryos, scientific progress in the study of embryo micromanipulation, embryological development, and cryopreservation of gametes could be hastened, and the resultant benefits could reach production agriculture more quickly.

Research to date has given emphasis to understanding the mechanisms regulating nuclear and cytoplasmic aspects of meiotic maturation, sperm penetration, and early development of oocytes and embryos in order to develop successful procedures for *in vitro* production of developmentally competent oocytes and embryos. The report of the birth of piglets from oocytes matured *in vivo* and fertilized *in vitro* (Cheng, 1985; Cheng et al., 1986) encouraged research on the utilization of immature follicular oocytes. Successful techniques for *in vitro* nuclear maturation and sperm penetration of porcine oocytes (Iritani et al., 1978; Nagai et al., 1984, 1988; Cheng et al., 1986; Wang et al., 1991), and *in vitro* culture of one-cell pig eggs to the blastocyst stage (Petters and Wells, 1993) have recently been developed. However, the progress toward efficient *in vitro* maturation (IVM) and *in vitro* fertilization (IVF) in the pig has been limited by a high incidence of polyspermy (Nagai et al., 1984; Cheng et al., 1986; Mattioli et al., 1988b; Funahashi and Day, 1993a), a reduced incidence of male pronuclear formation (Nagai et al., 1984; Mattioli et al., 1988b; Funahashi and Day, 1993a) and a low rate of early embryonic development (Nagai and Takahashi, 1992; Funahashi et al., 1994b, 1994e). As recently reviewed by us (Funahashi and Day, 1994) and others (Niwa, 1993; Mattioli, 1994; Nagai, 1994), continued progress in porcine IVM/IVF is being made. Piglets have been produced from oocytes matured and fertilized *in vitro* in a few laboratories (Mattioli et al., 1989; Yoshida et al., 1993c; Funahashi et al., unpublished observation). However, the high incidence of polyspermic penetration and the low rate of embryonic development *in vitro* still exist, although the former problem has been partially circumvented by modifying IVF conditions (Nagai and Moor, 1990; Funahashi and Day, 1993b). This review aims to present recent technological advancements and discuss characteristics of culture conditions for porcine IVM/IVF as compared with *in vivo* physiological environments.

Oocyte-Cumulus Complexes

Variation in Germinal Vesicle Oocytes

The oocyte-cumulus complexes (OCCs) used for IVM/IVF should be synchronized with respect to meiosis and high developmental competence and be of high quality. Only OCCs with uniform ooplasm and a compact cumulus cell mass should be used

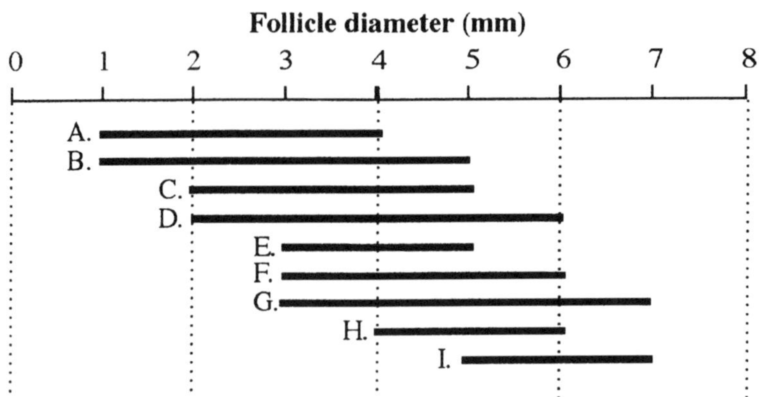

Figure 1. Variation in the size of follicles aspirated or dissected by various investigators for the collection of porcine OCCs. Letters within the figure indicate the following references, (A.) Kim and Menino (1995); (B.) Zheng and Sirard (1992); (C.) Mattioli et al. (1988a,b), Naito et al. (1988), Yoshida et al. (1992a,b, 1993c), Yoshida (1993), Reed et al. (1993), Laurincik et al. (1994), Sirtokin and Nitray (1994), Coskun and Lin (1994), Petr et al. (1994a, b), and Wang et al. (1994); (D.) Kano et al. (1994) and Ding et al. (1992a); (E.) Nagai and Moor (1990), Ocampo et al. (1993), Martinez et al. (1993), Kikuchi et al. (1995), and Schroeter and Meinecke (1995); (F.) Mattioli et al. (1989, 1991), Nagai et al. (1993), Nagai (1994), Funahashi and Day (1993a,b,c), and Funahashi et al. (1994a,b,c, 1995b); (G.) Ding and Foxcroft (1994c); (H.) Mattioli et al. (1990), Singh et al. (1993), Xia et al. (1994); (I.) Mattioli et al. (1994).

following their collection from selected antral follicles from ovaries of slaughtered prepubertal gilts. However, the size of follicles that are selected for *in vitro* production of porcine embryos differs among investigators (Figure 1). There is a large variation in the dictyate stage of the first meiotic prophase among oocytes collected from follicles of different sizes from slaughtered gilts (Table 1). In contrast, OCCs collected from gilts 72 hours after injection of equine chorionic gonadotropins (eCG) are closely synchronous (Funahashi and Day, unpublished data). Follicular heterogeneity in steroid concentration and gonadotropin binding ability has been observed in both cyclic (Grant et al., 1989; Hunter et al., 1989) and gonadotropin-treated gilts (Wiesak et al., 1990), and this heterogeneity has implications in the degree of oocyte maturation and early embryonic development (Foxcroft and Hunter, 1985; Hunter and Wiesak, 1990). A recent study demonstrated that the ability of follicles to secrete steroids and support cytoplasmic maturation of the oocyte is more dependent on the age than on the size of follicles (Ding and Foxcroft, 1994a). Therefore, an oocyte population that is harvested from follicles of different ages may be expected to have an increased range in quality of oocytes. Variation in oocyte quality may also be affected by the method (e.g., dissection vs. aspiration) used to collect OCCs (Nagai, 1994) and by the presence of corpora lutea (Ocampo et al., 1993).

Environment Before Meiotic Resumption

The number and size of follicles change dramatically during days 16 to 21 of the estrous cycle in gilts (Hunter and Baker, 1975; Grant et al., 1989; Hunter and Wiesak, 1990). Steroid concentrations and gonadotropin binding ability increase during follicular growth up to day 20 and then decrease on day 21. In contrast, the number of granulosa cells per follicle increases dramatically between day 20 and day 21 (Grant et al., 1989). Insulin-like growth factor I (IGF-I) has been found in porcine follicular fluid (Hammond et al., 1985), and concentrations increase in a time-dependent manner until 72 h after an eCG injection (Hammond et al., 1988). Follicle-stimulating hormone (FSH) appears to be an effective stimulator of IGF-I secretion by porcine granulosa cells, and luteinizing hormone (LH) and estradiol are facilitators of IGF secretion (Hsu and Hammond, 1986). The inclusion of IGF-I in the maturation medium does not affect nuclear maturation of porcine oocytes (Reed et al., 1993) but does stimulate the incidence of cleavage following IVF (Xia et al., 1994).

Table 1. Germinal vesicle stages in porcine oocytes collected from follicles of various sizes[a]

Diameter of follicles, mm	No. of oocytes examined	Percentage of germinal vesicle oocytes in the category of				
		GV-X	GV-I	GV-II	GV-III	GV-IV
Day < 2	53	69.7	18.9	3.8	5.7	1.9
$2 \leq$ Day < 4	48	16.7	56.3	8.3	8.3	10.4
$4 \leq$ Day < 6	42	4.7	40.5	31.0	2.4	21.4

[a]The germinal vesicle stage was determined by the categories GV-I to GV-IV as defined by Motlik and Fulka (1976). Oocytes regarded to be in a stage earlier than GV-I are shown as GV-X. Funahashi and Day (unpublished data).

Polyadenylation and deadenylation of maternal mRNAs during meiotic maturation and early embryogenesis are important mechanisms to regulate translational repression and induction (Wormington, 1994). The mRNAs associated with early development are polyadenylated in fully grown oocytes (c-mos mRNA) and during meiotic maturation (tissue-plasminogen activator mRNA and hypoxanthine phosphoribosyltransferase mRNA; Kinloch and Wassarman, 1993). Insulin-like growth factor I may affect cytoplasmic maturation at the translational level of maternal mRNA. Therefore, mimicking the hormonal environments of follicular maturation or regulating the exposure of OCCs to IGF-I before an artificial LH surge may synchronize and(or) improve the developmental competence of oocytes. Preincubation of OCCs in maturation medium without gonadotropins for 12 hours before exposing them to gonadotropins enhances the developmental competence of porcine oocytes following IVM/IVF (Funahashi and Day, 1995a). This is particularly true for OCCs derived from preantral follicles in which limited *in vitro* growth, maturation, and sperm penetration has been reported (Hirao et al., 1994). Although some physiological requirements for oocyte growth have been identified (Christmann et al., 1994; Lazzari et al., 1994; Petr et al., 1994a,b), further research is needed to reduce the variation in the quality of porcine OCCs used for *in vitro* embryo production.

Nuclear and Cytoplasmic Maturation

Nuclear Maturation

The continuation of the meiotic processes of oocyte chromatin is so-called nuclear maturation. The duration of nuclear maturation of porcine oocytes is almost twice that of bovine oocytes. Porcine oocytes remain in the dictyate stage of the first meiotic prophase until 18 hours after injection of human chorionic gonadotropin (hCG) and achieve first meiosis between 35 and 37 hours after hCG (Hunter and Polge, 1966). A recent study using ultrasonography indicated that the interval from the onset of the LH surge to ovulation averages 44 ± 3 hours (Soede et al., 1994). Although porcine OCCs have traditionally been matured *in vitro* without changes in the culture conditions, dramatic changes seem to occur *in vivo* in the nucleus and cytoplasm of oocytes, particularly during the period when the oocytes are at the germinal vesicle stage. Exposure of OCCs to gonadotropins for only the first 20-hour period of IVM is adequate (Funahashi and Day, 1993c), and removing cumulus cells 24 hours after the start of culture does not adversely affect nuclear maturation (Kameyama and Ishijima, 1994). Porcine oocytes are sensitive to a heterogenous nuclear RNA synthesis inhibitor (Meinecke and Meinecke-Tillmann, 1993), a protein phosphorylation inhibitor, and a protein synthesis inhibitor (Jung et al., 1993) during the first 12 hours of maturation *in vitro*. Therefore, transcriptional events, protein phosphorylation, and protein synthesis activities associated with nuclear maturation seem to occur during a relatively early period of IVM. A transient increase in cAMP levels is observed *in vivo* around 12 hours following hCG injection, and a similar increase is induced by LH during the first 12 hours of IVM in the presence, but not in the absence, of follicular cells (Mattioli et al., 1994). The transient increase of cAMP also may be associated with cytoplasmic maturation. Prolactin will stimulate cAMP output and inhibit estrogen secretion by porcine granulosa cells (Sirotkin and Nitray, 1994). It has been suggested that external calcium influx is not a direct requirement for germinal vesicle breakdown but may be required by porcine oocytes for progression beyond metaphase I of meiosis (Kaufman and Homa, 1993). Further, protein kinase A, protein kinase C, and calmodulin pathways modulate protein synthesis and phosphorylation activities in porcine oocytes and cumulus cells, and consequently seem to affect nuclear maturation (Jung et al., 1992). Stimulated protein kinase A pathway increases the activity of plasminogen activator (Kim and Menino, 1995) and the degree of intercellular metabolic coupling between the porcine oocyte and the cumulus cell mass (Racowsky, 1985).

Cytoplasmic Maturation

Cytoplasmic maturation is composed of the acquisition of factors that are needed for male pronuclear formation and, occasionally, early embryonic development. Cytoplasmic maturation (determined by the incidence of male pronuclear formation) abnormalities (Figure 2) have been reduced by modifications of IVM conditions. Early IVM protocols resulted in detrimental interactions among the culture medium, cumulus cells, and oocytes, which consequently affected low ability in oocytes to form a male pronucleus. In a recent study (Schroeter and Meinecke, 1995), it has been shown that polypeptides synthesized by porcine OCCs during IVM (in

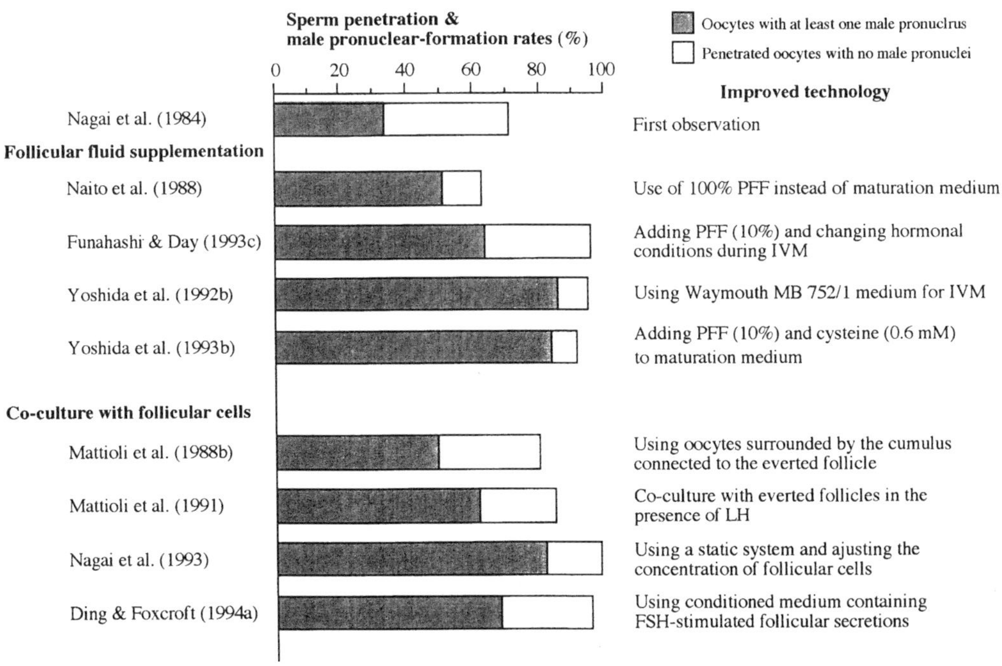

Figure 2. Recent progress in culture conditions during maturation of porcine oocytes as determined by the incidence of male pronuclear formation following IVF. Research is classified into two types, 1) culture systems supplemented with porcine follicular fluid (PFF) and(or) cysteine and 2) co-culture systems using follicular cells.

the presence of FSH and LH) correspond closely to those synthesized during *in vivo* maturation; however, there is a lag in the time of appearance and disappearance of polypeptides of OCCs matured *in vitro*. The ability of porcine oocytes to form a male pronucleus is affected by hormonal levels (Mattioli et al., 1988a; Funahashi and Day, 1993c; Funahashi et al., 1994a), follicular secretions (Mattioli et al., 1988b, 1991; Naito et al., 1988; Ding and Foxcroft, 1992, 1994c; Yoshida et al., 1992a; Zheng and Sirard, 1992; Funahashi and Day, 1993a; Nagai et al., 1993), intracellular ionic strength (Funahashi et al., 1994b,c), and supplementation with cysteine (Yoshida et al., 1992b, 1993b; Funahashi and Day, 1995b). In addition, recent evidence (Yoshida et al., 1993b; Funahashi et al., 1994c) has indicated that the incidence of male pronuclear formation is associated positively with oocyte glutathione content at the end of maturation. Yoshida et al. (1993b) showed that oocyte glutathione content increased during oocyte maturation when cysteine was added to the maturation medium but decreased without added cysteine. The presence of cumulus cells surrounding the oocyte is required throughout the 44-hour period of IVM to maintain a high oocyte glutathione level (Funahashi and Day, 1995b), whereas exposure of OCCs to a glutathione synthesis inhibitor inhibits male pronuclear formation only during the first 24 hours of IVM (Yoshida, 1993). Oocyte protein synthesis between 24 and 36 hours of IVM seems to be required for male pronuclear formation (Ding et al., 1992b).

Recent studies using follicular secretions suggest that the ability of porcine oocytes to form a male pronucleus seems to be related to the steroid environment (Nagai et al., 1993), especially the ratio of progesterone to estradiol (Ding and Foxcroft, 1994b). Differences in effectiveness of exogenous hormone supplementation in various IVM systems using follicular secretions (Naito et al., 1988; Mattioli et al., 1991; Ding and Foxcroft, 1994c; Funahashi et al., 1994a) may be associated with variable steroid concentrations in the follicular supplement. Although the mechanisms of oocyte glutathione regulation during maturation are not clear, follicular secretions with exogenous hormones may offer a more suitable environment for glutathione and protein synthesis in porcine OCCs.

Ionic Strength and Cytoplasmic Maturation

Although porcine follicular fluid contains a relatively high concentration of salts (Na^+, 128-145 m*M*; Cl^-, 97.3 m*M*; reviewed by Gosden et al., 1988), high NaCl concentration in the maturation medium detrimentally affects not only histone H1 kinase activity (Funahashi et al., 1994d), microfilament organization (Funahashi et al., unpublished data) and glutathione content of porcine oocytes at the end of IVM, but also the incidence of male pronuclear formation (Funahashi et al., 1994c) and *in vitro* development following IVM/IVF (Funahashi et al., 1994b). Supplementation of maturation media containing relatively high NaCl levels with organic osmolytes, such as taurine and sorbitol, reduces the severity of the detrimental effect (Funahashi et al., unpublished data). Follicular secretions may contain organic osmolytes since organic osmolytes exist universally in cells and physiological fluid (Garcia-Perez and Burg, 1991; Yancey, 1994). Therefore, a low NaCl concentration or the presence of organic osmolytes in the maturation medium seems to be required for normal porcine oocyte metabolism, and consequently for achievement of cytoplasmic maturation (Table 2 and Figure 3).

Growth Factors and Oocyte Maturation

Porcine follicular fluid contains active factors that promote cumulus expansion and male pronuclear formation (Yoshida et al., 1992a; Daen et al., 1994), but these factors do not seem to be exactly the same for each cellular process (Daen et al., 1994). Recent studies have increased our knowledge about the potential roles of growth factors found in porcine follicular fluid during oocyte maturation *in vivo*. Epidermal growth factor (EGF) is present in porcine follicular fluid (Feng et al., 1987; Hsu et al., 1987), and EGF (Sommer et al., 1992; Reed et al., 1993; Singh et al., 1993; Ding and Foxcroft, 1994b) and transforming growth-factor-α (TGF-alpha) (Coskun and Lin, 1994), which bind to the EGF receptor, enhance nuclear maturation of porcine oocytes. Further, gonadotropins (Ding and Foxcroft, 1994b), especially FSH (Singh et al., 1993), increase the stimulating effect of EGF because specific binding of EGF to granulosa cells is increased by FSH (Fujinaga et al., 1992). In contrast, TGF-beta detected in follicular fluid from small-size porcine follicles *in vitro* (Gangrade and May, 1990) inhibits porcine oocyte maturation (Coskun and Lin, 1994). Transforming

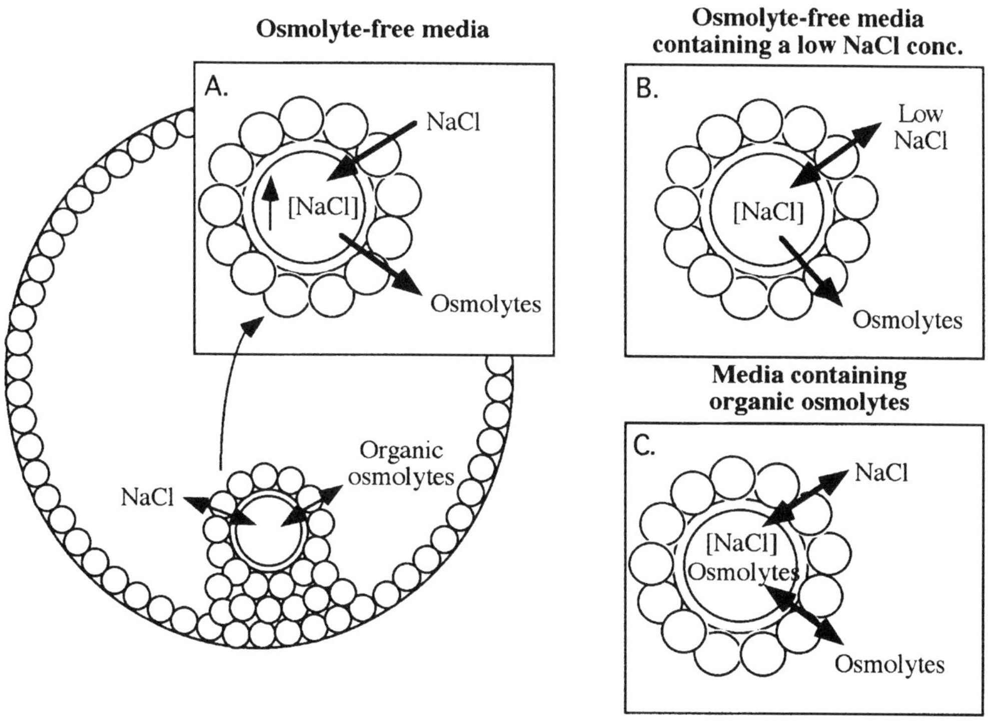

Figure 3. Proposed interactive effect of NaCl and organic osmolytes on intracellular ionic strength. When OCCs (in which intracellular ionic strength has been balanced with extracellular NaCl and organic osmolytes in follicles) are transferred to osmolyte-free media (A), intracellular ionic strength is increased by escape of the osmolytes and by influx of salts. However, the increase of intracellular ionic strength may not occur in the oocytes when OCCs are transferred to osmolyte-free media containing a low NaCl concentration (B) or when transferred to media containing organic osmolytes (C).

growth factor-beta has been found to produce dose- and time-dependent effects on FSH-induced progesterone production in porcine granulosa cells (Chang et al., 1993).

Sperm Penetration

Monospermic fertilization occurs in over 95% of *in vivo* matured and fertilized oocytes (Hunter, 1967, 1972). There seem to be two major mechanisms for the prevention of polyspermic fertilization in pigs. The first mechanism is via the female reproductive tract, which reduces the sperm population arriving simultaneously at the surface of oocytes (Hunter, 1990, 1991). The second mechanism is the direct preventive response of oocytes, such as exocytosis of cortical granule contents and secondary the reactions in perivitelline space and zona pellucida (Wassarman, 1988; Cran and Esper, 1990; Jones, 1990). However, the zone reaction *in vivo* is quite slow in pigs, as evidenced by the numerous sperm trapped in the zona pellucida.

Table 2. Selected base formulations of media used *in vitro* maturation of porcine oocytes

	Medium and compound concentration (m*M*)			
Compound	Modified Medium 199[a]	Waymouth MB 752/1[a]	Whitten's medium[b]	NCSU 23[c]
NaCl	116.36	102.65	68.49	108.73
KCl	5.37	2.01	4.78	4.78
$CaCl_2$	1.80	.81	—	1.70
KH_2PO_4	—	.59	1.19	1.19
NaH_2PO_4	1.01	—	—	—
Na_2HPO_4	—	2.11	—	—
$MgSO_4{\cdot}7H_2O$	.81	—	1.19	1.19
$MgCl_2$	—	1.18	—	—
$NaHCO_3$	26.19	26.67	25.07	25.07
Glucose	8.61	27.75	5.56	5.55
Na-lactate	—	—	25.20	—
Na-pyruvate	.91	—	.33	—
Ca-lactate	2.92	—	1.71	—
Cysteine	.00083	.83	—	—
Glutathione	.00016	.05	—	—
Glutamine	.68	2.40	—	1.00
Taurine	—	—	—	7.00
Hypotaurine	—	—	—	5.00
Inositol	.00028	.0055	—	—
Total Na^+	144.47	133.54	119.09	133.80
Total Ca^{2+}	4.72	.81	1.71	1.70
Total Cl^-	116.69	106.63	73.27	116.91
Total PO_4^{3-}	1.01	2.70	1.19	1.19
Total thiol (cysteine & glutathione)	.00099	.88	—	—
Total organic osmolytes	.68028	2.4055	—	13.00

[a]Calculated from the catalog of GIBCO BRL, Life Technology Inc.
[b]Beckmann and Day (1993); Petters and Wells (1993).
[c]Petters and Wells (1993).

Number of Spermatozoa Reaching Oocytes

Since an increased number of spermatozoa per oocyte is paralleled with a higher incidence of polyspermic penetration *in vitro* (Rath, 1992; Coy et al., 1993a), a reduced number of spermatozoa that become attached to the zona pellucida seem to be needed at the time of IVF. The population of capacitated spermatozoa *in vitro* is relatively small and does not seem to change during preincubation in modified M199 for 90 minutes (Funahashi and Day, unpublished observation). As mentioned previously (Mattioli, 1994), porcine oocytes may be exposed to an increased number of waves of capacitated spermatozoa in current IVF conditions. The duration of co-incubation of *in vivo* matured porcine oocytes with spermatozoa is reported to affect not only the incidence of sperm penetration but also the frequency of polyspermic penetration (Coy et al., 1993c).

The presence of macromolecules secreted from porcine oviductal epithelial cells (Nagai and Moor, 1990; Kano et al., 1994) and a low concentration of follicular fluid (Funahashi and Day, 1993b) in the fertilization media reduce the incidence of polyspermic penetration in porcine oocytes (Figure 4). Low molecular weight zona pellucida binding proteins (16 kDa to 23 kDa) on boar spermatozoa, which are derived from seminal fluid (Parry et al., 1992), have been detected in the acrosomal cap region (Mo et al., 1991). Proacrosin (53 kDa) in the acrosome of boar spermatozoa is also a secondary zona ligand (Jones and Brown, 1987). Binding of the above zona-binding glycoproteins of boar spermatozoa with zona pellucida glycoproteins (e.g., ZP3, 55 kDa) is inhibited by the presence of sulfated polymers such as dextran sulfate, fucoidan, polyvinylsulphate, and heparin (Jones, 1991; Parry et al., 1992). Therefore, oviductal and follicular sulfate proteoglycans in preincubation and(or) fertilization media likely stimulate the rate of sperm capacitation and acrosome reaction, compete with sperm receptors for binding to zona pellucida ligands, and consequently reduce the number of capacitated spermatozoa attaching to the surface of

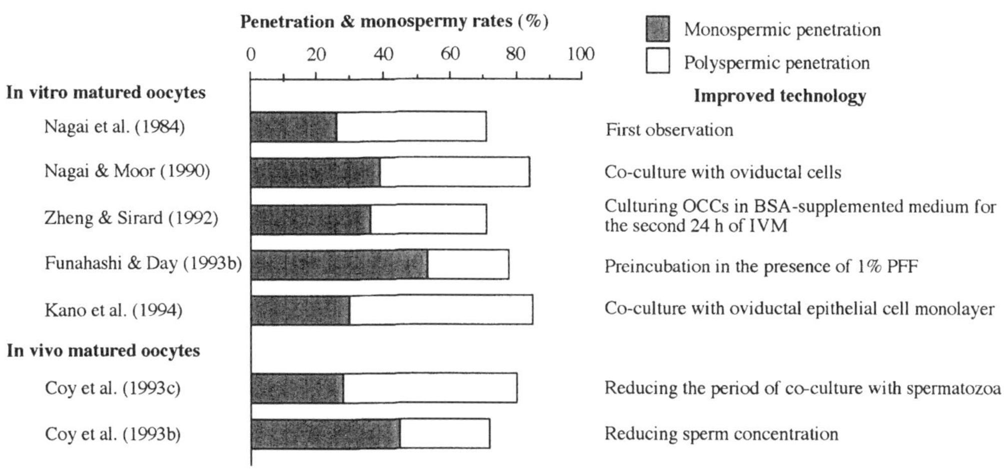

Figure 4. Recent progress in monospermic penetration of porcine oocytes following IVF. Progress has been achieved by modification of IVM or IVF.

porcine oocytes. Further, cumulus matrix co-cultured with spermatozoa seems to stimulate the incidence of sperm penetration and polyspermic penetration in both IVM (Kikuchi et al., 1993) and *in vivo* matured (Coy et al., 1993a) porcine oocytes.

Oocyte Reaction to Sperm Penetration

During fertilization in mammals, exocytosis of the cortical granule contents containing hydrolytic enzymes and saccharide components to the perivitelline space leads to the formation of a cortical granule envelope (Dandekar and Talbot, 1992), modification of zona pellucida proteins, and the inactivation of sperm receptors (Gulyas, 1980; Wassarman, 1988, 1990). A stabilized distribution of cortical granules in porcine oocytes requires an intercellular communication between the oocyte and cumulus cells (Galeati et al., 1991). In the pig, the timing of cortical granule migration in IVM oocytes is similar to that of oocytes matured *in vivo* (Yoshida et al., 1993a). Further, the time of reorganization of microfilament to the oocyte cortex area (Kim et al., 1996) corresponds with that of cortical granule migration in porcine oocytes (Cran and Cheng, 1985; Yoshida et al., 1993a). Even so, the cortical reaction mechanism may not be sufficiently effective in porcine IVF systems to prevent polyspermy. Although the sperm receptor activity of the zona pellucida of IVM oocytes is reduced during IVF (Hatanaka et al., 1992), sperm penetration is not blocked even after electrical stimulation (Funahashi et al., 1993, 1995a). Porcine IVM oocytes release all cortical granules following electrical stimulation (Sun et al., 1992) or microinjection of the G-protein stimulator guanosine-5'-O-(3'-thiotriphosphate) (Machaty et al., 1995). Cortical granule constituents in the perivitelline space of porcine IVM-IVF ova gradually disperse with time following sperm penetration (Yoshida et al., 1993a). It has been suggested that an increased volume of the perivitelline space may affect the incidence of polyspermic penetration and the number of spermatozoa in penetrated oocytes (Funahashi et al., 1994c). Further, oviductal secretory proteins detected within the zona pellucida, perivitelline space, and plasma membrane of oviductal porcine oocytes and embryos (Buhi et al., 1993) may delay the proteolytic digestion of zona pellucida (Broermann et al., 1989). It is unknown whether the "zona hardening" caused by interactions between oviductal secretions and the zona pellucida plays a role in the normal cortical reaction. Alternative techniques, such as gamete intrafallopian transfer (GIFT), may provide a means to assist in the production of embryos from IVM porcine oocytes (Rath et al., 1994a,b).

Male Pronuclear Formation

Glutathione and Male Pronuclear Formation

Male pronuclear formation in mammalian oocytes is completed following reduction of disulfides in the sperm protamine, replacement of sperm protamine by oocyte histones, and DNA synthesis (Zirkin et al., 1985, 1989; Perreault, 1990). Glutathione seems to be associated with the reduction of the disulfide bond cross-linking of sperm protamine (Perreault, 1990). As described above, synthesis of sufficient glutathione during IVM of porcine oocytes is required for sperm decondensation and male pronuclear formation (Yoshida et al., 1992b, 1993b; Yoshida, 1993; Funahashi et al., 1994c). Oocyte glutathione content is known to decrease following sperm penetration (Perreault et al., 1988; Yoshida et al., 1993b; Funahashi et al., 1995a), but not following electrical stimulation (Funahashi et al., 1995a) or microinjection of a G-protein stimulator (Funahashi et al., 1995b). The decrease of glutathione at fertilization seems to be due to gamma-glutamyltranspeptidase of spermatozoa (Funahashi et al., 1995b). Since IVF following microinjection of gamma-glutamyltranspeptidase into porcine IVM oocytes yielded a reduced incidence of male pronuclear formation (our unpublished observation), the ability of oocytes to form a male pronucleus seems to be dependent on oocyte glutathione level following sperm penetration (Figure 5). A low rate of male pronuclear formation in oocytes penetrated by a large number of spermatozoa (Coy et al., 1993b) may be due to an abnormally low level of oocyte glutathione induced by polyspermic penetration.

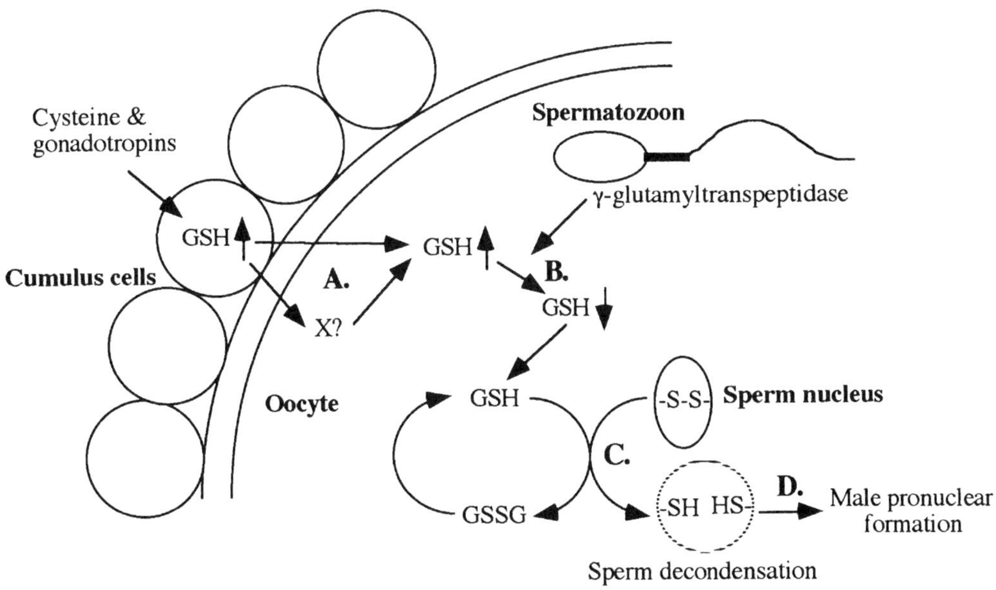

Figure 5. Hypothesis for changes of oocyte glutathione (GSH) content during maturation and fertilization and the potential role of GSH in sperm decondensation. During *in vitro* maturation, oocyte GSH content (A) is regulated by cumulus cells due to direct and(or) indirect (X) transfer of GSH to the oocyte. At the time of sperm penetration, oocyte glutathione content seems to be decreased by gamma-glutamyltranspeptidase of spermatozoa (B). However, sperm decondensation seems to be dependent on glutathione content after sperm penetration, and occurs (C) as described by Perreault (1990). Further, other factors, such as kinase and(or) phosphatase activities, are thought to contribute to male pronuclear formation following sperm decondensation (D). GSSG indicates glutathione disulfide.

Other Factor(s) Associated with Male Pronuclear Formation

Although sperm penetration in immature porcine oocytes is not different from that in matured oocytes if cumulus cells are removed before insemination (Martinez et al., 1993; Wang et al., 1994), sperm nuclei that penetrate oocytes at the germinal vesicle stage decondense 16 hours after insemination and then recondense or form metaphase chromatin 44 hours after insemination (W. H. Wang, K. Okuda, and K. Niwa, personal communication). Electrical activation of IVM porcine oocytes before or just after sperm penetration *in vitro* significantly decreases the incidence of male pronuclear formation (Funahashi et al., 1993). Therefore, decondensation of the sperm nucleus and male pronuclear formation seem to be affected not only by oocyte glutathione but also by cell cycle-dependent factor(s). Although a high positive correlation between histone H1 kinase activity in IVM porcine oocytes and the incidence of male pronuclear formation has been reported (Naito et al., 1992; Funahashi et al., 1994d), a high protein phosphatase activity is also required to dephosphorylate a 25-kDa polypeptide to a 22-kDa polypeptide (Ding et al., 1992a) necessary for pronuclear formation. The time of male pronuclear formation is delayed in porcine IVM oocytes as compared with those matured *in vivo* (Laurincik et al., 1994). Since asynchrony between male and female pronuclear formation occurs even in porcine oocytes with an improved glutathione content (Funahashi et al., 1995a), delayed male pronuclear formation may be due to unknown factor(s) regulating replacement of sperm protamine by oocyte histone and(or) other cell cycle-dependent factor(s).

Embryonic Development

In Vitro Embryonic Development

Successful *in vitro* development of one-cell porcine zygotes to the blastocyst stage has recently been achieved by culturing in modified Whitten's medium (Beckmann and Day, 1993; Youngs et al., 1993), modified Tyrode medium (Hagen et al., 1991), and NCSU-23 medium or NCSU-37 medium (Petters and Reed, 1991; Reed et al., 1992). Early embryonic development of IVM and IVF porcine embryos has also been achieved in amniotic fluid from chick embryos (Ocampo et al., 1994). Although IVM and IVF porcine oocytes develop to the blastocyst stage in these media or in a co-culture system (Figure 6), the efficiency is low, even with improved methods to promote male pronuclear formation and monospermic penetration.

As described above, the ability of IVM and IVF porcine embryos to cleave and develop to the blastocyst stage *in vitro* is improved not only by modification of IVM conditions (Funahashi et al., 1994b; Grupen et al., 1995), but also by preincubation in medium without added gonadotropins before IVM (Funahashi and Day, 1996). Further, developmental competence of porcine oocytes matured *in vivo* is lower following IVF than following *in vivo* fertilization (Rath et al., 1995b). Therefore, inadequacies existing not only in the IVM system but also in the IVF system impair the development of porcine oocytes. Further, there is a significantly lower number of cells in the trophectoderm of porcine IVF embryos as compared with those produced *in vivo* (Rath et al., 1995a).

Recently, adding IGF-I to the culture medium has been reported to enhance the cleavage rate of IVM and IVF porcine embryos (Xia et al., 1994). Binding of radiolabeled IGF-I to porcine trophectoderm has been demonstrated (Corps et al., 1990), and exogenous IGFs are known to have a stimulating effect on metabolism and cellular proliferation in porcine blastocysts (Lewis et al., 1992). Modification of culture conditions for IVM and IVF embryos will likely be improved by an increased understanding of the regulation of early embryonic development by growth factors.

Embryonic Development Following Embryo Transfer

As indicated previously, piglets from oocytes matured and fertilized *in vitro* have been produced by only a few laboratories (Mattioli et al., 1989; Yoshida et al., 1993c; Funahashi et al., unpublished observation). In our laboratory, transfer of IVM and IVF porcine embryos has resulted in extended estrous cycles and apparent failure of pregnancy during later stages of gestation. Further, a large variation in fetal size was observed in recipient gilts examined at day 30 of pregnancy. In polytocous animals such as the pig, it is proposed that asynchronous embryonic development reduces the number of embryos that develop to full term because estrogen secretion from the more advanced blastocysts adversely affects the development of smaller blastocysts (Pope et al., 1990; Greenwald and Roy, 1994). It is interesting to speculate that the quality control of oocytes before IVM may be an important factor for synchronized embryonic development and consequently the survival of IVM and IVF porcine embryos following embryo transfer. Inappropriate concentration or timing of administration of hormones during IVM may also be a major cause of flawed embryonic development and subsequent fetal mortality. This hypothesis is supported by *in vivo* results that indicate that embryo survival to day 35 of pregnancy is related to

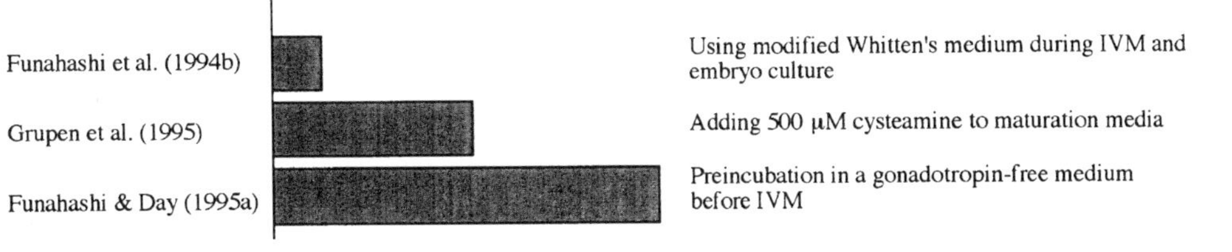

Figure 6. Recent progress in blastocyst formation of IVM and IVF porcine embryos. The development competence of IVM and IVF embryos has been improved by modification of culture conditions, such as use of a co-culture system or by addition of physiological fluids, during early development. Modifications in IVM conditions have also significantly improved developmental ability.

the interval between the peak concentrations of estradiol and LH; shorter intervals of estrogen exposure result in higher embryo survival (Blair et al., 1993; Soede et al., 1994).

Summary

Oocyte-cumulus complexes meiotically synchronized and of high developmental competence are desired for the *in vitro* production of embryos. However, a large variation exists in the germinal vesicle stage of oocytes collected from ovaries of slaughtered gilts. This variation may be due to the lack of exposure to a "normal" environment (induced during follicular maturation before meiotic resumption), and may be a cause of reduced developmental competence of IVM and IVF embryos. It is suggested that inappropriate concentrations and(or) timing of administration of hormones following OCC collection may lead to embryonic and(or) fetal mortality. Therefore, more precise mimicking of the hormonal environments of follicular maturation may be needed to improve the efficiency of *in vitro* production of porcine embryos. Although transcriptional events, protein phosphorylation, and protein synthesis that induce nuclear maturation seem to occur during a relatively early period of IVM, dynamic interactions among the culture medium, cumulus cells, and oocytes are required throughout the period of IVM for cytoplasmic maturation of oocytes. Further, a low NaCl concentration in the media, the addition of organic osmolytes, or the addition of follicular secretions is recommended to achieve cytoplasmic maturation in porcine oocytes. Oviductal and follicular sulfate proteoglycans added to culture medium likely stimulate the rate of sperm capacitation and acrosome reaction, compete with the sperm receptor for binding to zona pellucida glycoproteins, and consequently reduce the number of spermatozoa that attach to the zona pellucida. However, failure of the cortical reaction mechanism to function properly is still considered to be a problem in current porcine IVF systems. The ability of oocytes to form a male pronucleus seem to be dependent on oocyte glutathione level following sperm penetration. However, the delay in male pronuclear formation may be due to unknown factor(s) regulating replacement of sperm protamine by oocyte histone and(or) other cell cycle-dependent factor(s). Although an acceptable rate (10 to 20%) of blastocyst formation has been produced from IVM and IVF oocytes in some laboratories, culture conditions are still inadequate for satisfactory rates of embryonic and fetal development. It is anticipated that additional technological modifications of *in vitro* culture conditions will bring about not only an improved efficiency in *in vitro* production of porcine embryos but also an increased understanding of maturation, fertilization, and early embryonic development in the pig.

Acknowledgments

This manuscript was prepared while supported by NRI Competitive Grants Program/ USDA 93-37203-9186 and Farmland Industries and is a contribution from the Missouri Agricultural Experiment Station. Journal Series Number 12,318. We thank Betty Nichols for secretarial assistance with the preparation of this manuscript.

Literature Cited

Beckmann, L. S., and B. N. Day. 1993. Effect of media NaCl concentration and osmolarity on culture of the early stage porcine embryo and viability of embryos cultured in a selected superior medium. Theriogenology 39:611.

Blair, R. M., C. M. Coughlin, J. E. Minton, and D. L. Davis. 1993. Embryonic survival and variation in embryonic development in gilts and primiparous sows on day 11 of gestation. Proc. 4th Int. Conf. on Pig Reproduction, Columbia, MO. p 19. (Abstr.).

Broermann, D. M., S. Xie, K. P. Nephew, and W. F. Pope. 1989. Effects of the oviduct and wheat germ agglutinin on enzymatic digestion of porcine zona pellucidae. J. Anim. Sci. 67:1324.

Buhi, W. C., B. O'Brien, I. M. Alvarez, G. Erdos, and D. Dubois. 1993. Immunogold localization of porcine oviductal secretory proteins within the zona pellucida, perivitelline space, and plasma membrane of oviductal and uterine oocytes and early embryos. Biol. Reprod. 48:1274.

Chang, W. Y., H. Ohmura, S. K. Kulp, and Y. C. Lin. 1993. Transforming growth factor-ß1 regulates differentiation of porcine granulosa cells *in vitro*. Theriogenology 40:699.

Cheng, W.T.K. 1985. *In vitro* fertilization of farm animal oocytes. Ph.D. Dissertation. Council for National Academic Awards, Cambridge, U.K.

Cheng, W.T.K., C. Polge, and R. M. Moor. 1986. *In vitro* fertilization of pig and sheep oocytes. Theriogenology 25:146 (Abstr.).

Christmann, L., T. Jung, and R. M. Moor. 1994. MPF components and meiotic competence in growing pig oocytes. Mol. Reprod. Dev. 38:85.

Corps, A. N., D. R. Brigstock, C. J. Littlewood, and K. D. Brown. 1990. Receptors for epidermal growth factor and insulin-like growth factor-I on preimplantation trophectoderm. Development (Camb.) 110:221.

Coskun, S., and Y. C. Lin. 1994. Effects of transforming growth factors and activin-A on *in vitro* porcine oocyte maturation. Mol. Reprod. Dev. 38:153.

Coy, P., E. Martinez, S. Ruiz, J. M. Vazquez, J. Roca, and J. Gadea. 1993a. Environment and medium volume influence *in vitro* fertilization of pig oocytes. Zygote 1:209.

Coy, P., E. Martinez, S. Ruiz, J. M. Vazquez, J. Roca, and C. Matas. 1993b. Sperm concentration influences fertilization and male pronuclear formation *in vitro* in pigs. Theriogenology 40:539.

Coy, P., E. Martinez, B. Ruiz, J. M. Vazquez, J. Roca, C. Matas, and M. T. Pellicer. 1993c. *In vitro* fertilization of pig oocytes after different coincubation intervals. Theriogenology 39:1201.

Cran, D. G., and W.T.K. Cheng. 1985. Changes in cortical granules during porcine oocyte maturation. Gamete Res. 11:311.

Cran, D. G., and C. R. Esper. 1990. Cortical granules and the cortical reaction in mammals. J. Reprod. Fertil. Suppl. 42:177.

Daen, F. P., E. Sato, K. Naito, and Y. Toyoda. 1994. The effect of pig follicular fluid fractions on cumulus expansion and male pronucleus formation in porcine oocytes matured and fertilized *in vitro*. J. Reprod. Fertil. 101:667.

Dandekar, P., and P. Talbot. 1992. Perivitelline space of mammalian oocytes: Extracellular matrix of unfertilized oocytes and formation of a cortical granule envelope following fertilization. Mol. Reprod. Dev. 31:135.

Ding, J., N. Clarke, T. Nagai, and R. M. Moor. 1992a. Protein and nuclear changes in pig eggs at fertilization. Mol. Reprod. Dev. 31:287.

Ding, J., and G. R. Foxcroft. 1992. Follicular heterogeneity and oocyte maturation *in vitro* in pigs. Biol. Reprod. 47:648.

Ding, J., and G. R. Foxcroft. 1994a. Conditioned media produced by follicular shells of different maturity affect maturation of pig oocytes. Biol. Reprod. 50:1377.

Ding, J., and G. R. Foxcroft. 1994b. Epidermal growth factor enhances oocyte maturation in pigs. Mol. Reprod. Dev. 39:30.

Ding, J., and G. R. Foxcroft. 1994c. FSH-stimulated follicular secretions enhanced oocyte maturation in pigs. Theriogenology 41:1473.

Ding, J., R. M. Moor, and G. R. Foxcroft. 1992b. Effects of protein synthesis on maturation, sperm penetration, and pronuclear development in porcine oocytes. Mol. Reprod. Dev. 33:59.

Feng, P., M. Knecht, and K. Catt. 1987. Hormonal control of epidermal growth factor receptors by gonadotropins during granulosa cell differentiation. Endocrinology 120:1121.

Foxcroft, G. R., and M. G. Hunter. 1985. Basic physiology of follicular maturation in the pig. J. Reprod. Fertil. Suppl. 33:1.

Fujinaga, H., M. Yamoto, R. Nakano, and K. Shima. 1992. Epidermal growth factor binding site in porcine granulosa cells and their regulation by follicle-stimulating hormone. Biol. Reprod. 46:705.

Funahashi, H., T. C. Cantley, and B. N. Day. 1994a. Different hormonal requirement of porcine oocyte-complexes during maturation *in vitro*. J. Reprod. Fertil. 101:159.

Funahashi, H., T. C. Cantley, T. T. Stumpf, S. L. Terlouw, and B. N. Day. 1994b. *In vitro* development of *in vitro* matured porcine oocytes following chemical activation or *in vitro* fertilization. Biol. Reprod. 50:1072.

Funahashi, H., T. C. Cantley, T. T. Stumpf, S. L. Terlouw, and B. N. Day. 1994c. Use of low salt culture medium for *in vitro* maturation of porcine oocytes is associated with elevated oocyte glutathione levels and enhanced male pronuclear formation after *in vitro* fertilization. Biol. Reprod. 51:633.

Funahashi, H., and B. N. Day. 1993a. Effects of different serum supplements in maturation medium on meiotic and cytoplasmic maturation of pig oocytes. Theriogenology 39:965.

Funahashi, H., and B. N. Day. 1993b. Effects of follicular fluid at fertilization *in vitro* on sperm penetration in pig oocytes. J. Reprod. Fertil. 99:97.

Funahashi, H., and B. N. Day. 1993c. Effects of the duration of exposure to supplemental hormones on cytoplasmic maturation of pig oocytes *in vitro*. J. Reprod. Fertil. 98:179.

Funahashi, H., and B. N. Day. 1994. *In vitro* maturation/*in vitro* fertilization of porcine oocytes. Proc. Annu. Conf. of the Society for Theriogenology. pp 206-214.

Funahashi, H., and B. N. Day. 1995a. Effect of exposure of porcine oocyte-cumulus complexes to gonadotropins. Presented at Beltsville Symposia in Agricultural Research XX: Biotechnology's Role in the Genetic Improvement of Farm Animals, Beltsville, MD (Abstr.).

Funahashi, H., and B. N. Day. 1995b. Effects of cumulus cells on glutathione content of porcine oocyte during *in vitro* maturation. J. Anim. Sci. 73(Suppl. 1):90 (Abstr.).

Funahashi, H., Z. Machaty, R. S. Prather, and B. N. Day. 1995b. γ-Glutamyltranspeptidase (GGT) of spermatozoa may reduce oocyte glutathione (GSH) content at sperm penetration. Biol. Reprod. 52(Suppl. 1):140 (Abstr.).

Funahashi, H., T. T. Stumpf, T. C. Cantley, N.-H. Kim, and B. N. Day. 1995a. Pronuclear formation and intracellular glutathione content of *in vitro*-matured porcine oocytes following *in vitro* fertilization and/or electrical activation. Zygote 3:273.

Funahashi, H., T. T. Stumpf, N. H. Kim, and B. N. Day. 1994d. Effects of sodium chloride concentration in maturation media on histone H1 kinase (H1K) activity and cytoplasmic maturation of porcine oocytes. J. Reprod. Fertil. 13 (Abstr.):38.

Funahashi, H., T. T. Stumpf, S. L. Terlouw, T. C. Cantley, A. Rieke, and B. N. Day. 1994e. Developmental ability of porcine oocytes matured and fertilized *in vitro*. Theriogenology 41:1425.

Funahashi, H., T. T. Stumpf, S. L. Terlouw, and B. N. Day. 1993. Effects of electrical stimulation before or after *in vitro* fertilization on sperm penetration and pronuclear formation in pig oocytes. Mol. Reprod. Dev. 36:361.

Galeati, G., S. Modina, A. Lauria, and M. Mattioli. 1991. Follicle somatic cells influence pig oocyte penetrability and cortical granule distribution. Mol. Reprod. Dev. 29:40.

Gangrade, B. K., and J. V. May. 1990. The production of transforming growth factor-ß in the porcine ovary and its secretion *in vitro*. Endocrinology 127:2372.

Garcia-Perez, A., and M. B. Burg. 1991. Renal medullary organic osmolytes. Physiol. Rev. 71:1081.

Gosden, R. G., R.H.F. Hunter, E. Telfer, C. Torrance, and N. Brown. 1988. Physiological factors underlying the formation of ovarian follicular fluid. J. Reprod. Fertil. 82:813.

Grant, S. A., M. G. Hunter, and G. R. Foxcroft. 1989. Morphological and biochemical characteristics during ovarian follicular development in the pig. J. Reprod. Fertil. 86:171.

Greenwald, G. S., and S. K. Roy. 1994. Follicular development and its control. In: E. Knobil, J. D. Neill, G. S. Greenwald, C. L. Markert, and D. W. Pfaff (Ed.) The Physiology of Reproduction. pp 629-724. Raven Press, New York.

Grupen, C. G., H. Nagashima, and M. B. Nottle. 1995. Cysteamine enhances *in vitro* development of IVM-IVF porcine oocytes. Theriogenology 43:227 (Abstr.).

Gulyas, B. J. 1980. Cortical granules of mammalian eggs. Int. Rev. Cytol. 63:357.

Hagen, D. R., R. S. Prather, M. M. Sims, and N. L. First. 1991. Development of one-cell porcine embryos to the blastocyst stage in simple media. J. Anim. Sci. 69:1147.

Hammond, J. M., J.L.S. Baranao, D. Skaleris, A. B. Knight, J. A. Romanus, and M. M. Rechler. 1985. Production of insulin-like growth factors by ovarian granulosa cells. Endocrinology 117:2553.

Hammond, J. M., C.-J. Hsu, J. Klindt, B. K. Tsang, and B. R. Downey. 1988. Gonadotropins increase concentrations of immunoreactive insulin-like growth factor-I in porcine follicular fluid *in vivo*. Biol. Reprod. 38:304.

Hatanaka, Y., T. Nagai, T. Tobita, and M. Nakano. 1992. Changes in the properties and composition of zona pellucida of pigs during fertilization *in vitro*. J. Reprod. Fertil. 95:341.

Hirao, Y., T. Nagai, M. Kubo, T. Miyano, M. Miyake, and S. Kato. 1994. *In vitro* growth and maturation of pig oocytes. J. Reprod. Fertil. 100:333.

Hsu, C.-J., and J. M. Hammond. 1986. Gonadotropins and estradiol stimulate immunoreactive IGF-I production by porcine granulosa cells *in vitro*. Endocrinology 120:198.

Hsu, C.-J., S. D. Holmes, and J. M. Hammond. 1987. Ovarian epidermal growth factor-like activity. Concentrations in porcine follicular fluid during follicular enlargement. Biochem. Biophys. Res. Commun. 47:242.

Hunter, M. G., S. A. Grant, and G. R. Foxcroft. 1989. Histological evidence for heterogeneity in the development of preovulatory pig follicles. J. Reprod. Fertil. 86:165.

Hunter, M. G., and T. Wiesak. 1990. Evidence for and implications of follicular heterogeneity in pigs. J. Reprod. Fertil. Suppl. 40:163.

Hunter, R.H.F. 1967. The effects of delayed insemination on fertilization and cleavage in the pig. J. Reprod. Fertil. 13:133.

Hunter, R.H.F. 1972. Local action of progesterone leading to polyspermic fertilization in pigs. J. Reprod. Fertil. 31:433.

Hunter, R.H.F. 1990. Fertilization of pig eggs *in vitro* and *in vivo*. J. Reprod. Fertil. Suppl. 40:211.

Hunter, R.H.F. 1991. Oviduct function in pigs, with particular reference to the pathological condition of polyspermy. Mol. Reprod. Dev. 29:385.

Hunter, R.H.F., and T. G. Baker. 1975. Development and fate of porcine graafian follicles identified at different stages of the oestrous cycle. J. Reprod. Fertil. 43:193.

Hunter, R.H.F., and C. Polge. 1966. Maturation of follicular oocytes in the pig after injection of human chorionic gonadotrophin. J. Reprod. Fertil. 12:525.

Iritani, A., K. Niwa, and H. Imai. 1978. Sperm penetration *in vitro* of pig follicular oocytes matured in culture. J. Reprod. Fertil. 54:379.

Jones, R. 1990. Identification and functions of mammalian sperm-egg recognition molecules during fertilization. J. Reprod. Fertil. Suppl. 42:89.

Jones, R. 1991. Interaction of zona pellucida glycoproteins, sulphated carbohydrates and synthetic polymers with proacrosin, the putative egg-binding protein from mammalian spermatozoa. Development (Camb.) 111:1155.

Jones, R., and C. R. Brown. 1987. Identification of a zona-binding protein from boar spermatozoa as proacrosin. Exp. Cell Res. 171:505.

Jung, T., J. J. Fulka, C. Lee, and R. M. Moor. 1993. Effects of the protein phosphorylation inhibitor genistein on maturation pig oocytes *in vitro*. J. Reprod. Fertil. 98:529.

Jung, T., C. Lee, and R. M. Moor. 1992. Effects of protein kinase inhibitors on pig oocyte maturation *in vitro*. Reprod. Nutr. Dev. 32:461.

Kameyama, Y., and Y. Ishijima. 1994. Effect of cumulus cells on *in vitro* maturation of pig oocytes. Jpn. J. Fertil. Steril. 39:66.

Kano, K., T. Miyano, and S. Kato. 1994. Effect of oviductal epithelial cells on fertilization of pig oocytes *in vitro*. Theriogenology 42:1061.

Kaufman, M. L., and S. T. Homa. 1993. Defining a role for calcium in the resumption and progression of meiosis in the pig oocyte. J. Exp. Zool. 265:69.

Kikuchi, K., T. Nagai, J. Motlik, Y. Shioya, and Y. Izaike. 1993. Effect of follicle cells on *in vitro* fertilization of pig follicular oocytes. Theriogenology 39:593.

Kikuchi, K., K. Naito, F. P. Daen, Y. Izaike, and Y. Toyoda. 1995. Histone H1 kinase activity during *in vitro* fertilization of pig follicular oocytes matured *in vitro*. Theriogenology 43:523.

Kim, N.-H., H. Funahashi, R. S. Prather, G. Schatten, and B. N. Day. 1996. Microtubule and microfilament dynamics in porcine oocytes during meiotic maturation. Mol. Reprod. Dev. 43:248.

Kim, N.-H., and A.R.J. Menino. 1995. Effects of stimulators of protein kinase A and C and modulators of phosphorylation on plasminogen activator activity in porcine oocyte-cumulus cell complexes during *in vitro* maturation. Mol. Reprod. Dev. 40:364.

Kinloch, R. A., and P. M. Wassarman. 1993. Specific gene expression during oogenesis in mice. In: R.B.L. Gwatkin (Ed.) Genes in Mammalian Reproduction. pp 27-43. Wiley-Liss, New York.

Laurincik, J., D. Rath, and H. Niemann. 1994. Differences in pronucleus formation and first cleavage following *in vitro* fertilization between pig oocytes matured *in vivo* and *in vitro*. J. Reprod. Fertil. 102:277.

Lazzari, G., C. Galli, and R. M. Moor. 1994. Functional changes in the somatic and germinal compartments during follicle growth in pigs. Anim. Reprod. Sci. 35:119.

Lewis, A. M., P. L. Kaye, R. Lising, and R.D.A. Cameron. 1992. Stimulation of protein synthesis and expansion of pig blastocysts by insulin *in vitro*. Reprod. Fertil. Dev. 4:119.

Machaty, Z., M. A. Mayes, and R. S. Prather. 1995. Parthenogenetic activation of porcine oocytes with guanosine-5'-O-(3'-thiotriphosphate). Biol. Reprod. 52:753.

Martinez, E., J. M. Vazquez, C. Matas, J. Roca, P. Coy, and J. Gadea. 1993. Evaluation of boar spermatozoa penetrating capacity using pig oocytes at the germinal vesicle stage. Theriogenology 40:547.

Mattioli, M. 1994. Recent acquisitions in pig oocyte maturation and fertilization *in vitro*. Reprod. Domest. Anim. 29:346.

Mattioli, M., G. Galeati, M. L. Bacci, and E. Seren. 1988a. Follicular factors influence oocyte fertilizability by modulating the intercellular cooperation between cumulus cells and oocyte. Gamete Res. 21:223.

Mattioli, M., G. Galeati, and E. Seren. 1988b. Effect of follicle somatic cells during pig oocyte maturation on egg penetrability and male pronucleus formation. Gamete Res. 20:177.

Mattioli, M., M. L. Bacci, G. Galeati, and E. Seren. 1989. Developmental competence of pig oocytes matured and fertilized *in vitro*. Theriogenology 31:1201.

Mattioli, M., M. L. Bacci, G. Galeati, and E. Seren. 1991. Effects of LH and FSH on the maturation of pig oocytes *in vitro*. Theriogenology 36:95.

Mattioli, M., B. Barboni, M. L. Bacci, and E. Seren. 1990. Maturation of pig oocytes: Observations on membrane potential. Biol. Reprod. 43:318.

Mattioli, M., G. Galeati, B. Barboni, and E. Seren. 1994. Concentration of cyclic AMP during the maturation of pig oocytes *in vivo* and *in vitro*. J. Reprod. Fertil. 100:403.

Meinecke, B., and S. Meinecke-Tillmann. 1993. Effects of α-amanitin on nuclear maturation of porcine oocytes *in vitro*. J. Reprod. Fertil. 98:195.

Mo, H., T. Y. Yan, and Z. W. Shen. 1991. Isolation, characterization, and localization of the zona pellucida binding proteins of boar sperm. Mol. Reprod. Dev. 28:124.

Motlik, J., and J. Fulka. 1976. Breakdown of the germinal vesicle in pig oocytes *in vivo* and *in vitro*. J. Exp. Zool. 198:155.

Nagai, T. 1994. Current status and perspectives in IVM-IVF of porcine oocytes. Theriogenology 41:73.

Nagai, T., J. Ding, and R. M. Moor. 1993. Effect of follicle cells and steroidogenesis on maturation and fertilization *in vitro* of pig oocytes. J. Exp. Zool. 266:146.

Nagai, T., and R. M. Moor. 1990. Effect of oviduct cells on the incidence of polyspermy in pig eggs fertilized *in vitro*. Mol. Reprod. Dev. 26:377.

Nagai, T., K. Niwa, and A. Iritani. 1984. Effects of sperm concentration during preincubation in a defined medium on fertilization *in vitro* of pig follicular oocytes. J. Reprod. Fertil. 70:271.

Nagai, T., and M. Takahashi. 1992. Culture of *in vitro* matured and fertilized pig oocytes. 12th Int. Congr. Anim. Reprod. 3:1324.

Nagai, T., T. Takahashi, H. Masuda, Y. Shioya, M. Kuwayama, M. Fukushima, S. Iwasaki, and A. Hanada. 1988. In-vitro fertilization of pig oocytes by frozen boar spermatozoa. J. Reprod. Fertil. 84:585.

Naito, K., F. P. Daen, and Y. Toyoda. 1992. Comparison of histone H1 kinase activity during meiotic maturation between two types of porcine oocytes matured in different media *in vitro*. Biol. Reprod. 47:43.

Naito, K., Y. Fukuda, and Y. Toyoda. 1988. Effects of porcine follicular fluid on male pronucleus formation in porcine oocytes matured *in vitro*. Gamete Res. 21:289.

Niwa, K. 1993. Effectiveness of *in vitro* maturation and *in vitro* fertilization techniques in the pig. J. Reprod. Fertil. Suppl. 48:49.

Ocampo, M. B., L. C. Ocampo, T. Mori, J. Ueda, H. Shimizu, and H. Kanagawa. 1994. Blastocyst formation of pig embryos derived from *in vitro* fertilization of *in vitro* matured pig oocytes in the amniotic fluid of a developing chick embryo. Anim. Reprod. Sci. 37:65.

Ocampo, M. B., L. C. Ocampo, I. S. Ryu, T. Mori, J. Ueda, and H. Kanagawa. 1993. Effects of culture time, ovarian activity, cumulus cells and sera on the nuclear and cytoplasmic maturation of pig oocytes *in vitro*. Anim. Reprod. Sci. 34:135.

Parry, R. V., P. J. Barker, and R. Jones. 1992. Characterization of low Mr zona pellucida binding proteins from boar spermatozoa and seminal plasma. Mol. Reprod. Dev. 33:108.

Perreault, S. D. 1990. Regulation of sperm nuclear reactivation during fertilization. In: B. D. Bavister, J. Cummins, and E.R.S. Roldan (Ed.). Fertilization in Mammals. pp 285-296. Serono Symposia, Norwell, MA.

Perreault, S. D., R. R. Barbee, and V. L. Slott. 1988. Importance of glutathione in the acquisition and maintenance of sperm nuclear decondensing activity in maturing hamster oocytes. Dev. Biol. 125:181.

Petr, J., J. Rozinek, J. J. Fulka, and F. Jilek. 1994a. Influence of cytoplasmic microinjection on meiotic competence in growing pig oocytes. Reprod. Nutr. Dev. 34:81.

Petr, J., O. Tepla, R. Grocholova, and F. Jilek. 1994b. Inhibition of meiotic maturation in growing pig oocytes by factor(s) from cumulus cells. Reprod. Nutr. Dev. 34:149.

Petters, R. M., and M. L. Reed. 1991. Addition of taurine or hypotaurine to culture medium improves development of one- and two-cell pig embryos. Theriogenology 35:253 (Abstr.).

Petters, R. M., and K. D. Wells. 1993. Culture of pig embryos. J. Reprod. Fertil. Suppl. 48:61.

Pope, W. F., S. Xie, D. M. Broermann, and K. P. Nephew. 1990. Causes and consequences of early embryonic diversity in pigs. J. Reprod. Fertil. Suppl. 40:251.

Racowsky, C. 1985. Effect of forskolin on maintenance of meiotic arrest and stimulation of cumulus expansion, progesterone and cyclic AMP production by pig oocyte-cumulus complexes. J. Reprod. Fertil. 74:9.

Rath, D. 1992. Experiments to improve *in vitro* fertilization techniques for *in vivo*-matured porcine oocytes. Theriogenology 37:885.

Rath, D., L. A. Johnson, G. R. Welch, and H. Niemann. 1994a. Successful gamete intrafallopian transfer (GIFT) in the porcine. Theriogenology 41:1173.

Rath, D., H. Niemann, and L. A. Johnson. 1994b. Gamete intrafallopian transfer (GIFT), an alternative to *in vitro* fertilization procedures for special applications. Reprod. Domest. Anim. 29:349.

Rath, D., H. Niemann, T. Tao, and M. Boerjan. 1995a. Ratio and number of inner cell mass and trophoblast cells of *in vitro* and *in vivo* produced porcine embryos. Theriogenology 43:304 (Abstr.).

Rath, D., H. Niemann, and C.R.L. Torres. 1995b. *In vitro* development to blastocysts of early porcine embryos produced *in vivo* or *in vitro*. Theriogenology 43:913.

Reed, M. L., J. L. Estrada, M. J. Illera, and R. M. Petters. 1993. Effects of epidermal growth factor, insulin-like growth factor-I, and dialyzed porcine follicular fluid on porcine oocyte maturation *in vitro*. J. Exp. Zool. 266:74.

Reed, M. L., M. J. Illera, and R. M. Petters. 1992. *In vitro* culture of pig embryos. Theriogenology 37:95.

Schroeter, D., and B. Meinecke. 1995. Comparative analysis of the polypeptide pattern of cumulus cells during maturation of porcine cumulus oocyte complexes *in vivo* and *in vitro*. Reprod. Nutr. Dev. 35:85.

Singh, B., G. J. Barbe, and D. T. Armstrong. 1993. Factors influencing resumption of meiotic maturation and cumulus expansion of porcine oocyte-cumulus cell complexes *in vitro*. Mol. Reprod. Dev. 36:113.

Sirotkin, A. V., and J. Nitray. 1994. Effects of prolactin on estrogen, cAMP and oxytocin secretion by porcine granulosa cells *in vitro*. Reprod. Nutr. Dev. 34:141.

Soede, N. M., F. A. Helmond, and B. Kemp. 1994. Periovulatory profiles of oestradiol, LH and progesterone in relation to oestrus and embryo mortality in multiparous sows using transrectal ultrasonography to detect ovulation. J. Reprod. Fertil. 101:633.

Sommer, P., D. Rath, and H. Niemann. 1992. *In vitro* maturation of porcine oocytes in the presence of follicular granulosa cells, FSH and/or EGF. Proc. 12th Int. Congr. Anim. Reprod. 1:378.

Sun, F. Z., J. Hoyland, X. Huang, W. Mason, and R. M. Moor. 1992. A comparison of intercellular changes in porcine eggs after fertilization and electroactivation. Development (Camb.) 115:947.

Wang, W. H., L. R. Abeydeera, K. Okuda, and K. Niwa. 1994. Penetration of porcine oocytes during maturation *in vitro* by cryopreserved, ejaculated spermatozoa. Biol. Reprod. 50:510.

Wang, W. H., K. Niwa, and K. Okuda. 1991. In-vitro penetration of pig oocytes matured in culture by frozen-thawed ejaculated spermatozoa. J. Reprod. Fertil. 93:491.

Wassarman, P. M. 1988. Zona pellucida glycoproteins. Ann. Rev. Biochem. 57:415.

Wassarman, P. M. 1990. Regulation of mammalian fertilization by zona pellucida glycoproteins. J. Reprod. Fertil. Suppl. 42:79.

Wiesak, T., M. G. Hunter, and G. R. Foxcroft. 1990. Differences in follicular morphology, steroidogenesis and oocyte maturation in naturally cyclic and PMSG/hCG-treated prepubertal gilts. J. Reprod. Fertil. 89:633.

Wormington, M. 1994. Unmasking the role of the 3' UTR in the cytoplasmic polyadenylation and translational regulation of maternal mRNA. BioEssays 16:533.

Xia, P., F. R. Tekpetey, and D. T. Armstrong. 1994. Effect of IGF-I on pig oocyte maturation, fertilization, and early embryonic development *in vitro*, and on granulosa and cumulus cell biosynthetic activity. Mol. Reprod. Dev. 38:373.

Yancey, P. H. 1994. Compatible and counteracting solutes. In: K. Strange (Ed.). Cellular and Molecular Physiology of Cell Volume Regulation. pp 81-109. CRC Press, Boca Raton, FL.

Yoshida, M. 1993. Role of glutathione in the maturation and fertilization of pig oocytes *in vitro*. Mol. Reprod. Dev. 35:76.

Yoshida, M., D. G. Cran, and V. G. Pursel. 1993a. Confocal and fluorescence microscopic study using lectins of the distribution of cortical granules during the maturation and fertilization of pig oocytes. Mol. Reprod. Dev. 36:462.

Yoshida, M., Y. Ishizaki, H. Kawagishi, K. Bamba, and Y. Kojima. 1992a. Effects of pig follicular fluid on maturation of pig oocytes *in vitro* and on their subsequent fertilizing and developmental capacity *in vitro*. J. Reprod. Fertil. 95:481.

Yoshida, M., K. Ishigaki, T. Nagai, M. Chikyu, and V. G. Pursel. 1993b. Glutathione concentration during maturation and after fertilization in pig oocytes: Relevance to the ability of oocytes to form male pronucleus. Biol. Reprod. 49:89.

Yoshida, M., K. Ishigaki, and V. G. Pursel. 1992b. Effect of maturation media on male pronucleus formation in pig oocytes matured *in vitro*. Mol. Reprod. Dev. 31:68.

Yoshida, M., Y. Mizoguchi, K. Ishigaki, T. Kojima, and T. Nagai. 1993c. Birth of piglets derived from *in vitro* fertilization of pig oocytes matured *in vitro*. Theriogenology 39:1303.

Youngs, C. R., S. P. Ford, L. K. McGinnis, and L. H. Anderson. 1993. Investigations into the control of litter size in swine: I. Comparative studies on *in vitro* development of Meishan and Yorkshire preimplantation embryos. J. Anim. Sci. 71:1561.

Zheng, Y. S., and M. A. Sirard. 1992. The effect of sera, bovine serum albumin and follicular cells on *in vitro* maturation and fertilization of porcine oocytes. Theriogenology 37:779.

Zirkin, B. R., S. D. Perreault, and S. J. Naish. 1989. Formation and function of the male pronucleus during mammalian fertilization. In: H. Schatten, and G. Schatten (Ed.). The Molecular Biology of Fertilization. pp 91-114. Academic Press, San Diego, CA.

Zirkin, B. R., D. A. Soucek, T.S.K. Chang, and S. D. Perreault. 1985. *In vitro* and *in vivo* studies of mammalian sperm nuclear decondensation. Gamete Res. 11:349.

Recent Observations on Effects of Cell Cycle in Nuclear Transfer

Ian Wilmut, Pedro Otaegui, and Keith Campbell

Roslin Institute, Roslin, Midlothian, EH25 9PS, U.K.

Abstract Many factors influence development of embryos produced by nuclear transfer, including species, donor embryo stage, choice of recipient cytoplast, method of activation of the oocyte, and the culture system for the embryo. However, one other factor that has been shown to have a great impact is the coordination of cycles of donor and recipient cells. If a nucleus is transferred into an oocyte at metaphase II then DNA replication is induced regardless of the stage of the donor cell cycle. Hence, it is essential that donor cells be in G1 phase of the cell cycle. By contrast, if the oocyte is activated and cultured to allow the decay of Maturation Promoting Factor activity, then the nucleus determines whether replication should occur. Such a cytoplast is suitable for nuclei at any stage of the cell cycle. By use of a "universal recipient" it has been possible to increase the proportion of embryos that develop and to obtain development to term with nuclei from later stages of development, including cultured embryonic cells. Nuclear transfer from cultured embryonic cells would permit dissemination of genetic improvement by embryo multiplication and the introduction of precise genetic changes by gene targeting.

Key Words: Nuclear Transfer, Cell Cycle, Livestock, Genetic Improvement

Introduction

Revolutionary opportunities in animal breeding and in research were created by the establishment of procedures for "nuclear transfer" in mammals (McGrath and Solter, 1983). It is now almost a decade since these procedures were first applied in farm animals (Willadsen, 1986) and there has been considerable research activity directed to their improvement. During that time many factors have been shown to influence development following nuclear transfer, including species, donor embryo stage of development, source of recipient cytoplast, and method of activation of oocytes. One factor with a critical influence on the developmental potential of the reconstituted embryo is the regulation of the cell cycle of the donor and recipient cells. This paper will describe the method of nuclear transfer and review recent investigations of the effects of cell cycle before describing our recent experiments to establish novel methods for embryo multiplication and genetic modification through nuclear transfer from established cell lines.

The Method of Nuclear Transfer

Embryo multiplication in mammals is achieved by fusion of a donor cell (or karyoplast) to an oocyte or zygote from which the genetic material has been removed. The recipient cell is usually a secondary oocyte. The method of enucleation that has been used most often is to remove the second metaphase either by aspiration of that portion of the cytoplasm that is adjacent to the polar body or by use of DNA-specific fluorochromes to identify the chromosomes. In livestock, fusion and activation are usually induced by exposure to direct current (Willadsen, 1986). Development of the reconstituted embryo to the blastocyst stage is usually in the oviduct of a temporary sheep recipient; *in vitro* culture is believed to be less effective.

Regulation of Cell Cycles

In the first uses of nuclear transfer, which were with mouse embryos, no account was paid to the stage of the cell cycle of the donor cells and recipient cytoplasts were prepared from zygotes (McGrath and Solter, 1983). Subsequently, it was noted that when transferring between zygotes or from 2-cell embryos to zygotes, the stage of the cell cycle of both donor and recipient cells had a profound effect on the proportion of embryos that developed (Smith et al., 1988). These effects associated with cell cycle have not been fully accounted for by subsequent experiments.

Since then increasing attention has been paid to the regulation of cell cycle stage during nuclear transfer, and it has become clear that donor cell cycle stage can have a great influence on subsequent embryo development. Our research suggests that inappropriate choice of cell cycle stage may lead to aneuploidy, because of inappropriate regulation of DNA replication, and suggests several approaches to nuclear transfer. The critical mechanisms governing the cell cycle will be considered in the context of nuclear transfer.

The investigations with oocytes and embryos have drawn heavily upon earlier research in which cell cycle stage was shown to have profound effects on the nuclear response following the fusion of two cells (see review by Rao and Sunkara, 1978). Two Hela cells were fused by exposure to Sendia virus and the response of both nuclei was observed directly. It was found that cytoplasm from cells in G2 phase induced nuclear envelope breakdown and chromosome condensation in nuclei in G1 phase or S phase. It is now believed that complex groups of molecules ensure that cells normally replicate their DNA once, but only once, during each cell cycle (see Murray and Hunt, 1993) and they have a critical influence on the events after nuclear transfer. One important activity causes nuclear membrane breakdown during mitosis and meiosis and is known as MPF, for maturation/meiosis/mitosis promoting factor. It was this activity that was present in cells at G2 phase in the experiments with Hela cells.

A series of changes is observed when a cell is fused to an oocyte at metaphase II: first, there is breakdown of the nuclear membrane and condensation of the chromosomes, before the membrane reforms, and over a period of several hours the nucleus swells in a manner similar to that in pronuclei (Czolowska et al., 1984). These changes are believed to be brought about by the activity of MPF.

In a recent experiment with cattle oocytes and embryos, we examined the effect on subsequent DNA replication of donor and recipient cell cycle stages (Campbell et al., 1993). Nuclei were transferred in G1, S or G2 phases. Two groups of oocytes were used as recipient cells: one with high levels of MPF activity and one with low levels. Metaphase II oocytes were expected to have high MPF activities. By contrast, when oocytes that had been induced to activate were cultured for 10 hours before nuclear transfer MPF activity was found to be low.

There was a profound effect of recipient cell cycle stage on events in the first cell cycle. When nuclei were transferred to metaphase II oocytes replication of DNA was observed in almost all cases, regardless of the stage of the cell cycle of donor cells. Direct observation of rabbit nuclei after transfer shows that those in S phase also attempt to re-replicate their DNA and form abnormal chromosomes (Collas et al., 1992b), extending our observation with nuclei in G1 or G2 to the other stage of the cell cycle. As a result of these responses, the only stage of the cell cycle that would be expected to support normal development if transferred to a metaphase II cytoplast is G1 phase, as replication of DNA is required.

By contrast, following transfer 10 hours after activation, when MPF activity had disappeared, the stage of the cell cycle of the nucleus determined whether replication of DNA occurred: DNA replication was observed in G1 nuclei and S-phase nuclei, but not in G2 nuclei. In this case it seems that the pre-activated oocytes function as "universal recipients" and are suitable for nuclei at any stage of the cell cycle.

Effects on Development

These observations suggest two possible approaches to the maintenance of normal ploidy after nuclear transfer, and it is not yet clear which will be more effective in supporting normal development. Two experimental systems have been used to investigate these possibilities. In the mouse nuclei have been transferred following synchronization of the donor cells by treatment with a cytoskeletal inhibitor, whereas sheep nuclei have been transferred at unknown stages of the cell cycle to universal recipient oocytes.

It would be expected that nuclei transferred in the G1 phase of the cell cycle would replicate normally if transferred at the time of activation to oocytes at metaphase, and development to blastocyst was observed more frequently if the donor cell was in G1 than in S phase (rabbit, Collas et al, 1992a). However, routine use of this approach depends on a method to synchronize the cell cycle stage, with the attendant risk that this may reduce developmental potential. Methods have been established for obtaining mouse embryonic cells early in the cell cycle by exposing embryos to nocodazole to hold the cells in mitosis, before culture for a short period (Otaegui et al., 1994). The effect on subsequent development of the treated embryos depended on the concentration of the drug and the stage of development of the embryos. Development of 4-cell stage embryos to term was not reduced provided that an exposure time of 9 hours was used. Development of mouse embryos is relatively synchronous, so it is possible to obtain groups of cells from 8-cell embryos of which over 90% will complete mitosis between 60 and 90 minutes after the end of treatment. A comparison was made of development to blastocyst following transfer to metaphase II oocyte of nuclei from cells apparently either in G1 phase or in S phase. Whereas 27% of embryos in the G1-phase group developed to blastocysts there were none in the S-phase group, a very dramatic difference. The effects of transfer of nuclei in other phases of the cell cycle is under investigation at present.

In contrast to the satisfactory methods for synchronization of murine embryonic cells, we know of no published reports of reliable methods for cell cycle synchronization of cells from the embryos of any livestock species. Several factors may account for this apparent difference. By comparison with livestock species the timing of development of embryos recovered from inbred strains of donor mice is relatively predictable and consistent. As a result, it is possible to recover embryos shortly before a specific cleavage and successfully control mitosis in all of the cells by exposure to a drug for a relatively short period. This apparent difference between the species may be exaggerated by the fact that studies with bovine embryos have been with morulae, by which stage there is great variation both within and between embryos in the time of mitosis. Hence it would be necessary to expose livestock embryos to drugs for a period of time that has been shown to be toxic in the mouse (Otaegui et al., 1994).

The universal recipient approach to maintaining normal ploidy was first tested in the sheep with nuclei from 16-cell embryos (Campbell et al., 1994). Nuclei taken at unknown stages of the cell cycle were transferred either at the time of activation or to pre-activated oocytes at 5 to 18 hours after activation. Because embryos progress rapidly through the cell cycle it is likely that the great majority of cells were in S phase. It was found that 22% of the oocytes developed to blastocysts in the control group, in which fusion and activation were simultaneous. It was observed that activation occurred spontaneously in a similar proportion of unactivated oocytes, so it is probable that, in some of these embryos, activation also occurred before nuclear transfer, so creating a universal recipient inadvertently. In this way normal development might be expected even if S-phase nuclei were being transferred. By contrast, 55% of embryos formed from universal recipients developed to blastocysts, a significantly different proportion.

More recently, this approach has been used to establish the developmental potential of cultured cells derived from sheep embryos (Campbell et al., 1995). Day 9 blastocysts were cultured as though for the isolation of mouse embryonic stem cells and cells used as nuclear donors for transfer to universal recipients. Lambs have been born following transfer of nuclei from cells after one, two, or three passages.

Although there were no published reports of previous attempt to exploit this approach, it seems probable that the procedure used commercially in cattle may have done so unintentionally (Barnes et al., 1990). When the methods were first described it was a surprise that the recipient oocytes were far older than those used for *in vitro* fertilization (Willadsen, 1986). Whereas bovine oocytes are fertilized around 24 hours after the onset of maturation (e.g., Gordon and Lu, 1990), recipient cytoplasts have commonly been used some 16 to 24 hours later (Bondioli et al., 1990). It seems that at that age the oocytes are activated by minor changes in the environment, such as changes in temperature or exposure to the fusion medium, and the rate of nuclear membrane formation is faster than in younger oocytes (Powell and Barnes, 1992). As a result, by the time of fusion the oocyte was no longer equivalent to one at second metaphase, but rather was similar to the universal recipient advocated as a result of our analyses.

Together these observations show that by gaining an understanding of the events that follow transfer of a nucleus it has been possible to increase the proportion of embryos that develop to term and to obtain development with cells derived from embryos at later stages of development. It is not known which of the two basic approaches for regulation of cell cycle will provide more normal embryos. Despite this progress much remains to be learned of other factors that influence development of embryos produced by nuclear transfer. It may be argued that normal development of an embryo produced by nuclear transfer depends on reprogramming gene expression. Although there is indirect and direct evidence of changes in gene expression (Prather et al., 1992), the mechanisms are not understood. Nor is the efficiency of the process known. It may be that the efficiency of this reprogramming varies with stage of the cell cycle. Whereas the chromosomes are exposed to the cytoplasm when nuclear membrane breakdown occurs after nuclear transfer in the presence of MPF activity, this does not happen if the recipient cytoplast is a universal recipient. Do the active molecules pass through the membrane or reach the chromosomes during the next mitosis? Assessment of the importance of many of these factors will be addressed when it becomes possible to transfer nuclei from cultured cells at known specific stages of the cycle.

Application

The availability of cell lines from which it is possible to obtain normal development following nuclear transfer would provide two new opportunities for animal breeding: embryo multiplication and genetic modification.

Embryo multiplication has the potential to enhance the rate of genetic improvement in national herds. Genetic improvement by selection can be considered in two stages: improvement achieved in specialist, elite herds and dissemination throughout the national herd. Embryo multiplication is unlikely to assist selection in the elite herd but has the theoretical potential to accelerate the second phase. This difference reflects the fact that progress within herds depends on maintaining variation.

The ability to introduce precise genetic changes into cells before their use as nuclear donors will provide an opportunity to modify many aspects of animal performance. At present the techniques for gene transfer by direct injection have serious disadvantages that are well documented (Wilmut and Clark, 1991); however, many of these would be overcome by techniques for gene targeting in cultured cells before nuclear transfer. The great potential advantage of gene targeting is that it will permit precise modification to existing genes to either to change the pattern of gene expression or to alter the protein that is produced. Existing variation in productivity reflects exactly such variation that has arisen by spontaneous mutation. Research to investigate physiological mechanisms will reveal other potential sites for targeted change. These may be at many different sites in the production or transport of hormones or in the response to those hormones in different tissues. Direct injection can only induce change by overriding such homeostatic mechanisms, with the risk of undesirable consequences, such as those observed after increasing the level of growth hormone in pigs. By gene targeting there will be the opportunity for measured changes. While there are only a small number of candidate genes at present, it is anticipated that more will be identified by the gene mapping projects. Experience also suggests that the full potential of a new technique is only appreciated some time after it becomes available.

Conclusions

This review suggests that considerable progress has been made in the improvement of methods for embryo multiplication. As a result, development to term has been obtained with nuclei from later stages of development than before. The biological limits to the potential of nuclear transfer in livestock remain to be discovered. However, the techniques used at present are laborious and expensive. Further research is required, not only to improve upon the present methods, but also to provide simple methods of manipulation that can be automated.

Acknowledgments

We acknowledge financial support from Ministry of Agriculture, Fisheries and Food, Biotechnology, Biological Sciences Research Council and the European Commission.

Literature Cited

Barnes, F. L., M. E. Westhusin, and C. R. Looney. 1990. Embryo cloning: Principles and progress. Proc. 4th World Congr. on Genet. Appl. Livest. Prod. Edinburgh, Scotland, 16:323.

Bondioli, K. R., M. E. Westhusin, and C. R. Looney. 1990. Production of identical bovine offspring by nuclear transfer. Theriogenology 33:165.

Campbell, K., P. Loi, P. Cappia, and I. Wilmut. 1994. Improved development to blastocyst of ovine nuclear transfer embryos reconstructed during the presumptive S-phase of enucleated activated oocytes. Biol. Reprod. 50:1385.

Campbell, K., J. McWhir, W. A. Ritchie, and I. Wilmut. 1995. Production of live lambs following nuclear transfer of cultured embryonic disc cells. Theriogenology 43:181 (Abstr.).

Campbell, K., W. A. Ritchie, and I. Wilmut. 1993. Nucleocytoplasmic interactions during the first cell cycle of nuclear transfer reconstituted bovine embryos. Biol. Reprod. 49:933.

Collas, P., J. J. Balise, and J. M. Robl. 1992a. Influence of the cell cycle stage of the donor nucleus on development of nuclear transplant rabbit embryos. Biol. Reprod. 46:492.

Collas, P., C. Pinto-correia, F. A. Ponce de Leon, and J. M. Robl. 1992b. Effect of donor cell cycle stage on chromatin and spindle morphology in nuclear transplant rabbit embryos. Biol. Reprod. 46:501.

Czolowska, R., J. A. Modlinski, and A. K. Tarkowski. 1984. Behavior of thymocyte nuclei in nonactivated and activated mouse oocytes. J. Cell Sci. 69:19.

Gordon, I., and K. H. Lu. 1990. Production of embryos in vitro and its impact upon livestock production. Theriogenology 33:77.

McGrath, J., and D. Solter. 1983. Nuclear transplantation in the mouse embryo by microsurgery and cell fusion. Science (Wash DC) 220:1300.

Murray, A., and T. Hunt. 1993. The Cell Cycle, An Introduction. pp 251. Oxford University Press, Oxford, U.K.

Otaegui, P. J., G. T. O'Neill, K.H.S. Campbell, and I. Wilmut. 1994. Transfer of nuclei from 8-cell stage mouse embryos following use of nocodazole to control the cell cycle. Mol. Reprod. Dev. 39:147.

Powell, R., and F. L. Barnes. 1992. The kinetics of oocyte activation and polar body formation in bovine embryo clones. Mol. Reprod. Dev. 33:53.

Prather, R. S., T. T. Stumpf, and L. F. Rickords. 1992. Reprogramming the nucleus and synchronizing it with the cytoplasm. In: G. E. Seidel, Jr. (Ed.) Symposium on Cloning Mammals by Nuclear Transplantation. p 26. Colorado State University, Boulder.

Rao, P. N., and P. S. Sunkara. 1978. Cell fusion and regulation of DNA synthesis. In: J. Jeter, I. Cameron, and A. Zimmerman (Ed.) Cell Cycle Regulation. Academic Press, New York.

Smith, L. C., I. Wilmut, and R.H.F. Hunter. 1988. Influence of cell cycle stage at nuclear transplantation on the development *in vitro* of mouse embryos. J. Reprod. Fertil. 84:619.

Willadsen, S. M. 1986. Nuclear transplantation in sheep embryos. Nature (Lond.) 320:63.

Wilmut, I., and A. J. Clark. 1991. Basic techniques for transgenesis. In: G. E. Lamming, A.P.F. Flint, and B. J. Weir (Ed.) Reproduction in Domestic Ruminants. J. Reprod. Fertil. Suppl. 43:265.

Gender Preselection in Mammals

L. A. Johnson[*,1], D. G. Cran[†], G. R. Welch[*], and C. Polge[†]

[*]*Germplasm & Gamete Physiology Laboratory, Agricultural Research Service, U.S. Department of Agriculture, Beltsville, MD 20705, U.S.A. and* [†]*Mastercalf, Ltd. Parkhead, Bucksburn, Aberdeen AB2 95X, Scotland*

Abstract Gender preselection through the use of separated X- and Y-chromosome-bearing spermatozoa based on DNA content is now a reality in farm animals. The preselection for gender using DNA technology in clinical trials is underway in humans. Semen from most livestock can now be successfully separated into predominantly X or Y sperm populations prior to their use for intra-tubal insemination or for *in vitro* fertilization (IVF) to produce sexed embryos for transfer. Birth of progeny of the desired sex in cattle, sheep, and swine under semi-practical conditions has successfully validated the technology. Sperm are separated based on the inherent difference in DNA content in the X- and Y-chromosome-bearing sperm using flow cytometry/cell sorting technology. Sperm are stained with Hoechst 33342, which binds to the DNA helix in an amount proportional to the amount of DNA, thus forming the basis for the method. Forty-five calves of the predicted sex have been born using sorted bull sperm in conjunction with IVF resulting in embryos for transfer. Swine, rabbit, and sheep progeny have been produced using surgical techniques to inseminate sorted sperm intra-tubally. At the present time, standard artificial insemination techniques are not suitable with flow cytometrically separated X or Y sperm populations. This is due to slow throughput of the sorting system that is not sufficient to produce enough sperm per hour for a normal insemination dose in any species. Several hundred offspring have been born using the Beltsville Sperm Sexing Technology with no evidence of any morphological abnormality. Cattle, swine, and rabbit offspring have been reproduced through the second generation with normal morphology and reproductive function.

Key Words: Embryos, X-Y Sperm, Cattle, Swine, *In Vitro* Fertilization, Flow Cytometer

Introduction

Application of sexed semen for production of livestock of predetermined sex is dependent on economics, effectiveness, and ease of use. The most cost-effective approach for achieving sex preselection would involve separating adequate numbers of the X from the Y sperm followed by use of the individual sperm populations for artificial insemination (AI). One would, however, need fewer sperm if in vitro fertilization (IVF) and subsequent embryo transfer (ET) were used. Because one is predetermining the outcome of conception there are obvious advantages over embryo sexing. Livestock management schemes can benefit from sex preselection because of the ability to plan matings for a specific sex. In addition to faster genetic progress from sexed sperm or embryos, there are additional advantages of management efficiency and facility flexibility.

[1]Author to whom correspondence should be addressed: Bldg. 200, BARC-East.

Historical Aspects of Sex Preselection Until the report of Johnson et al. (1989), virtually all the efforts directed toward sexing viable semen fell under the category of "physical separation" methods. Although the literature contains a vast number of reports that claim success in achieving sex pre-determination using physical separation methods (particularly in the human and in cattle) none have proven to be repeatable. The reason for this lack of success can generally be attributed to the inability of a particular technique or process to detect very small differences (either perceived or actual) between X- and Y-bearing sperm. In addition, the lack of reliable procedures for determining the success of sperm separation has caused confusion. This review will deal primarily with the well-established DNA difference between X and Y sperm and how it has been developed and used as a functional gender preselection method. Readers interested in the various physical separation methods are directed to the following papers that give excellent summaries of those methods (Kiddy and Hafs, 1971; Amann and Seidel, 1982; Gledhill, 1988; Johnson, 1992, 1994; Windsor et al., 1993). The remainder of this review will deal with the methodology associated with DNA based separation of X and Y sperm.

DNA Measurement in Sperm

The DNA content of individual sperm is the only known and scientifically validated, measurable difference between X- and Y-chromosome-bearing sperm. The Y-chromosome is smaller and contains less DNA than the larger X-chromosome, while the autosomes carried by X- or Y-bearing sperm are identical in DNA content. The difference in X and Y sperm DNA has formed the basis for flow cytometric analysis of X- and Y-bearing sperm that was initially attempted by Gledhill et al. (1976). Since then, advances made in sperm analysis and sorting for DNA have paralleled advances made in improving flow cytometry instrumentation. Cellular staining and the utilization of fluorochromes to identify various organelles have also advanced. The successful adaptation of flow cytometry for DNA analysis and sorting can be traced to the need to orient sperm heads to the laser beam (Dean et al., 1978), resulting in the differentiation of X and Y sperm in a sample of semen (Pinkel et al., 1982; Johnson and Pinkel, 1986). These reports were followed by improved stain and preparation protocols (Johnson et al., 1987a) that resulted in fertilization using intracytoplasmic sperm injection (ICSI) of sorted sperm nuclei (Johnson and Clarke, 1988) and sorting of viable X and Y sperm capable of producing offspring (Johnson et al., 1989; Johnson, 1991; Cran et al., 1993, 1995).

Unassisted fertilization with viable and intact X or Y sperm flow cytometrically sorted on the basis of DNA content difference was the final confirmation of the validity of the DNA-based Beltsville Sperm Sexing Technology for sex preselection (Johnson et al., 1989). Several factors proved central to this success: incubation of sperm at 35°C with a vital fluorochrome (Hoechst 33342) to get uniform stain penetration (Johnson et al., 1989), collecting the sorted sperm in TEST-Yolk extender to preserve viability (Johnson et al., 1989; Johnson 1991), and the use of flow cytometric reanalysis of the sorted sperm for DNA content in order to validate sort purities and predict the ultimate sex of the offspring (Johnson et al., 1987a, 1989).

DNA-Based Flow Cytometric Sperm Sorting

The resolution of X- and Y-bearing sperm using flow cytometric DNA analysis of sperm is difficult compared to sorting other cell types. The inordinate compactness of chromatin in the morphologically flat, paddle, or ovoid shaped sperm head that is characteristic for livestock causes a high index of refraction. The difference in refractive index between the sperm head and the surrounding medium, coupled with the flat shape of the sperm head, results in preferential emission of light in the plane of the cell (from the edge of the sperm head). Because of these properties, the orientation of the sperm head with respect to the excitation laser beam and the optical detectors (0°) is critical for resolution. Figure 1 illustrates a flow cytometer/cell sorter system that has been modified specifically for sperm. The modification consists of a forward detector in place of the normal light scatter detector and a beveled sample injection needle in place of a cylindrical needle (Johnson and Pinkel, 1986). The modified flow cytometer and cell sorter is essential for attaining separate populations of X- and Y-chromosome-bearing sperm based on DNA content difference and for reanalyzing sorted sperm to determine the proportions of X or Y sperm in a given sorted sample (Johnson et al., 1987b, 1989). Typical histograms from flow cytometric analyses of sperm from several species are shown in Figure 2.

Sorting of sperm for the differentiation of the amount of DNA present in the X and Y chromosome can be accomplished on virtually all commercial flow cytometer/cell sorters that have been manufactured in the past 15 years. Flow cytometer/cell sorter systems manufactured by the Coulter Corporation, Miami, FL (EPICS V; EPICS 750 series, ELITE) and the Becton Dickinson Co., San Jose, CA (FACS Vantage; FACStar Plus; FACStar; FACS 440) have been modified for sperm sorting. Each system was modified to provide for a beveled sample injection needle and a second fluorescence detector located 0° to the laser. Recently an Ortho Cytofluorgraf cell sorter has also been modified for sperm (J. Aten, personal communication). Altogether 12 instruments worldwide have been modified for sperm sorting as described by Johnson and Pinkel (1986).

Production of Sorted X and Y Sperm

The rate at which viable X and Y sperm can be produced approximates 3×10^5 to 5×10^5 sperm per hour. The rate is slow because increased flow rate of the sperm through the system tends to increase the coefficient of variation of the analysis. It is essential that the sperm be properly oriented to the laser beam to effectively differentiate their X and Y DNA content. Depending on the orientation, with respect to the forward detector, only 10 to 20% of the sperm put through the system can be resolved into X and Y populations. The lower rate is based on the need to achieve a high purity of the individual sperm populations (the greater the purity the narrower the sorting windows and the slower the sorting rate). Although one cannot achieve 100% purity (under normal sorting conditions) of a livestock sperm population with this flow sorting system, one can approach 95% purity (Johnson, 1992). Virtual 100% purity is possible with some species in which the DNA difference is much greater, such as in the creeping vole (*Microtus oregoni*; 12.5%; Johnson and Clarke, 1990) and the *Chinchilla langier* (7.5%; Johnson et al., 1987b). Therefore, the highest purity

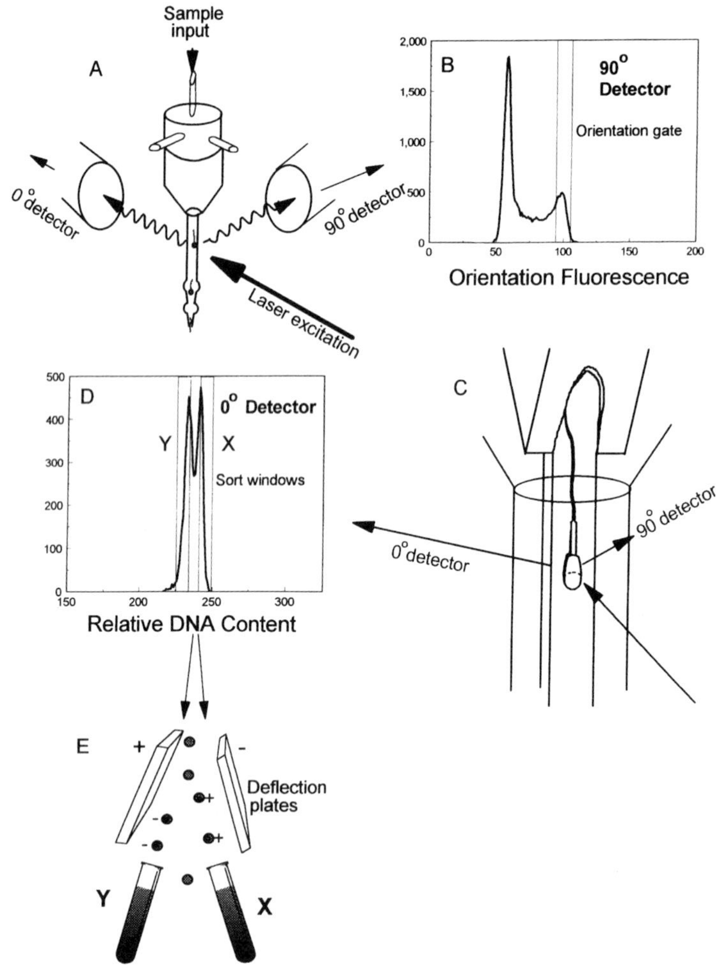

Figure 1. Schematic of the modified sperm sorting system used to carry out the Beltsville Sperm Sexing Technology. Section A illustrates the configuration of the system with two optical detectors. The two optical detectors are situated in the 0° and 90° positions relative to the laser beam. Section B depicts a histogram of the response of the system to all the sperm that pass through it; only the sperm falling within the gate are properly oriented and that population of sperm are seen by the 0° detector. The sample stream is drawn into the flow cell through a beveled needle illustrated in section C. The purpose of the bevel is to create a fluid stream that is more ribbon-shaped than cylindrical. The flat ovoid sperm head takes up the plane of the stream and thus one can control the orientation of the sperm relative to the laser and the optical detectors that collect the fluorescent light given off in respective directions. Section D shows the histogram produced from the 0° detector illustrating the differential measurement (bimodal peaks) of DNA content from the sperm. Section E illustrates the actual sorting system whereby sperm that fit the proper criteria in the amount of DNA they contain are enclosed in droplets that fall through the electrostatic field and the sperm are then drawn to their respective tubes depending on the charge and fall within the appropriate collection tube. The sperm are selected for sorting by whether or not they fall within the particular sort window that is electronically set (section D) (Johnson, 1992, 1994).

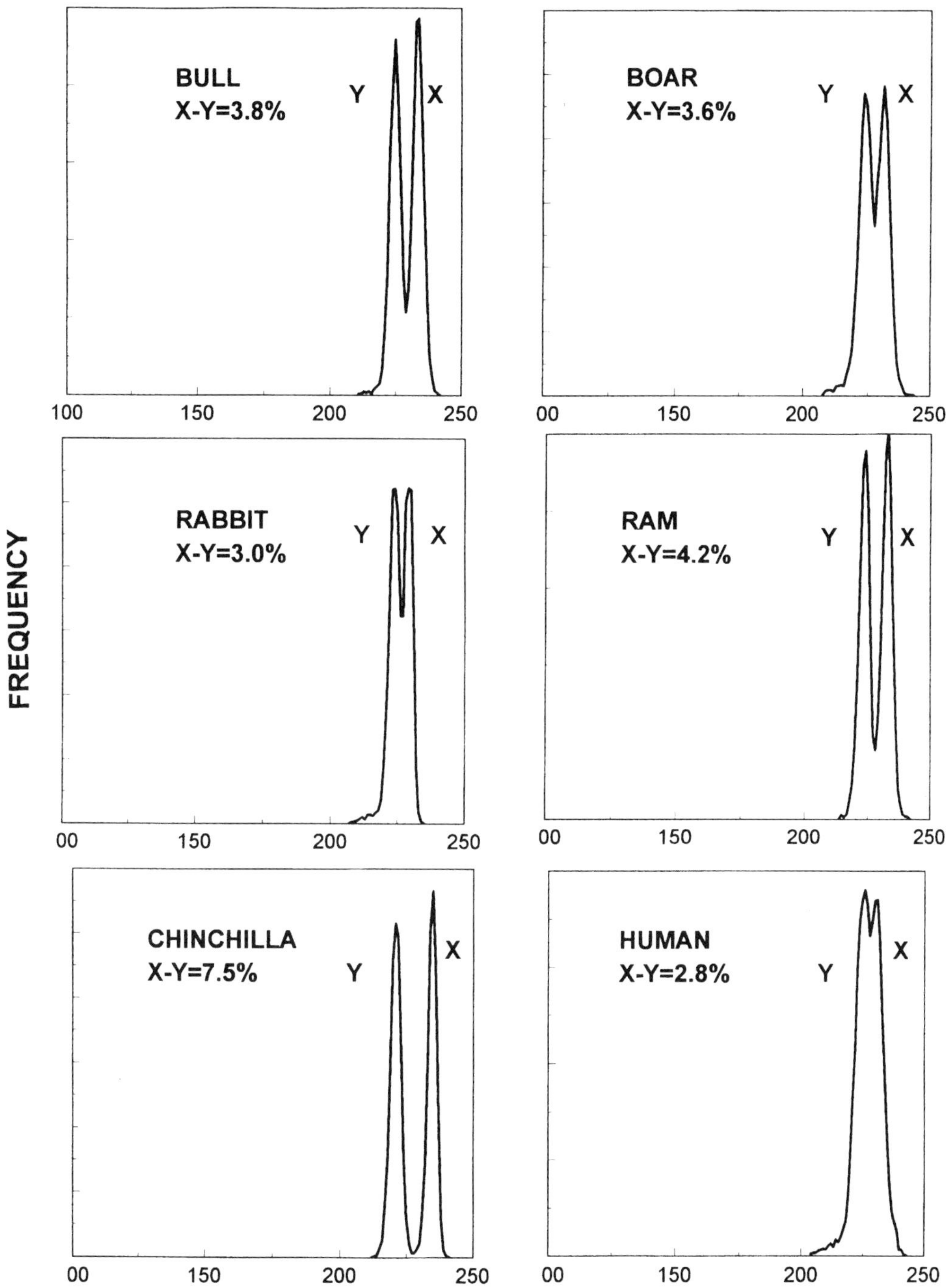

Figure 2. Histograms of DNA content from six different species are illustrated. The histograms represent typical profiles for the species shown along with the percentage difference in DNA content for X and Y sperm for each species. Note that the greater the difference in DNA content (7.5%; chinchilla) the easier it is to sort a high purity sample from the respective X or Y peak as compared to the human (2.8%) (Data from Johnson, 1992).

attainable in using flow sorting is dependent on four factors: 1) the genetically determined DNA content difference between the X- and Y-chromosome-bearing sperm, 2) proper configuration and sensitivity of the flow system to allow selection of a properly oriented population of sperm, 3) uniform and consistent binding of the fluorochrome to the DNA in the sperm head, and 4) the width of the selected sort windows.

At the rate of 3×10^5 to 5×10^5 sperm per hour, one cannot economically produce enough sperm for AI. Application of the Beltsville Sperm Sexing Technology in practice requires that it be used either for intra-tubal (ITI) or intra-uterine (IUI) insemination, intra-cytoplasmic sperm injection (ICSI), or for IVF. Use with IVF requires optimal procedures for those species for which IVF is developed. This is true for cattle (Cran et al., 1993, 1995) but not for swine (Rath et al., 1994). For the human IVF, ITI and ICSI are being used (Levinson et al., 1995). At the present time, it is clear that IVF offers the most feasible means of applying the current sexing technology to gender preselection in cattle. However, twofold improvement in sorted sperm production could increase the applicability for ultrasound-guided insemination in cattle by which smaller numbers of sperm can be used. Recent results with low numbers of liquid stored sperm and artificial insemination in cattle suggest that even at the current sexed sperm production rates, there may be applicability for AI of 1 to 2×10^5 per insemination (Seidel et al., 1995).

Factors Affecting Quality and Viability of Sorted X and Y Sperm

Embryo and Fetal Development Reduced litter size in does and sows inseminated with flow-sorted X or Y sperm (Johnson et al., 1989; Johnson, 1991) led us to evaluate early embryo development to determine the impact of flow-sorted sperm. McNutt and Johnson (1996) found an influence of sorting on development of embryos flushed at 42 h after surgical insemination. Embryos had reached the 8- to 16-cell stage, whereas embryos from the controls (unstained, unsorted sperm) had already progressed to the early morula stage ($P<.01$). The impact seen early in development was carried through pregnancy; there were significantly fewer fetuses from flow-sorted sperm inseminations than from control unsorted-unstained sperm inseminations. Results from cattle indicated that most embryos produced from IVF using non-stained, non-sorted sperm generate fully expanded blastocysts by 7 days after fertilization. When stained, sorted sperm were used for IVF, there was a delay of some 12 hours in blastocyst development. Overall, both of these studies confirmed that flow-sorted sperm have a definite impact on the early and subsequent fetal development when such sperm are used for fertilization.

Effects of Laser Power on Sperm Separation and Fertilization Flow cytometric separation of X and Y sperm based on DNA requires the illumination of the Hoechst-stained sperm by UV laser. The duration of illumination is determined by the sperm's flow velocity as it traverses the laser beam (about 2 to 3 microseconds). The intensity of illumination is determined by the output power of the laser and is readily adjustable. We have investigated the impact of laser power on both the ability to resolve X from Y sperm and their fertilizing capability.

Sperm were first analyzed starting with 175 mW of UV laser power and subsequently at 25 mW reductions in power. Resolution decreased modestly between 175 and 125 mW. At power levels below 125 mW the loss of resolution became more pronounced. This affected the resulting purities of the Y sperm population to the greatest extent. Even at low laser power (below 125 mW), where the resolution of X and Y sperm appeared diminished, the X sperm population could still be sorted to 90% purity but Y sperm purity dropped to 70%.

Fertilization rates in rabbits with sperm sorted using varying laser powers also showed a more pronounced change at the 125 mW level. The percentage of unfertilized eggs decreased from 60% at 200 mW of power to 25% at 125 mW. Similarly, the percentage of 8- to 16-cell and early morula embryos increased from 30% at 200 mW to 70% at 125 mW. Our current protocols utilize 125 mW of power to minimize the possible effects of laser power in the presence of the Hoechst stain (unpublished data). We have also tested a newly developed membrane-permeant nucleic acid-specific stain (SYBR-14; Molecular Probes, Eugene, OR) in an effort to avoid the use of UV lasers. However, lack of stability in the staining of living sperm prevents clear resolution of the bimodal X and Y sperm peaks. Consequently, the variability proved to be too large for use in conjunction with X/Y sperm separation (unpublished data).

Fertility Results Achieved Using Embryos Produced from Sexed Sperm and IVF

Our experience has shown that it is important to locate the sperm sorting facility in close proximity to the IVF and embryo culture laboratory. This is based on the fact that survivability of the sorted sperm tends to be shorter than that of non-sorted sperm. Our standard rule is to sort the sperm over a 2-hour period, and depending on the level of capacity required, achieving approximately 8×10^5 sperm for use in IVF. The sperm are concentrated and coincubated with the eggs within 1 hour of the end of the sort (Lu et al., 1987; Cran et al., 1993, 1995). This represents a time period of about 5 to 6 hours from the time of semen collection to the time of fertilization. After the sorted sperm are collected, they are centrifuged to concentrate and wash the sperm prior to placement with the eggs. Protocols for sperm preparation are straightforward and easy to carry out (Johnson et al., 1989; Johnson, 1991). Only ova recovered from slaughterhouse ovaries have been used for IVF by flow-sorted sperm. Standard IVF techniques have been used for in vitro maturation of ova prior to coincubation with flow-sorted sperm (Lu et al., 1987; Cran et al., 1993, 1995).

The protocol that we have developed for sorting sperm and subsequent IVF involves the processing of 20 million sperm by staining and diluting to 2 mL in extender. Sorting over a 2-hour period will utilize about 65% of the sperm prepared. This results in the accumulation of approximately 8×10^5 sorted X sperm and the same number of sorted Y sperm with sort windows set to achieve 80 to 90% purity of either X or Y sperm (i.e., about 8% of all droplets formed contain a sperm and are available for sorting). Decreasing the sort window by 15% results in a 10% increase in purity (>90%). This can result in about a 20% reduction in the throughput of sperm that will be sorted over a 2-hour period. Viability remains unaffected by manipulation of the sort windows. Sperm motility and acrosomal integrity of sorted sperm are similar to the original sample. Cleavage rates range from 60 to 80% when using approximately 2,000 to 4,000 sperm per egg (Cran et al., 1993).

Sperm Sort Validation in the Laboratory

Reanalysis of Sorted Sperm for DNA Content Aliquots of sperm are removed from the sorting tube, sonicated to remove the tails, and Hoechst 33342 added to maintain staining uniformity and then flow cytometrically analyzed but not resorted. The proportion of X and Y sperm is determined based on the DNA difference and histograms are analyzed by computer fitting to double gaussian peaks (Figure 2; Johnson et al., 1987a).

Molecular Genetic Analysis of Single Sorted Sperm A qualitative determination of the presence of the X- or Y-chromosome in an individual sperm can be made by using the allele-specific polymerase chain reaction (PCR). By analyzing several hundred individual sperm, a quantitative determination of the proportion of X and Y sperm in a given population can be made (Welch et al., 1995).

Single sperm that were determined to be X or Y by flow cytometric analysis were deposited into each well of a 48-well plate. Microscopic verification of each well's contents after single sperm sorting can easily be done. The wells also served as the first-round amplification vessel for PCR. Specific X and Y products from a second amplification were separated by agarose gel electrophoresis, product bands identified and counted, and resulting purities determined. Results from 100 amplifications of each X and Y sort showed 94% purity for X and 90% purity for Y. These purities are consistent with results obtained by flow cytometric reanalysis of sorted sperm described above.

Embryo Sexing by PCR Embryos produced from IVF with sorted sperm can be frozen by standard methods and stored for later analysis of gender by PCR. After thawing and incubation with Proteinase K, the DNA is probed using primers derived from a multiple repeat bovine Y-chromosome specific sequence (Cran et al., 1993).

Fluorescence In Situ Hybridization (FISH) to Validate Proportions of X and Y Sperm Application of FISH to the determination of X and Y proportions is fairly straightforward with human sperm because of the availability of commercial microsatellite centromeric DNA probes. Our initial work on the human was validated using FISH (Johnson et al., 1993). Sperm were sorted on to slides, fixed with methanol and acetic acid, exposed to biotinylated DNA probes for X- or Y-chromosomes, and incubated; the signals indicating an X- or Y-chromosome were then counted using fluorescence microscopy. This technique is particularly helpful when the X/Y DNA difference is small and DNA reanalysis by flow cytometry is inconsistent. Preliminary work has been conducted in this area with animal sperm, but there are fewer probes available for domestic animals.

Fertility Using Flow-Sorted X and Y Sperm in Livestock

The initial studies to validate the efficacy of the Beltsville Sperm Sexing Technology were conducted in rabbits and swine. The sex ratios of the litters produced from females surgically inseminated with sorted X or Y sperm supported the results of reanalyzing aliquots of sorted sperm populations (Johnson et al., 1989; Johnson, 1991). A summary of the results obtained in these studies is presented in Table 1. After this initial proof that the technology was valid and consistent across species, we sought to establish its usefulness for cattle. Application in cattle required the adaptation of the sorting process to IVF due to the problematic nature of surgical insemination. The preliminary study (Cran et al., 1993) was conducted at Cambridge and resulted in nine twin transfers of sexed embryos produced from IVF. Four cows calved, producing two sets of twins and two single calves. All were of the predicted sex based on DNA reanalysis and embryo sexing using PCR.

Following the success of the initial study, a field trial was designed in which only Y sperm were sorted with a minimum 90% purity based on reanalysis of sorted sperm populations for DNA. This criterion is intended to ensure that the resulting sex of the embryo and subsequent calves born is greater than 85% of the desired sex. Sexed embryos were produced and cultured to days 7 and 8. Grade-one blastocysts were frozen for later direct transfer to recipient cows. Twin transfers were conducted on 11 farms in Scotland, resulting in 41 calves from 35 calvings. Of the total born, 37 were males and four were females, for a 9:1 ratio (Cran et al., 1995). A more detailed summary of fertility results is included in Table 1. These results along with those with swine and rabbits (Table 1) confirm the efficacy of the sorting technology as an effective means of skewing the sex ratio of livestock progeny.

Table 1. Fertility data. Progeny produced from intratubal or intrauterine insemination or transfer of sexed embryos produced from sorted X and Y sperm based on DNA content

Treatment	Number of females inseminated	Number kindled farrowed/ calved	Number born	Actual %		Significance[a]
				Male	Female	
Rabbit[b]						
Sorted Y	16	5	21	81	19	**
Sorted X	14	3	16	6	94	**
Sorted (X+Y recomb.)	17	5	14	43	53	NS
Swine[c]						
Sorted Y	8	4	37	68	32	*
Sorted X	10	5	34	26	74	**
Unsort/stain	11	5	40	43	57	NS
Unsort/unstained	7	5	46	42	48	NS
Cattle[d]						
Sorted X	4	2	3	0	100	
Sorted Y	5	2	3	100	0	
Sorted Y	106	35	37	90	10	**

[a]Deviation from 50:50: **P<.01, *P<.04.
[b]Johnson et al. (1989).
[c]Johnson (1991).
[d]Cran et al. (1993, 1995).

Application of Sexing Technology to Human Sperm

As indicated earlier there is wide interest in the application of sexed semen to livestock production for various reasons, most of which are economic. In the human, however, the reasons for such interest are more varied and often controversial. Nevertheless, three reasons predominate. The first is sex-linked disease, another is family balance, and also there are cultural considerations. The least controversial of the three is sex selection for the avoidance of sex-linked disease. With this purpose in mind, we sought to determine the applicability of the sexing technology to human sperm.

Human sperm differ in DNA content by about 2.8% (Johnson et al., 1993). As indicated earlier, the smaller the DNA difference the more difficult the task of separating two nearly pure populations. A project to determine the feasibility of sorting human sperm was undertaken with the collaboration of the Genetics and IVF Institute of Fairfax, VA. Even though the human sperm head has a more angular dimension than that of domestic animals, it still has an "edge" and a "flat side," thus giving differential fluorescence when treated in the same way. Because of the small difference in DNA our usual reanalysis of sorted populations could not be accomplished with consistency. We therefore adapted FISH technology for validating the proportions of X and Y sperm in the sorted semen. The results are shown in Figure 3. In summary, we were successful in skewing the sperm sex ratio of human semen by 80 to 85% (Johnson et al., 1993). The succeeding step in the development of the process, that of testing the technology in a clinical setting has begun. This work is being done at the Genetics and IVF Institute, Fairfax, VA under appropriate ethics board supervision. Pregnancies are in progress following insemination with sorted sperm, in clients qualifying on the basis of genetic history for sex-linked disease (Levinson et al., 1995).

Adaptation of Flow-Sorting to Cryopreservation Parameters

Cryopreservation of Sorted X and Y Sperm Freshly ejaculated semen has been the source of sperm for all the sexing protocols developed. Increasing the flexibility with which the sorted sperm can be used would be advantageous, as would the use of a broader source of semen. Cryopreservation of sperm after the sorting process would increase the flexibility with which separated X and Y sperm could be used for fertilization. Our experience to date with freezing sorted sperm show that modifications to standard equilibration and staining protocols are required. With these adjustments, it is possible to achieve around 30% sperm motility with 40% acrosomal integrity after thawing. Fertilization testing with sperm that have been frozen after sorting is currently underway.

Sorting of Cryopreserved Semen Using cryopreserved semen to begin the sorting process would give one access to the best genetics available worldwide. Studies have been initiated to examine the potential use of previously frozen sperm for sorting. Due to cell damage during freezing, the number of sperm available for sorting is very limited. More than 50% of the cells may be dead or incapable of fertilizing. The diluent within which the sperm are frozen must be removed by density gradient

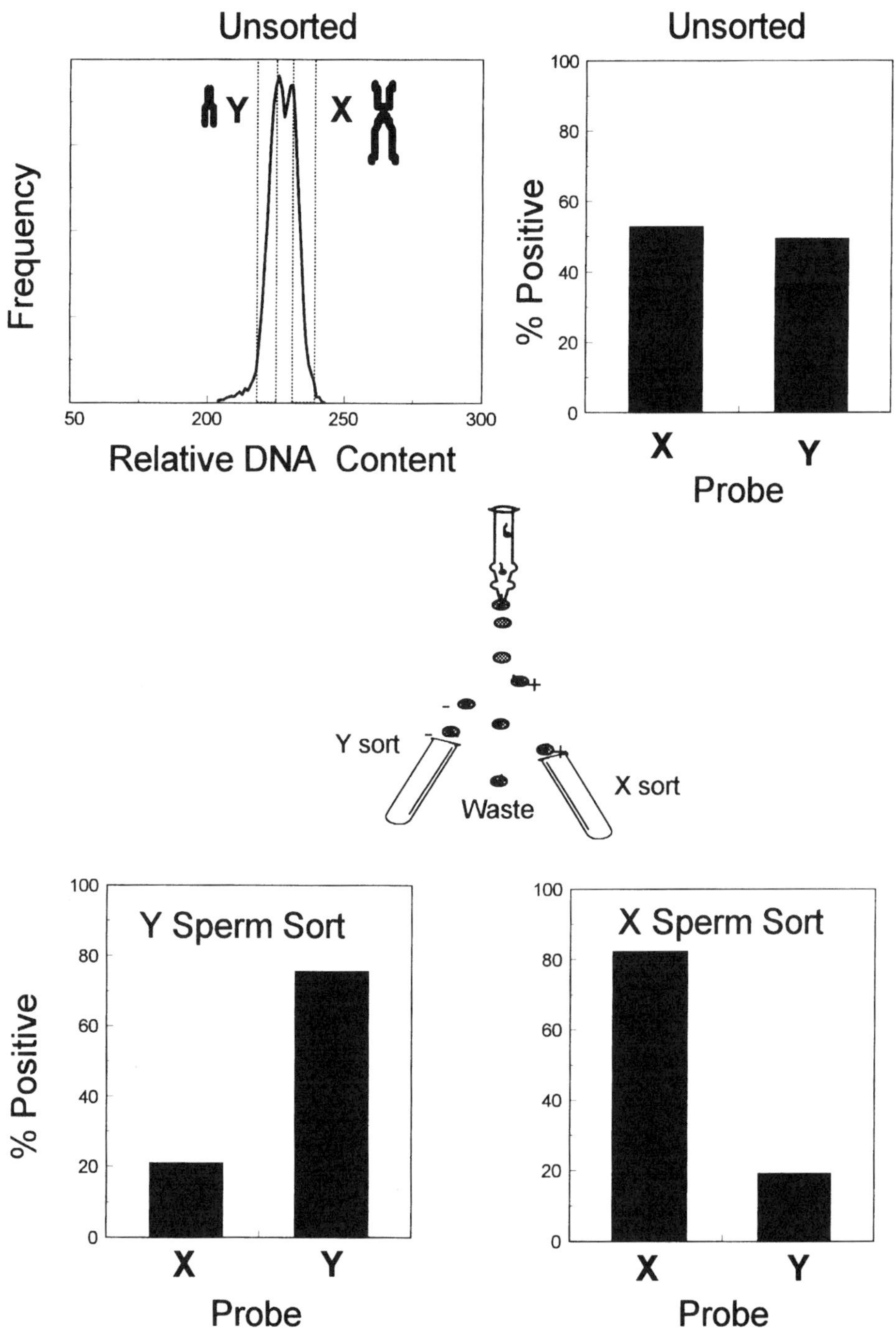

Figure 3. Histogram and accompanying bar graphs showing the purity of sorts received when human sperm were sorted into respective X and Y populations. Purity was determined using fluorescence *in situ* hybridization (FISH) on the sorted sperm. Sperm were sorted directly onto slides, and the slides were stained with DNA probes specific for X or Y chromosomes. The bar graphs show nearly 82% purity for the X- and 75% for the Y-chromosome-bearing sperm (Data from Johnson et al., 1993).

centrifugation, as must the components of the gradient. The freezing process causes changes in the plasma membrane, necessitating a different staining protocol. Preliminary results indicate that X and Y sperm purities in excess of 80% can be achieved. However, results are unavailable for the fertilizing capacity of such sperm. Indications are that considerable variation will occur between samples. Nevertheless, cleavage rates of up to 50% have been recorded (Cran et al., unpublished observation).

Implications

The ability to preselect the sex of progeny in livestock has been validated on live births, reanalysis of sorted sperm for DNA content, and on the basis of determining the sex of the resulting embryos. The flow cytometric sorting method for separation of X- and Y-bearing sperm can be used in conjunction with IVF to produce sexed embryos that can be subsequently transferred to eligible recipients for the duration of gestation. This semi-practical sexing method, though impractical for some production systems, could be used to provide a more flexible option for many livestock production operations. The use of this technology for gender preselection in humans is currently being clinically tested. Although we would predict that the outcome would be similar to that for livestock, we must await the completion of numerous pregnancies before making a clear assessment of application in the human.

Future Prospects

Flow cytometric sorting of sperm for reproductive purposes requires significant commitment of expertise and resources in order to apply the systems commercially or in research situations. Research continues to improve the efficiency of sorting in terms of number of sperm per hour. Current research involving the removal of dead sperm from the sorting process through the addition of the dead membrane permeant propidium iodide in combination with Hoechst 33342 may result in a more efficiently sorted population (Johnson et al., 1994). Slit-scan optics developed for chromosome sorting (Rens et al., 1993) applied to the sperm sexing technology are showing potential for enhancing the efficiency of the sorting process (W. Rens, G. R. Welch, and L. A. Johnson, unpublished observations). High-speed sorting as developed for mammalian chromosomes (Van den Engh and Stokdijk, 1989) may have application to the sperm sorting process if the need to orient sperm to the laser beam can be minimized. Such a development would allow the sperm to flow faster through the system. Finally, the search continues for a sperm (surface protein) sex chromosome specific marker that can be exploited in the development of an antibody to tag sperm under affinity chromatography protocols or for use with micromagnetic bead separation systems. Initial results have not proven the validity of this approach (Fenner et al., 1992; Hendriksen et al., 1993; Hendriksen et al., unpublished data). Obviously an antibody-based X/Y sperm separation procedure would be the preferred process simply for economy and practicality. Recently, Seidel et al. (1995) showed that fertility can be achieved inseminating fresh sperm numbers far below the norm in cattle ($<2 \times 10^5$). Success has been achieved producing sexed calves from inseminating small numbers of flow-sorted X or Y bull sperm (Seidel et al., 1996).

Literature Cited

Amann, R. P., and G. E. Seidel (Ed.). 1982. Prospects for Sexing Mammalian Sperm. Colorado University Press, Boulder.

Cran, D. G., L. A. Johnson, N.G.A. Miller, D. Cochrane, and C. Polge. 1993. Production of bovine calves following separation of X- and Y-chromosome bearing sperm and in vitro fertilisation. Vet. Rec. 132:40.

Cran, D. G., L. A. Johnson, and C. Polge. 1995. Sex preselection in cattle: A field trial. Vet. Rec. 136:495.

Dean, P. N., D. Pinkel, and M. L. Mendelsohn. 1978. Hydrodynamic orientation of sperm heads for flow cytometry. Biophys. J. 23:7.

Fenner, G. P., L. A. Johnson, W. R. Hruschka, and D. J. Bolt. 1992. Two-dimensional electrophoresis and densitometric analysis of solubilized bovine sperm plasma membrane proteins detected by silver staining and radioiodination. Arch. Androl. 29:21.

Gledhill, B. L. 1988. Gender preselection: Historical, technical, and ethical perspectives. Semin. Reprod. Endocrinol. 6:385.

Gledhill, B. L., S. Lake, L. L. Steinmetz, J. W. Gray, J. R. Crawford, P. N. Dean, and M. A. Van Dilla. 1976. Flow microfluorometric analysis of sperm DNA content: Effect of cell shape on the fluorescence distribution. J. Cell. Physiol. 87:367.

Hendriksen, P.J.M., M. Tieman, T. Van der Lende, and L. A. Johnson. 1993. Binding of anti-H-Y monoclonal antibodies to separated X and Y chromosome bearing porcine spermatozoa. Mol. Reprod. Dev. 35:189.

Johnson, L. A. 1991. Sex preselection in swine: Altered sex ratios in offspring following surgical insemination of flow sorted X- and Y-bearing sperm. Reprod. Domest. Anim. 26:309.

Johnson, L. A. 1992. Gender preselection in domestic animals using flow cytometrically sorted sperm. J. Anim. Sci. 70(Suppl. 2):8.

Johnson, L. A. 1994. Isolation of X- and Y-bearing sperm for sex preselection. In: H. H. Charlton (Ed.). Oxford Reviews of Reproductive Biology. Vol. 16. pp 303-326. Oxford University Press, Oxford, UK.

Johnson, L. A., and R. N. Clarke. 1988. Flow sorting of X and Y chromosome-bearing mammalian sperm: Activation and pronuclear development of sorted bull, boar and ram sperm microinjected into hamster oocytes. Gamete Res. 21:335.

Johnson, L. A., and R. N. Clarke. 1990. Sperm DNA and sex chromosome differences between two geographical populations of the creeping vole, Microtus oregoni. Mol. Reprod. Dev. 27:159.

Johnson, L. A., J. P. Flook, and H. W. Hawk. 1989. Sex preselection in rabbits: Live births from X and Y sperm separated by DNA and cell sorting. Biol. Reprod. 41:199.

Johnson, L. A., J. P. Flook, and M. V. Look. 1987a. Flow cytometry of X and Y chromosome-bearing sperm for DNA using an improved preparation method and staining with Hoechst 33342. Gamete Res. 17:203.

Johnson, L. A., J. P. Flook, M. V. Look, and D. Pinkel. 1987b. Flow sorting of X and Y chromosome-bearing spermatozoa into two populations. Gamete Res. 16:1.

Johnson, L. A., and D. Pinkel. 1986. Modification of a laser-based flow cytometer for high resolution DNA analysis of mammalian spermatozoa. Cytometry 7:268.

Johnson, L. A., G. R. Welch, and D. L. Garner. 1994. Improved flow sorting resolution of X and Y chromosomes bearing viable sperm separation using dual staining and dead cell gating. Cytometry 17(Suppl. 7):83.

Johnson, L. A., G. R. Welch, K. Keyvanfar, A. Dorfmann, E. F. Fugger, and J. D. Schulman. 1993. Gender preselection in humans? Flow cytometric separation of X and Y spermatozoa for the prevention of X-linked diseases. Hum. Reprod. 8:1733.

Kiddy, C. A., and H. D. Hafs (Ed.). 1971. Sex Ratio at Birth–Prospects for Control. J. Anim. Sci. Symp. Suppl. Am. Soc. Anim. Sci., Champaign, IL.

Levinson, G., K. Keyvanfar, J. C. Wu, E. F. Fugger, R. A. Fields, G. L. Harton, F. T. Palmer, M. E. Sisson, K. M. Starr, L. Dennison-Lagos, L. Calvo, R. J. Sherins, D. Bick, J. D. Schulman, and S. H. Black. 1995. DNA-based X-enriched sperm separation as an adjunct to preimplantation genetic testing for the prevention of X-linked disease. Hum. Reprod. 10:979.

Lu, K. H., L. Gordon, M. Gallagher, and H. McGovern. 1987. Pregnancy established in cattle by transfer of embryos derived from in vitro fertilization of oocytes matured in vitro. Vet. Rec. 121:259.

McNutt, T. L., and L. A. Johnson. 1996. Flow cytometric sorting of sperm: Influence on fertilization and embryo/fetal development in the rabbit. Mol. Reprod. Dev. 43:261.

Pinkel, D., B. L. Gledhill, M. A. Van Dilla, D. Stephenson, and G. Watchmaker. 1982. High resolution DNA measurements of mammalian sperm. Cytometry 3:1.

Rath, D., L. A. Johnson, G. R. Welch, and H. Niemann. 1994. Successful gamete intrafallopian transfer (GIFT) in the porcine. Theriogenology 41:1173.

Rens, W., C. H. van Oven, J. Staap, and J. A. Aten. 1993. Effectiveness of pulse-shape criteria for the selection of dicentric chromosomes by slit-scan flow cytometry and sorting. Anal. Cell. Pathol. 5:147.

Seidel, G. E., C. H. Allen, Z. Brink, J. K. Graham, and M. B. Cattell. 1995. Insemination of Holstein heifers with very low numbers of unfrozen spermatozoa. J. Anim. Sci. 73(Suppl. 1):232 (Abstr.).

Seidel, G. E., Jr., L. A. Johnson, C. A. Allen, G. R. Welch, M. D. Holland, Z. Brink, and M. B. Cattell. 1996. Artificial insemination with X- and Y-bearing bovine sperm. Theriogenology 45:309.

Van den Engh, G., and W. Stokdijk. 1989. Parallel processing data acquisition system for multilaser flow cytometry and cell sorting. Cytometry 10:282.

Welch, G. R., G. C. Waldbieser, R. J. Wall, and L. A. Johnson. 1995. Flow cytometric sperm sorting and PCR to confirm separation of X- and Y-chromosome bearing bovine sperm. Anim. Biotechnol. 6:131.

Windsor, D. P., G. Evans, and I. G. White. 1993. Sex predetermination by separation of X and Y chromosome bearing sperm: A review. Reprod. Fertil. Dev. 5:155.

Modification of Milk Composition in Transgenic Animals

R. J. Wall

Beltsville Agricultural Research Center, Livestock and Poultry Sciences Institute, Gene Evaluation and Mapping Laboratory, ARS, USDA, Beltsville, MD 20705, U.S.A.

Abstract Amid the explosion of fundamental knowledge generated from transgenic animal models, a small group of scientists has been producing transgenic livestock to test the feasibility of improving animal production efficiency and generating new products. Modifying the genetic control of mammary glands provides an opportunity to pursue several distinctly different goals. The objective of the emerging "gene pharming" industry is to produce drugs for treating human diseases. It is argued that mammary glands are an ideal site for producing complex bioactive proteins that can be harvested and purified in a cost-effective manner. Several products produced in this way are approaching clinical trials. A second strategy that has been widely discussed but has received less attention in the laboratory focuses on altering the ratio of endogenous milk proteins to improve the nutritional quality of milk and to reduce processing costs of dairy products. Proposals include increasing protein, decreasing fat, altering physical properties, and "humanizing" cow's milk. The potential profitability of gene pharming seems clear; however, cost-benefits of genetically engineered dairy products will remain marginal until inefficiencies of transgenic technology such as low gene integration rates, poor embryo survival, and unpredictable transgene behavior are overcome.

Key Words: Transgenic Livestock, Bioreactors, Milk, Mammary Gland

Introduction

In 1987 Lothar Hennighausen and Heiner Westphal at the NIH in collaboration with Katy Gordon and her colleagues at Integrated Genetics started a mini-revolution when they reported that a pharmaceutical could be produced in mammary glands of transgenic animals (Gordon et al., 1987). Within 5 years of that report a new industry sponsored by venture capital was formed. The remarkable speed with which transgenic animal bioreactor technology moved from laboratories to industry attests to the perceived potential value of this approach for producing pharmaceuticals. However, it is not surprising that no products derived from transgenic animals are yet in physicians' hands, given the time required to put a new drug on the market. But it is rumored that at least three potential products are currently being tested in animal models (preclinical trials).

General Goals for Transgenic Livestock In addition to altering milk composition, a variety of agricultural applications have been proposed that utilize transgenic livestock. Approaches to improving animal production efficiency include improving growth rate, feed efficiency, and carcass composition (Pursel and Rexroad, 1993), enhancing disease resistance (Lo et al., 1991; Weidle et al., 1991; Brem, 1993; Clements et al., 1994), increasing rate and quality of wool production (Rogers, 1990; Ward and Nancarrow, 1991; Bawden et al., 1995), and endowing livestock with the ability to utilize high-fiber diets (Wells and Petters, 1994). Projects for transgenic livestock not related to agriculture include producing human blood substitutes (Swanson et al., 1992), producing human antibodies (Brüggemann et al., 1993; Geller et al., 1994; Wagner et al., 1994), and developing genetic disease models (Mehtali et al., 1992; Breslow, 1993; Fabry, 1993; Fukuchi et al., 1993; Huang, 1993; Rothnagel et al., 1993; Shear et al., 1993; Thompson et al., 1993; Yamamura et al., 1993). Disease models have only been evaluated in mice. The topic now receiving most industry attention, only rivaled by mammary gland bioreactors, is an attempt to engineer pigs as a source of organs for human use (Cozzi et al., 1994; Fodor et al., 1994; Martin et al., 1993; Rosengard et al., 1995; Yannoutsos et al., 1995).

Goals for Genetic Engineering of Mammary Glands Commercial interests have focused primarily on modifying the genetic control of mammary glands for the purpose of producing pharmaceutical proteins. However, transgenic technology also offers the opportunity to reformulate endogenous protein production of the mammary gland in a variety of other ways. Several excellent reviews have been written on the topic (Richardson, 1985; Bremel et al., 1989; Hoeschele, 1990; Wilmut et al., 1990; Gibson, 1991; Houdebine, 1994). These reviewers and others have suggested strategies for changing milk composition to provide a source of milk for lactose-intolerant individuals, to provide milk that is more efficiently processed into cheese, and of course to provide "natural" low-fat milk. One of the most ambitious schemes proposed is to "humanize" milk. Some bovine milk proteins would be eliminated and others would be replaced with coding sequences derived from human genes. The resulting milk would more closely resemble human breast milk and would be used to supplement or replace infant formula.

Economic Considerations From an agricultural perspective, genetic improvements in livestock will be judged solely on their total economic merit. Assessing cost-benefit ratios of a genetic alteration is a complex and possibly inaccurate process at best. Assessing the economics of projects directed toward solving human disease problems, which involve ethical considerations, adds to the complexity of the equation. However, the economics of even those types of projects can and should be evaluated. Assessment of the merits of a given strategy for improving livestock traits is more straightforward. Economic value can be increased by expanding the market for a product, improving the efficiency of producing the product, or increasing the intrinsic value of the product. The economics of a particular approach may be neutral or even of negative value for the livestock industry, but if overall economic value to society is positive, pursuit of the goal may be justified. For example, a strategy might be devised to significantly decrease carcass fat and increased feed efficiency. However, if it took longer for animals to reach finished weight, the producer might not realize a direct economic benefit. However, if consumers were willing to pay a premium for thinner carcasses and consumption of meat increased, the industry as a whole would realize an economic advantage. The increase in demand would eventually benefit producers. Society as a whole would benefit from the improvement in feed utilization, by the sparing of natural resources.

Except for the mammary gland bioreactor arena, there are few examples of proven transgenic technologies that seem to pass the economic test. But policy makers should not view this situation as a reason to abandon the development of these technologies. Rather, they should make additional research resources available to overcome the impediments to progress.

Before mammary gland gene modifications are discussed in detail, it is appropriate to consider the current status of transgenic animal technology. Some of the hurdles will be addressed that continue to impede use of this technology to solve agricultural problems.

Transgenic Animal Model System

Most of the goals currently proposed for genetically altering milk composition, or improving production efficiency, rely on "gain of function" strategies (introduction of a new gene by means of pronuclear microinjection). However, it is currently popular to consider that "loss of function" approaches (eliminating or altering gene function) can best be achieved through stem cell-dependent gene knock-out technology. The status of stem cell technology is discussed elsewhere in these symposium proceedings and will not be dealt with in this review. However, eliminating or reducing the concentration of milk proteins through genetic manipulation can, at least in theory, be achieved by adding new genetic information. By introducing ribozyme or anti-sense genes, translation of specific milk protein genes can be blocked, or at least reduced.

While gain of function experiments have predominated as the genetic strategy, pronuclear microinjection has predominated as the means of introducing new genetic material. Other methods are under development (Petters et al., 1987; Francolini et al., 1993; Lauria and Gandolfi, 1993; Schellander et al., 1995) but have not been widely applied. Production of transgenic animals by pronuclear microinjection is practiced today essentially as described in Jon Gordon's pioneering first report (Gordon et al., 1980) and as further characterized by Brinster and colleagues (Brinster et al., 1985) and as adapted to livestock by our laboratory (Hammer et al., 1985; Wall et al., 1985). The milestones from Jon Gordon's pioneering effort to production of transgenic bioreactors are listed in Table 1.

Interestingly, while the technique has changed little, the definition of the word "transgenic" has taken on new meanings. Jon Gordon and Frank Ruddle originally proposed the term to describe animals in which new genetic material had been introduced by pronuclear microinjection. The term has been broadened considerably to include genetically engineered plants, animals produced with the aid of stem-cells, and those that have been subjected to somatic cell genetic engineering. Investigators have also been referring to laboratory animals subjected to *in vivo* somatic cell genetic engineering (DNA injected directly into living animals) as transgenic animals. One wonders whether humans treated by gene therapy will one day be referred to as transgenics.

Efficiency of Producing Transgenic Animals The efficiency of producing transgenic animals is still low, ranging from about one transgenic animal per 40 eggs injected (mice) to one in 1,600 eggs injected (cattle). Low efficiency is not of particular concern to those working with mice, but it is a major impediment to those attempting to produce transgenic livestock. Three parameters account for the low efficiency of the process: gene integration rate; embryo survival; and transgene behavior. The last two parameters influence laboratory animals and livestock equally, while transgene integration rates are significantly lower in farm animals than in laboratory animals

(Brem, 1985; Hammer et al., 1985; Table 2). The data presented in Table 2 are somewhat biased. The numbers of manipulated embryos for the different species are unequal, and a week-long culture was imposed on the cow eggs before transfer to recipients. Nevertheless, there seems to be two observable trends. The proportions of transferred embryos that survive to term in the livestock species and laboratory animals are similar (11% vs. 15%). Integration frequency, as measured by proportion of transgenic animals born, for livestock species is about half of that for laboratory animals (9% vs. 17%). That difference in integration rate may indicate important biological differences between laboratory animal and livestock zygotes.

Causes of Low Efficiency Genetic diversity in livestock species may play a role in the low integration frequency observed. Laboratory animals employed to produce transgenic animals are derived from highly inbred lines, and investigators often chose to use specific strains whose embryos "culture easily." Scientists who work with livestock embryos do not have the same inbred resources. However, examples can be found in the livestock industry where selection of animals is based partly on gamete quality. In the artificial insemination industry, selection is not only based on a bull's genetic merit but also on the "freezability" of his sperm. However, very few in the research community have the resources to perform the same kind of selection on embryo donors. Even though breeds of livestock are genetically more diverse than inbred strains of mice, there may be some value in comparing transgene integration efficiency among livestock breeds.

Another cause of low transgene integration rate may be related to procedural differences between microinjection of livestock and laboratory animal zygotes. Livestock eggs are more challenging to microinject than eggs from laboratory species in

Table 1. Milestones in transgenic animal bioreactor technology

Year	Accomplishment	Reference
1980	Transgenic animal produced by pronuclear microinjection	Gordon et al., 1980
1981	Transmission of transgene to progeny	Gordon and Ruddle, 1981
1982	Fusion gene introduced into transgenic mice	Brinster et al., 1982
1987	Expression of pharmaceutical directed to the mammary gland of transgenic mice	Gordon et al., 1987
1988	First transgenic bioreactor farm animals produced (sheep)	Simons et al., 1988
1990	Mammary gland specific transgene expressed in rabbits	Bulher et al., 1990
1991	Mammary gland specific transgene expressed in goats	Ebert et al., 1991
1991	Mammary gland specific transgene expressed in pigs	Wall et al., 1991
1991	Transgenic bull produced with mammary gland specific gene construct	Krimpenfort et al., 1991
1992	Mammary gland specific transgene expressed in rats	Hochi et al., 1992
1994	Bioreactor products being tested in animal models (pre-clinical testing)	Henryk Lubon, personal communication

part because livestock pronuclei are more difficult to visualize. Cow and sow eggs must be centrifuged before microinjection, though there is little evidence that embryo survival is compromised by centrifugation (Wall et al., 1985). But the procedure may influence integration rate. Within livestock species, the lower integration rate observed in cattle embryos may also be attributable to the mobility of their pronuclei. They tend to move away when touched by an injection needle. But that is not likely to account for the 50% or greater reduction in integration efficiency between livestock and laboratory animals.

It is also possible that transgenes become integrated at the similar rates in all species, but the development of injected embryos is disproportionately compromised in livestock. However, the similar poor survival rate of injected embryos of both laboratory and livestock embryos does not support that hypothesis. With the data available it is not possible to determine to what extent the injection of transgenes accounts for poor livestock embryo survival, but it is probably not a major component (Hawk et al., 1989).

A more compelling argument could be made for an association of integration failure with inappropriate timing of microinjection. It is well known that cell transfection is most efficient in dividing populations. It has been inferred from that observation that DNA replication is required for integration of foreign genes into the genome (Bishop and Smith, 1989). If that is the case, then the timing of pronuclear microinjection should be synchronized with onset of the DNA synthetic phase (S-phase) of the first cell cycle to ensure the maximum likelihood of an integration event. As depicted in the stylized graph of Figure 1, mouse eggs are microinjected about 8 hours after mating. This results in DNA being introduced into zygotes during the beginning of S-phase (Krishna and Generoso, 1977). Sow and cow eggs are injected toward the end of the S-phase (Kay et al., 1988; Eyestone and First, 1989; Wall et al., 1991; Eid et al., 1994; Laurincik et al., 1995), possibly reducing the

Table 2. Embryo survival and gene integration efficiencies in transgenic animal experiments

Species	No. of injected and transferred embryos (n)[a]	Fetuses or born		Transgenic animals produced			Reference[c]
		No.	% of Transferred	No.	% of Born	% of Transferred	
Mice	12,314 (18)	1,847	15.0	321	17.3	2.61	1
Rabbit	1,907 (1)	218	11.4	28	12.8	1.47	2
Rat	1,403 (5)	353	29.6	62	17.6	4.42	3
Cattle[b]	1,018 (7)	193	19.0	7	3.6	.68	4
Pig	19,397 (20)	1,920	9.9	177	9.2	.91	5
Sheep	5,242 (10)	556	10.6	46	8.3	.88	5

[a]Number of experiments, which in most cases was equivalent to different gene constructs, is shown in parentheses.

[b]Eleven thousand two hundred six eggs were microinjected and cultured, of which 1,018 developed to morula or blastocysts and were transferred into recipient cows.

[c]1, R. J. Wall, unpublished data; 2, Hammer et al., 1985; 3, Ninomiya et al., 1994; 4, Hill et al., 1992; 5, Pursel and Rexroad, 1993.

probability of an integration event. Characterization of pronuclear development in Figure 1 was determined by non-vital staining techniques. Pronuclei in pig and cattle zygotes form well before the usual timing of microinjection (Eid et al., 1994; Laurincik et al., 1995). However, the time of microinjection, in those species, was established to coincide with the time pronuclei can be visualized in a non-destructive manner (differential interference contrast microscopy). At earlier times pronuclei cannot be seen, even though they are present. Therefore, it does not seem possible to microinject transgenes into pronuclei until DNA synthesis is nearly completed in cow and sow eggs. Experimental evidence on the influence of time of microinjection on transgene integration frequency is lacking and may be difficult to achieve. Efforts to inject *in vitro* fertilized bovine zygotes earlier than 13 hours after insemination have failed because of difficulties in visualizing pronuclei (K. Bondioli, personal communication and unpublished data).

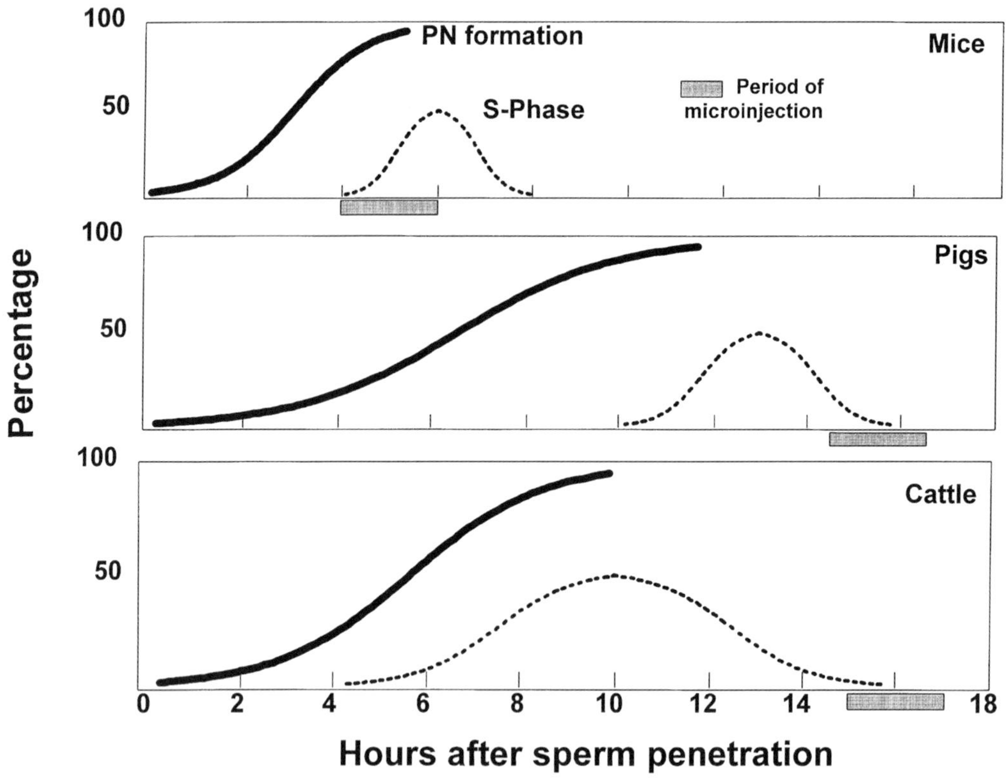

Figure 1. Timing of S-Phase with respect to pronuclear (PN) formation in mice, pigs and cattle. Data from several sources (see text) were combined to generate this stylized graph showing the relative relationship of pronuclear development, DNA synthetic phase (S-phase) of the first cell cycle and typical timing of pronuclear microinjection. Pronuclear formations was characterized using non-vital DNA staining techniques. Timing of pronuclear microinjection is based on the ability to visualize pronuclei with differential interference microscopy.

Aside from the differences in efficiency of producing transgenic animals, phenotypic characteristics of transgenic goats, mice, pigs, rabbits, rats and sheep seem to be similar. There are insufficient data to make the same claim for transgenic cattle, but there is no reason to expect that transgenes will behave differently in that species. Transgenes seem to integrate randomly into the genome. Approximately half of transgenic animal lines express their transgenes, though some specific transgenes are expressed in a higher proportion of transgenic animals (Grosveld et al., 1987; Harris et al., 1991; Dale et al., 1992; Bonifer et al., 1994). Adding matrix attachment region sequences to transgene constructs may "rescue" expression in lines that otherwise might carry nonfunctional transgenes (McKnight et al., 1992). About 70% of transgenic founder mice transmit the transgene to offspring (Wilkie et al., 1986). Though few transgenic livestock experiments have produced enough offspring to precisely estimate transmission rates, it would seem that transgenes will be inherited in livestock as they are in mice (Carver et al., 1993). Significant exceptions to predicted transmission frequency have been observed, but those exceptions were a consequence of specific transgene products (Grosveld et al., 1987; Bayna and Rosen, 1990; Bonifer et al., 1990; Whitelaw et al., 1992).

Transgenic Animal Bioreactors

The New Industry At least nine companies are involved in developing transgenic animal bioreactor products (Table 3). Most of those organizations were formed within the last decade specifically to exploit transgenic technology. In the last few years several companies have created subsidiaries to develop the technology (Galagen, GenePharming Europe, and DNX), and several larger organizations have acquired smaller biotechnology companies (Astra bought Symbicon, Genzyme bought Integrated Genetics, and Pharmaceutical Protein Limited bought TransPharm) to give them the technical base to enhance their developmental efforts.

This new industry has been created because of the perception that production of pharmaceuticals in transgenic animals will be more cost-effective than by more conventional means. The world market, for the products thus far identified as appropriate for this technology, exceeds three billion dollars a year (Table 4). It is noteworthy

Table 3. Industrial organizations engaged in bioreactor research and development projects

Company	Location
American Red Cross	Rockville, Maryland, USA
AltraBio (Galagen)	Arden Hills, Minnesota, USA
FinnGene (GenePharming)	Kupio, Finland
GenePharming Europe B.V.	Leiden, The Netherlands
Genzyme	Framingham, Massachusetts, USA
NexTran (DNX)	Princeton, New Jersey, USA
PPL Therapeutics PLC	Roslin, Scotland, UK
Symbicon (Astra)	Umea, Sweden
YS New Technology Institute	Tochigi, Japan

that all but one of the products (glucocerebrosidase) listed in Table 4 are currently derived from human blood. That may account for the fact that the American Red Cross's Blood Derivatives Laboratory is a central player in the field.

However, there are clearly other approaches available for the synthesis of the products listed in Table 4. Why have these organizations chosen to utilize what can legitimately be characterized as an emerging technology rather than employ more well-established approaches?

Are Mammals the Right Bioreactor Organism? Almost any living organism, or part thereof, that can be "domesticated" could serve as a bioreactor. Bacteria, yeast, insect cells, mammalian cells in culture, plants, and chicken eggs are all potential competing production systems. Each system has specific advantages and disadvantages. In general, prokaryotic systems and plants can be genetically engineered rapidly, at relatively low cost and can be propagated rapidly but they may lack the appropriate enzymes to perform some of the critical post-translational modifications, such as signal peptide cleavage, glycosylation, amidation, acetylation, carboxylation, and phosphorylation required by complex mammalian proteins (Houdebine, 1994). Not withstanding some deficiencies, production of pharmaceuticals in transgenic plants may be a formidable competitor in the future. Baculovirus-insect cell expression systems (Luckow and Summers, 1988) and stably transfected mammalian cells have the capacity to perform authentic post-translational modifications. However, yields in those systems are often an order of magnitude lower than has already been achieved in transgenic animal bioreactors. Furthermore, proteins generated in those systems must be subjected to costly purification schemes (Buckholtz and Gleeson, 1991; Blum et al., 1992).

Table 4. Estimated annual worldwide requirements and costs of potential bioreactor products[a]

	Pharmaceutical						
Item	F-VIII[b]	F-IX[c]	GC[d]	Protein C	AT III[e]	Fibrin[f]	Albumin[g]
Estimated quantity needed, kg	.3	4	10	10	21	150	315×10^3
Current cost per gram, $	2.9×10^6	40,000	100,000	10,000	7,000	1,000	3.56
Annual market, $10^6 \times$ $	.87	160	1,000	100	150	150	1,120

[a]Information from William Drohan and Henryk Lubon, American Red Cross; Edward Ginns, NIH, personal communication; Paleyanda et al., 1991; Lee and de Boer, 1994.
[b]Blood coagulating Factor VIII.
[c]Blood coagulating Factor IX.
[d]Glucocerebrosidase.
[e]AntiThrombin III.
[f]Fibrinogen.
[g]Human serum albumin.

Is the Mammary Gland the Right Organ? The mammary gland is the defining organ of mammals. It provides nutrients for neonates and is the focus of the developmental biologist because it is the only organ in adults, aside from the uterus, to undergo dramatic cycles of differentiation, development, and apoptosis. A complex, horizontally integrated industry, the dairy industry, is based on bovine mammary glands.

The mammary gland is a prodigious production system, capable of generating between 23 g (dairy cattle) and 205 g (rat) of protein per kilogram of body weight during peak lactation (Oftedal, 1984). In ruminants the mammary gland is also responsible for *de novo* synthesis of 50% of the lipid found in milk (Bremel et al., 1989), which can exceed 40 g per liter. Milk is clearly the least complicated bodily fluid to collect, especially from ruminants. Milking of pigs can even be mechanized (Grun et al., 1993). The ample production capacity of the mammary gland coupled with the relative ease of harvesting milk in a non-invasive manner recommends the mammary gland as the organ of choice for producing pharmaceutical products in animals.

Another often cited advantage of producing biologically active products in the mammary gland is the isolation the gland provides from the circulation. It is argued that bioreactor animals would be protected from the potentially untoward effects of biologically active compounds because those compounds would be sequestered in the milk and therefore unavailable to the circulatory system. However, endogenous products produced in the mammary gland (milk proteins) are indeed found in the circulation in cattle, especially during late gestation and at the time of parturition (McFadden et al., 1987). Also, mammary gland-specific transgenes and milk protein genes are transiently expressed during estrus, even in virgin mice (Robinson et al., 1995). Therefore, to safeguard bioreactor animals, it may be appropriate to consider designing gene constructs in such a way that their product is converted to an active form after it is isolated from milk.

One of the more serious and obvious constraints imposed by a mammary gland-based production system is that only half of the transgenic animals produced, females, can be used to generate product. Furthermore, product can only be obtained after females have successfully produced offspring. There have been attempts to induce lactation in virgins, though the results have been mixed (Shamay et al., 1992; Ebert et al., 1994). Targeting production of bioreactor products to almost any other organ would eliminate both of those problems.

Post-translational Modifications A less obvious, but possibly more serious, constraint imposed by expression of foreign proteins in mammary epithelial cells is their ability to provide appropriate post-translational modifications such as N- or O-glycosylation, and γ-carboxylation. It can be assumed that lack of faithful post-translational modification of a protein will affect its biological activity, as has been the case for at least one bioreactor protein (Henryk Lubon, personnel communication). Furthermore, carbohydrate residues of glycoproteins can serve as antigens. Those residues can also influence secretion of a protein and affect a protein's half-life in the circulation (James et al., 1995).

When transgenes for interferon-γ were expressed in Chinese hamster ovary cells, baculovirus-infected Sf9 insect cells and mammary glands of transgenic mice, N-glycosylation patterns (sites glycosylated) and composition of the sugar residues differed significantly (James et al., 1995). It has also been demonstrated that glycosylation patterns of Protein C produced in the mammary glands of mice differ from Protein C isolated from human serum (Drohan et al., 1994). The glycosylation patterns of Protein C isolated from transgenic pig milk also differed from that observed in human serum (Morcol et al., 1994). Furthermore, the Protein C carbohydrate patterns in mouse milk differed from those in pig milk.

No additional data are available to assess the significance of these observations on the transgenic bioreactor approach to producing pharmaceuticals. However, at least one prominent laboratory in the field is beginning to address the issue by creating transgenic mice containing two mammary gland-directed transgenes, one for the protein of interest, the other to increase the gland's post-translational modification capacity (Drews et al., 1995).

Other Organs May Have Advantages Other bodily fluids and excretions such as saliva, blood, urine, and feces would be more difficult to harvest than milk but may have some potential advantages over mammary glands. Saliva, which is produced in prodigious amounts by ruminants, would be difficult to collect in large quantities. Feces would contain bacteria that would complicate extraction and purification processes. Blood and urine seem to be the best alternatives.

Blood serves as a collection reservoir for secreted products of many organs. Therefore, if it were necessary to produce the desired compound in a specific organ (e.g., in order to achieve the appropriate post-translational modifications), blood could serve as a convenient carrier that could be harvested by catheterization or at slaughter (Logan, 1993). The technology, though costly, is already well developed for purifying many of the most commonly proposed bioreactor products from blood.

Using urine as the carrier for bioreactor products would, like blood, present collection difficulties, but unlike milk and blood, urine of healthy animals contains relatively little protein (approximately 250 mg/L) and therefore purification of bioreactor products from urine might be quite cost-effective. Analogous to blood, many secreted proteins find their way into the urine, after filtration through the kidneys. If the desired product can be protected from degradation by the liver and is small enough to avoid being trapped by glomerular filtration, the product could be harvested from urine. Ideally, urine might be perceived as a better carrier if hepatic and renal processing could be bypassed. However, urothelial (epithelium lining the ureters and bladder) specific gene regulatory elements have not been available until the recent identification of the uroplakin II gene (Lin et al., 1995). Regulatory sequences of the uromodulin gene might also direct proteins to the urine; however, sequence data for this gene's promoter is not currently available (Pennica et al., 1987).

Production Capacity of Mammary Glands Many mammary gland-specific fusion genes have been tested in transgenic animals (Table 5). However, it is difficult to predict the behavior of specific regulatory elements when ligated to different protein coding sequences. To obtain a rough estimate, an average has been calculated from the available expression data for several species (Table 6). Such a simple arithmetic mean, which neither distinguishes between genomic and cDNA-based constructs nor distinguishes among which regulatory elements were used, at least provides a crude "snapshot" of the current state of expression efficiencies across species. Average concentrations of transgene products measured in the milk of rabbits, rats, pigs, sheep, and goats are 350 μg/mL, 1 mg/mL, 1 mg/mL, 11 mg/mL, and 1 mg/mL, respectively. The mean value for sheep (11 g/L) is disproportionately high because of one exceptional line (Wright et al., 1991) and therefore inclusion of data from that line may compromise the predictive value of the mean. However, at present, there is no compelling reason to expect other fusion genes will not behave as well as the β-lactoglobulin/α-antitrypsin transgene has in sheep. Even if achievable, it may not be desirable to strive for high transgene product concentrations such as the 35 g/L achieved in sheep, because excessive protein expression may exceed cells' capacity to perform appropriate post-translational modifications, as has been demonstrated in cell culture (Rehemtulla and Kaufman, 1992) and in transgenic mice (Drews et al., 1995).

To put these mammary gland transgene expression values in perspective, they can be compared with the endogenous protein production capacity of mammary glands. Production efficiency (100 × maximum transgene product concentration/endogenous milk protein concentration) has been highest in the mouse (Table 6). In one experiment transgenic mice produced 47% more transgene protein (growth hormone) than is typical for endogenous mouse milk proteins (Devinoy et al., 1994). Rabbits and pigs have exhibited the lowest efficiency, producing .7 and 2% of endogenous milk protein concentrations, respectively. It is impossible to know whether these values reflect future trends, but it is clear that, so far, most transgene products have not taxed the production capacity of mammary glands of any species, with the exception of mice and possibly sheep.

It would be especially useful to have a means of predicting the transgene product production capacity of cattle, given the significant expenditures in time and money involved. Unfortunately, no expression data have been reported for transgenic cattle bioreactors, though the data from one project (Krimpenfort et al., 1991) should be available shortly. An estimate of transgene production efficiency for cows can be calculated by applying data from other species, ruminants. Goats and sheep have produced transgene products at 10 and 20% of endogenous milk protein levels, respectively. Is it reasonable to assume that the transgene production efficiency of cattle will fall within that range? For the purposes of this discussion a conservative estimate of 1 g/L of transgene products (1.8% of endogenous protein production capacity) will be used for cows. That concentration is double the minimum said to be commercially useful for many of the proteins currently being targeted by the bioreactor industry (W. N. Drohan, personal communication).

Number of Bioreactor Animals Needed Based on the values presented in Table 6, one can calculate the number of animals that would be required to produce some of the proposed pharmaceutical bioreactor products. The results of those calculations are presented in Table 7. At first inspection it seems unreasonable to think an organization would consider generating over 100,000 rabbits necessary to produce 150 kg of fibrinogen. The labor required to maintain and milk those animals would be enormous, especially in light of the fact that 17 cows might be capable of producing all of the required fibrinogen to satisfy current world needs (Table 7). However, the required number of rabbits could be produced in a little over a year, by using homozygous males and artificial insemination, whereas it would take over four years to produce the 17 cows (Table 8). This is an exaggerated comparison, but it points out the need to consider generation interval as well as production capacity when choosing a species for a bioreactor project.

Table 5. Summary of transgenes with mammary gland specific regulatory elements introduced into animals of various species[a]

5′ Regulatory elements		Coding sequences						
Gene	Source	Gene	Type	Source	3′ or other sequences	TG species	Amount, mL^{-1}	Refs.[c]
α_{s1}-Casein	Bovine	IGF-I	cDNA	Human		Rabbits	mg	1
α_{s1}-Casein	Bovine	Growth hormone	genomic	Human	hGH	Rat	mg	2
α_{s1}-Casein	Bovine	Lactoferrin	cDNA	Human		Cattle	?	3
α_{s1}-Casein	Bovine	Lactoferrin	cDNA	Human		Mice	mg	4
α_{s1}-Casein	Bovine	Lysozyme	cDNA	Human	α_{s1}-Casein	Mice	?	5
α_{s1}-Casein	Bovine	t-PA	cDNA	Human		Mice	μg	6
α_{s1}-Casein	Bovine	t-PA	cDNA	Human		Rabbit	μg	6
α_{s1}-Casein	Bovine	Urokinase	cDNA	Human	α_{s1}-Casein	Mice	mg	7
β-Casein	Bovine	Lysozyme	cDNA	Human	β-Casein	Mice	116% of WAP	5
β-Casein	Bovine	Growth hormone	genomic	Human	hGH	Rat	mg	2
β-Casein*	Goat	β-Casein	genomic	Goat		Mice	mg	8, 9
β-Casein	Goat	CFTR	cDNA	Human		Mice	μg	10
β-Casein	Rabbit	Interleukin-2	cDNA ?	Human	Interleukin-2	Rabbits	ng	11
β-Casein	Rat	β-Casein	genomic	Rat	β-Casein (rat)	Mice	mg	12
β-Casein	Rat	FSH	cDNA	Human		Mice	μg	13
κ-Casein	Bovine	Growth hormone	genomic	Human	hGH	Rat	ng	2
α-Lactalbumin	Bovine	β-Casein	genomic	Bovine	β-Casein	Mice	mg	14
α-Lactalbumin*	Bovine	α-Lactalbumin	genomic	Bovine	α-Lactalbumin	Mice	mg	15, 16
α-Lactalbumin	Bovine	Trophoblastin	cDNA	Ovine	α-Lactalbumin	Mice	μg	17
α-Lactalbumin	Bovine	Growth hormone	genomic	Human	hGH	Rat	mg	2
α-Lactalbumin*	Guinea-pig	α-Lactalbumin	genomic	Guinea pig	α-Lactalbumin	Mice	?	46
β-Lactoglobulin	Ovine	α_1 Antitrypsin	genomic	Human		Mice	mg	18
β-Lactoglobulin	Ovine	α_1 Antitrypsin	genomic	Human		Sheep	mg	19
β-Lactoglobulin	Ovine	Factor VIII	cDNA	Human		Sheep	?	20
β-Lactoglobulin	Ovine	Factor IX	cDNA	Human	β-Lactoglobulin + β-lac	Mice	?	21
β-Lactoglobulin	Ovine	Factor IX	cDNA	Human	β-Lactoglobulin	Sheep	ND	22
β-Lactoglobulin	Ovine	γ-Interferon	cDNA	Human	β-Lactoglobulin	Mice	ng	23
β-Lactoglobulin*	Ovine	β-Lactoglobulin	Genomic	Ovine	β-Lactoglobulin	Mice	mg	24, 25
β-Lactoglobulin	Ovine	Serum albumin	cDNA	Human	β-Lactoglobulin	Mice	mg	26
β-Lactoglobulin	Ovine	Superoxide dismutase	cDNA	Human		Mice	ng?	27

WAP	Mouse	Growth hormone	Genomic	Human		Mice	μg	28, 29
WAP	Mouse	Growth hormone	Genomic	Human	hGH	Rat	ug	2
WAP	Mouse	β-Interferon	cDNA	Human		Mice	ND	30
WAP	Mouse	Protein C	cDNA	Human		Mice	ng	31
WAP	Mouse	Protein C	cDNA	Human		Pigs	mg	32
WAP	Mouse	PS2	Genomic	Human		Mice	μg	33
WAP	Mouse	Superoxide dismutase	cDNA	Human		Mice	mg	27
WAP	Mouse	t-PA	cDNA	Human	SV40	Mice	ng	34, 35
WAP	Mouse	t-PA	cDNA	Human		Goats	μg	36
WAP*	Mouse	WAP	Genomic	Mouse	WAP	Mice	ND	37
WAP*	Mouse	WAP	Genomic	Mouse	WAP + MAR	Mice	ng	38
WAP*	Mouse	WAP	Genomic	Mouse	WAP	Pigs	ng	39, 40
WAP*	Mouse	WAP	Genomic	Mouse	WAP	Sheep	μg	41
WAP	Rabbit	α_1 Antitrypsin	Genomic	Mouse		Mice	mg	42
WAP	Rabbit	Growth hormone	Genomic	Human		Mice	mg	43
WAP*	Rat	WAP	Genomic	Rat		Mice	mg	44, 45

[a]Rows marked with an asterisk contain intact milk protein genes.
[b]ND = not determined.
[c]1, Brem et al., 1994; 2, Ninomiya et al., 1994; 3, Krimpenfort et al., 1991; 4, Platenburg et al., 1994; 5, Maga et al., 1994; 6, Riego et al., 1993; 7, Meade et al., 1990; 8, Persuy et al., 1992; 9, Roberts et al., 1992; 10, DiTullio et al., 1992; 11, Buhler et al., 1990; 12, Lee et al., 1988; 13, Greenberg et al., 1991; 14, Bleck et al., 1995; 15, Soulier et al., 1992; 16, Vilotte et al., 1989; 17, Stinnakre et al., 1991; 18, Archibald et al., 1990; 19, Wright et al., 1991; 20, Halter et al., 1993; 21, Clark, 1989; 22, Simons et al., 1988; 23, Dobrovolsky et al., 1993; 24, Simons et al., 1987; 25, Harris et al., 1990; 26, Shani et al., 1992; 27, Hansson et al., 1994; 28, Reddy et al., 1991; 29, Gunzburg et al., 1991; 30, Schellander and Peli, 1992; 31, Velander et al., 1992b; 32, Velander et al., 1992a; 33, Tomasetto et al., 1989; 34, Gordon et al., 1987; 35, Pittius et al., 1988; 36, Ebert et al., 1991; 37, Burdon et al., 1991; 38, McKnight et al., 1992; 39, Shamay et al., 1991; 40, Wall et al., 1991; 41, Wall et al., 1996; 42, Bischoff et al., 1992; 43, Devinoy et al., 1994; 44, Bayna and Rosen, 1990; 45, Dale et al., 1992; 46, Maschio et al., 1991.

The data presented in Tables 7 and 8, admittedly based on preliminary findings, bode well for the genetic engineering of mammary glands to produce commercially useful concentrations of foreign proteins. But from a biological perspective transgene protein production efficiency values are low for all species with the possible exception of sheep. Are the relatively modest concentrations of transgene proteins observed the result of inept transgene design or are mammary glands already producing close to their maximum capacity? Will milk protein genes have to be "knocked-out" to provide additional capacity for transgene proteins? Will higher transgene product levels be disruptive to mammary gland function as they were in one study (Bleck et al., 1995)? These questions cannot be answered at this time.

Table 6. Milk production capacity of various species of animals

	Transgenic animal species						
Item	Mice	Rat	Rabbit	Pig	Sheep	Goat	Cattle
Length of lactation, days	21	21	35	28	250	250	305
Milk protein concentration, %[a]	1.5	8.9	14	4.8	5.5	2.9	3.5
Milk volume per day at peak lactation[b]	1.6 mL	40 mL	262 mL	12.8 L	2.4 L	4.7 L	35 L
Harvestable volume per lactation[c]	9.0 mL	80 mL	1.4 L	260 L	306 L	600 L	9,000 L
Max. transgene protein produced[d]	22 g/L	11 g/L	1 g/L	1 g/L	11(35)[g] g/L	3 g/L	NDA[h]
Transgene protein production capacity[e]	198 mg	880 mg	1.4 g	260 g	3.4 kg	1.8 kg	?
Transgene protein production efficiency[f], %	147	12.4	.7	2.1	20.0	10.3	?

[a]Milk protein concentrations (Dils, 1986).
[b]Milk volume per day at peak lactation was calculated from Oftedal (1984).
[c]Values for mice and rats were estimated based on collecting every third day (mice, Hanrahan and Eisen, 1970) and every other day (rats). Milk yield data for rabbits (Duby et al., 1993) were based on daily milking and pig yields were calculated from once a week milkings (Grun et al., 1993).
[d]Maximum concentration of transgene proteins reported to date.
[e](Maximum transgene protein concentration) × (Harvestable milk volume per lactation).
[f](Maximum concentration of transgene protein production reported)/(Endogenous milk protein concentration) × 100.
[g]One exceptional line of sheep expressed α-1-antitrypsin at 35 g/L. However, in five additional lines expression ranged from 1.1 g/L to 16 g/L (Carver et al., 1993). An average (11 g/L) of the six lines was used to calculate the transgene protein production capacity and transgene protein production efficiency for sheep.
[h]No data available for cattle.

Modification of Milk to Improve Nutrition and Processing

As mentioned at the outset, many reviewers have suggested a variety of ways to modify milk to improve its nutritional quality and improve the efficiency of manufacturing milk products, such as cheese, ice cream, and yogurt. Unfortunately, this area of investigation has received less attention in laboratories than it has in review articles. To date only five papers containing original research findings on this topic have been published. Two papers report efforts to express human lactoferrin in mice (Platenburg et al., 1994) and cattle (Krimpenfort et al., 1991). Two other nutraceutical (orally administered biologics that provide some health benefit) projects describe successful attempts to express lysozyme (Maga et al., 1994) and superoxide dismutase (Hansson et al., 1994) in the mammary glands of transgenic mice. The fifth report (not yet published as a full-length paper) describes a landmark experiment conducted by a team of researchers at the University of California at Davis led by Jim Murray and Gary Anderson (Anderson et al., 1995). In that study bovine κ-casein was expressed in milk of transgenic mice. The milk of those mice had decreased micelle size and increased curd strength, an important parameter in cheesemaking. These results, which support the predictions of dairy scientists (Richardson, 1985), are a significant first step toward validating the concept that genetic manipulation of mammary gland-specific genes can alter milk composition in beneficial ways.

Why Has Progress Been Slow? The limited efforts in this area of research, in comparison to pharmaceutical production, may reflect the perceived rate of financial return on food products vs. biomedical products. However, a less obvious factor may also have influenced the differences in emphasis of the two fields of mammary gland genetic engineering. A significant segment of the pharmaceutical industry has embraced the transgenic bioreactor approach to making drugs, whereas there is little evidence that the dairy industry has been as enthusiastic about modifying the type

Table 7. Estimated number of transgenic animals needed to satisfy the annual world market for selected pharmaceuticals[a]

	Pharmaceutical						
Species	F-VIII[b]	F-IX[c]	GC[d]	Protein C	AT III[e]	Fibrin	Albumin[f]
Rat	345	4,454	11×10^3	11×10^3	24×10^3	171×10^3	358×10^6
Rabbit	217	2,857	7,143	7,143	15,000	107×10^3	225×10^6
Pig	2	15	38	38	81	577	$1{,}212 \times 10^3$
Goat	1	3	6	6	12	83	175,000
Sheep	1	1	3	3	6	45	93,000
Cattle[g]	1	1	2	2	3	17	35,000

[a]Based on estimated quantity needed (Table 4) and transgene protein production capacity (Table 5).
[b]Blood coagulating Factor VIII.
[c]Blood coagulating Factor IX.
[d]Glucocerebrosidase.
[e]AntiThrombin III.
[f]Human serum albumin.
[g]Assumes 1 g/L production of transgenic protein in cattle.

of milk that cows produce. The pharmaceutical industry has a history of obtaining its products from a variety of sources, including extracting drugs from plants, synthesizing them from chemical constituents, and isolating them from animal and human tissues. Isolating their products from the milk of genetically engineered animals does not require a significant change in mind set. On the other hand, the proposition of creating cows that produce specialized milk may have momentous consequences for the structure of the dairy industry.

The goal of the dairy industry, at least in the United States, has been to create a feed-efficient, healthy cow that produces copious amounts of milk that can serve all the needs of the industry. Genetic engineering offers the opportunity for a paradigm shift, a reshaping of the industry from producers to the processing plants. Dairy farmers may have the opportunity to choose to produce high-protein milk, milk that has accelerated curd clotting time destined for cheese manufacturing, or milk containing nutraceuticals to supplement or replace infant formula. Such a scenario would be a radical change for the dairy industry. But it is by no means without precedence in agricultural production systems. One has only to look to grain, fruit, and vegetable crops to see numerous examples of varieties that have been propagated to fill, or create, specific niche markets. Whether such a paradigm shift will sweep through the entire dairy industry or become a subset of the industry remains to be seen. There is, however, little question that products of genetic engineering will become part of the dairy industry in the next century.

Proposed Changes in Composition of Milk Many of the proposed changes in milk structure are listed in Table 9. It is readily apparent that there may be a number of ways to achieve the same or similar goals. It is also clear that altering the characteristics of one component to enhance a particular processing feature may make milk unsuitable for other uses. It will be up to dairy scientists to determine during the next decade which approach is most efficacious to achieve the desired goal.

Reducing Milk Fat Gibson recently published a thorough analysis of the economic potential of altering milk fat composition (Gibson, 1991) and pointed out several

Table 8. Estimated months required to produce transgenic animals needed to satisfy the annual world market for blood coagulating Factor IX and Fibrinogen[a]

	Pharmaceutical	
Species	Factor IX	Fibrinogen
Rabbit	15	27
Sheep	18	26
Pig	37	59
Cattle	33	54

[a]Based on number of animals needed (Table 7) and utilizing the most time-efficient breeding schemes available for each species.

potential limitations to that strategy. One consequence of altering milk fat composition would be a concomitant change in the melting point of milk. Increasing the proportion of unsaturated fats (and as a consequence decreasing saturated fats) would decrease the melting point of milk, which could impair whipping of cream and decrease the hardness of butter. Gibson concluded that the interactive consequences of altering fat content of milk on processing of other milk products would not be economically advantageous to large segments of the dairy industry, but would only be suitable for niche markets.

However, if the goal is to decrease total fat content rather than to alter the ratios of various fat components, a different economic picture emerges. It is likely if cows produced 2.0% fat milk that milk would not be used to produce cream or butter, but would rather be directed to the fluid milk market.

On average, dairy cows in this country produce milk that contains 3.8% fat. Approximately 50% of that fat is synthesized in the mammary gland. Therefore, if that synthetic machinery in the mammary gland could be disrupted, milk containing 2.0% fat (a 40% reduction) might be achievable (Yom and Bremel, 1993).

Robert Bremel at the University of Wisconsin has suggested that *de novo* fat synthesis might be reduced by blocking acetyl CoA carboxylase gene expression (Bremel et al., 1989). This might be achieved through stem-cell (knock-out), anti-sense, or ribozyme technology. A recent study confirmed the potential of this concept by

Table 9. Some proposed modifications of milk constituents[a]

Change	Consequence
Increase α- and β-caseins	Enhanced curd firmness for cheese-making, improved thermal stability, increased calcium content
Increase phosphorylation sites in caseins	Increased calcium content, improved emulsification
Introduce proteolytic sites in caseins	Increased rate of textural development (improved cheese ripening)
Increase κ-casein concentration	Enhanced stability of casein aggregates, decreased micelle size, decreased gelation and coagulation
Eliminate β-lactoglobulin	Decreased high temperature gelation, improved digestibility, decreased allergenic response, decreased primary source of cysteine in milk
Decrease α-lactalbumin	Decreased lactose, increase market potential of fluid milk, decreased ice crystal formation, compromise osmotic regulation of mammary gland
Add human lactoferrin	Enhanced iron absorption, protect against gut infections
Add proteolytic sites to κ-casein	Increased rate of cheese ripening
Decrease expression of acetyl CoA carboxylase	Decreased fat content, improved nutritional quality, reduce milk production costs
Express immunoglobulin genes	Protection against pathogens such as salmonella and listeria
Replace bovine milk proteins genes with human equivalents	Mimic human breast milk

[a]Adapted from Clark, 1992; Yom and Bremel, 1993; Houdebine, 1994.

transfecting preadipocyte cells with a ribozyme gene directed against acetyl CoA carboxylase RNA (Ha and Kim, 1994). Acetyl CoA carboxylase RNA was reduced in comparison to non-ribozyme-expressing cells. The decrease in acetyl CoA carboxylase mRNA was associated with a significant decrease in acetyl CoA carboxylase enzyme activity, and the rate of fatty acid synthesis fell to about 30 to 70% of controls. This strategy would also have collateral consequences. But in this scenario the attendant economic consequences might be very positive.

Fats are energetically expensive molecules to make. Thus, reducing fat content of milk to 2.0% could reduce feed energy requirements. Typical dairy rations consist of about 60% forage and 40% concentrate. The cost of forage accounts for about 40% of the cost of feed and the grain component accounts for about 50% of the feed's cost. It has been estimated that engineering cows to produce 2.0% fat milk would allow the forage component of rations to be increased to 83% (Yom and Bremel, 1993). The consequence of that change in feed composition would be a reduction in feed cost per kilogram of milk by 22%. Notwithstanding the significant savings to the dairy farmer, there are further ramifications, which could have a positive environmental impact. The amount of forage production would have to be increased by 33%, but the amount of corn produced for dairy rations could be decreased by 50%. Based on corn and forage yields here at Beltsville, an acre of land could be taken out of corn production for every four cows engineered to produce 2.0% fat milk. Over two million acres used for grain production could be reclaimed if all 10 million dairy cows in the United States were genetically engineered to produce 2% fat milk. This is admittedly an unlikely occurrence.

Conclusions

The most easily identified constraint that is impeding progress in this field is the low efficiency of producing useful transgenic animals. As mentioned earlier, low embryo survival and transgene integration rates have clearly been documented. Because of those two factors, producing a transgenic sheep or goat can easily cost $60,000 and producing a transgenic cow or bull can exceed $300,000. The situation is further exacerbated by the unpredictable nature of fusion gene constructs. Currently there is no foolproof way to assess the function of a transgene short of using it to make transgenic animals of the species of interest. This is a particularly acute problem for mammary gland-targeted transgenes because there are no authentic cell culture systems for that tissue. Most organizations test transgene constructs in mice before introducing them into livestock. However, transgenic mice have proven to be poor predictors of transgene function in another species. A way must be found to accurately evaluate transgene constructs in the tissue of the animal of choice without having to invest both time and money to produce transgenic livestock.

Increasing efficiency will not only have the obvious benefit of reducing costs, but will also encourage participation in this field by investigators who cannot currently afford to use this technology. The participation of other researchers will be necessary, especially in the area of modifying milk for post-harvest processing purposes, because a great deal of fundamental research is needed to understand mammary gland biology.

Transgenic animal bioreactor organizations have demonstrated that it is possible to produce almost any desired protein in mammary glands, mostly through empirical studies. The field has now reached the stage of maturity where the focus of attention will be shifting to concerns of post-translational modifications of those proteins, efficacious purification schemes, and product safety.

Fifteen years ago transgenic animal technology was invented, and since then an industry has formed to exploit that technology. In the next 15 years products created by that technology will probably be in the hands of consumers.

Literature Cited

Anderson, G. B., A. Gutierrez, E. A. Maga, J. F. Medrano, and J. D. Murray. 1995. The use of transgenes to alter the physical and functional properties of milk. In: Proc. Beltsville Symposia on Agricultural Animals xx. p 3 (Abstr.).

Archibald, A. L., M. McClenaghan, V. Hornsey, J. P. Simons, and A. J. Clark. 1990. High-level expression of biologically active human alpha 1-antitrypsin in the milk of transgenic mice. Proc. Natl. Acad. Sci. USA 87:5178.

Bawden, C. S., A. V. Sivaprasad, P. J. Verma, S. K. Walker, and G. E. Rogers. 1995. Expression of bacterial cysteine biosynthesis genes in transgenic mice and sheep: Toward a new *in vivo* amino acid biosynthesis pathway and improved wool growth. Transgenic Res. 4:87.

Bayna, E. M., and J. M. Rosen. 1990. Tissue-specific, high level expression of the rat whey acidic protein gene in transgenic mice. Nucleic Acids Res. 18:2977.

Bischoff, R., E. Degryse, F. Perraud, W. Dalemans, D. Ali-Hadji, D. Thëpot, E. Devinoy, L. M. Houdebine, and A. Pavirani. 1992. A 17.6 kbp region located upstream of the rabbit WAP gene directs high level expression of a functional human protein variant in transgenic mouse milk. FEBS Lett. 305:265.

Bishop, J. O., and P. Smith. 1989. Mechanism of chromosomal integration of microinjected DNA. Mol. Biol. Med. 6:283.

Bleck, G. T., R. Jimenez-Florest, and R. D. Bremel. 1995. Abnormal properties of milk from transgenic mice expressing bovine b-casein under control of the bovine a-lactalbumin 5' flanking region. Int. Dairy J. 5:619.

Blum, P., M. Velligan, N. Lin, and A. Martin. 1992. DNA K-mediated alteration in human growth hormone protein inclusion bodies. Bio/Technology 10:301.

Bonifer, C., M. Vidal, F. Grosveld, and A. E. Sippel. 1990. Tissue specific and position independent expression of the complete gene domain for chicken lysozyme in transgenic mice. EMBO J. 9:2843.

Bonifer, C., N. Yannoutsos, G. Krüger, F. Grosveld, and A. E. Sippel. 1994. Dissection of the locus control function located on the chicken lysozyme gene domain in transgenic mice. Nucleic Acids Res. 22:4202.

Brem, G. 1993. Inheritance and tissue-specific expression of transgenes in rabbits and pigs. Mol. Reprod. Dev. 36:242.

Brem, G., P. Hartl, U. Besenfelder, E. Wolf, N. Zinovieva, and R. Pfaller. 1994. Expression of synthetic cDNA sequences encoding human insulin-like growth factor-1 (IGF-1) in the mammary gland of transgenic rabbits. Gene 149:351.

Brem, G. B. 1985. Production of transgenic mice, rabbits and pigs by microinjection into pronuclei. Zuchthygiene 20:251.

Bremel, R. D., H. C. Yom, and G. T. Bleck. 1989. Alteration of milk composition using molecular genetics. J. Dairy Sci. 72:2826.

Breslow, J. L. 1993. Transgenic mouse models of lipoprotein metabolism and atherosclerosis. Proc. Natl. Acad. Sci. USA 90:8314.

Brinster, R. L., H. Y. Chen, and R. D. Palmiter. 1982. Regulation of metallothionein-thymidine kinase fusion plasmids injected into mouse eggs. Nature (Lond.) 296:39.

Brinster, R. L., H. Y. Chen, M. E. Trumbauer, M. K. Yagle, and R. D. Palmiter. 1985. Factors affecting the efficiency of introducing foreign DNA into mice by microinjecting eggs. Proc. Natl. Acad. Sci. USA 82:4438.

Brüggemann, M., N. P. Davies, and I. R. Rosewell. 1993. Designer mice: The production of human antibody repertoires in transgenic animals. Year Immunol. 7:33.

Buckholtz, R. G., and M.A.G. Gleeson. 1991. Yeast systems for the commercial production of heterologous proteins. Bio/Technology 9:1067.

Buhler, T. A., T. Bruyere, D. F. Went, G. Stranzinger, and K. Burki. 1990. Rabbit β-casein promoter directs secretion of human interleukin-2 into the milk of transgenic rabbits. Bio/Technology 8:140.

Burdon, T., L. Sankaran, R. J. Wall, M. Spencer, and L. Hennighausen. 1991. Expression of a whey acidic protein transgene during mammary development. Evidence for different mechanisms of regulation during pregnancy and lactation. J. Biol. Chem. 266:6909.

Carver, A. S., M. A. Dalrymple, G. Wright, D. S. Cottom, D. B. Reeves, Y. H. Gibson, J. L. Keenan, J. D. Darrass, A. R. Scott, A. Colman, and I. Garner. 1993. Transgenic livestock as bioreactors: Stable expression of human alpha-1-antitrypsin by a flock of sheep. Bio/Technology 11:1263.

Clark, A. J. 1989. Expression of human anti-hemophilic factor IX in the milk of transgenic sheep. Bio/Technology 7:487.

Clark, A. J. 1992. Prospects for the genetic engineering of milk. J. Cell. Biochem. 49:121.

Clements, J. E., R. J. Wall, O. Narayan, D. Hauer, R. Schoborg, D. Sheffer, A. Powell, L. M. Carruth, M. C. Zink, and C. E. Rexroad. 1994. Development of transgenic sheep that express the visna virus envelope gene. Virology 200:370.

Cozzi, E., G. A. Langford, A. Richards, K. Elsome, R. Lancaster, P. Chen, N. Yannoutsos, and D.J.G. White. 1994. Expression of human decay accelerating factor in transgenic pigs. Transplant. Proc. 26:1402.

Dale, T. C., M. J. Krnacik, C. Schmidhauser, C. L. Yang, M. J. Bissell, and J. M. Rosen. 1992. High-level expression of the rat whey acidic protein gene is mediated by elements in the promoter and 3' untranslated region. Mol. Cell Biol. 12:905.

Devinoy, E., D. Thepot, M. G. Stinnakre, M. L. Fontaine, H. Grabowski, C. Puissant, A. Pavirani, and L. M. Houdebine. 1994. High level production of human growth hormone in the milk of transgenic mice: The upstream region of the rabbit whey acidic protein (WAP) gene targets transgene expression to the mammary gland. Transgenic Res. 3:79.

Dils, R. R. 1986. Comparative aspects of milk fat synthesis. J. Dairy Sci. 69:904.

DiTullio, P., S. H. Cheng, J. Marshall, R. J. Gregory, K. M. Ebert, H. M. Meade, and A. E. Smith. 1992. Production of cystic fibrosis transmembrane conductance regulator in the milk of transgenic mice. Bio/Technology 10:74.

Dobrovolsky, V. N., O. V. Lagutin, T. V. Vinogradova, I. S. Frolova, V. Kuznetsov, and O. A. Larionov. 1993. Human gamma-interferon expression in the mammary gland of transgenic mice. FEBS Lett. 319:181.

Drews, R., R. K. Paleyanda, T. K. Lee, R. R. Chang, A. Rehemtulla, R. J. Kaufman, W. N. Drohan, and H. Lubon. 1995. Engineering the protein processing capacity of the mammary gland. Proc. Natl. Acad. Sci. USA 92:10462.

Drohan, W. N., D. W. Zhang, R. K. Paleyanda, R. Chang, M. Wroble, W. Velander, and H. Lubon. 1994. Inefficient processing of human protein C in the mouse mammary gland. Transgenic Res. 3:355.

Duby, R. T., M. B. Cunniff, J. M. Belak, J. J. Balise, and J. M. Robl. 1993. Effect of milking frequency on collection of milk from nursing New Zealand white rabbits. Anim. Biotechnol. 4:31.

Ebert, K. M., P. DiTullio, C. A. Barry, J. E. Schindler, S. L. Ayres, T. E. Smith, L. J. Pellerin, H. M. Meade, J. Denman, and B. Roberts. 1994. Induction of human tissue plasminogen activator in the mammary gland of transgenic goats. Bio/Technology 12:699.

Ebert, K. M., J. P. Selgrath, P. DiTullio, J. Denman, T. E. Smith, M. A. Memon, J. E. Schindler, G. M. Monastersky, J. A. Vitale, and K. Gordon. 1991. Transgenic production of a variant of human tissue-type plasminogen activator in goat milk: Generation of transgenic goats and analysis of expression. Bio/Technology 9:835.

Eid, L. N., S. P. Lorton, and J. J. Parrish. 1994. Paternal influence on S-phase in the first cell cycle of the bovine embryo. Biol. Reprod. 51:1232.

Eyestone, W. H., and N. L. First. 1989. Cell cycle analysis of early bovine embryos. Theriogenology 29:243.

Fabry, M. E. 1993. Transgenic animal models of sickle cell disease. Experientia 49:28.

Fodor, W. L., B. L. Williams, L. A. Matis, J. A. Madri, S. A. Rollins, J. W. Knight, W. Velander, and S. P. Squinto. 1994. Expression of a functional human complement inhibitor in a transgenic pig as a model for the prevention of xenogeneic hyperacute organ rejection. Proc. Natl. Acad. Sci. USA 91:11153.

Francolini, M., M. Lavitrano, C. L. Lamia, D. French, L. Frati, F. Cotelli, and C. Spadafora. 1993. Evidence for nuclear internalization of exogenous DNA into mammalian sperm cells. Mol. Reprod. Dev. 34:133.

Fukuchi, K.-I., C. E. Ogburn, A. C. Smith, D. D. Kunkel, C. E. Furlong, S. S. Deeb, D. Nochlin, S. M. Sumi, and G. M. Martin. 1993. Transgenic animal models for Alzheimer's disease. Ann. NY Acad. Sci. 695:217.

Geller, S. A., W. S. Nichols, S. Kim, T. Tolmachoff, S. Lee, M. J. Dycaico, K. Felts, and J. A. Sorge. 1994. Hepatocarcinogenesis is the sequel to hepatitis in Z#2 a_1-antitrypsin transgenic mice: Histopathological and DNA ploidy studies. Hepatology 19:389.

Gibson, J. P. 1991. The potential for genetic change in milk fat composition. J. Dairy Sci. 74:3258.

Gordon, K., E. Lee, J. A. Vitale, A. E. Smith, H. Westphal, and L. Hennighausen. 1987. Production of human tissue plasminogen activator in transgenic mouse milk. Bio/Technology 5:1183.

Gordon, J. W., and F. H. Ruddle. 1981. Integration and stable germ line transmission of genes injected into mouse pronuclei. Science (Wash DC) 214:1244.

Gordon, J. W., G. A. Scangos, D. J. Plotkin, J. A. Barbosa, and F. H. Ruddle. 1980. Genetic transformation of mouse embryos by microinjection of purified DNA. Proc. Natl. Acad. Sci. USA 77:7380.

Greenberg, N. M., J. W. Anderson, A. W. Hsueh, K. Nishimori, J. J. Reeves, D. M. de Avila, D. N. Ward, and J. M. Rosen. 1991. Expression of biologically active heterodimeric bovine follicle-stimulating hormone in milk of transgenic mice. Proc. Natl. Acad. Sci. USA 88:8327.

Grosveld, F., A. van, D. R. Greaves, and G. Kollias. 1987. Position-independent, high-level expression of the human beta-globin gene in transgenic mice. Cell 51:975.

Grun, D., G. Reiner, and V. Dzapo. 1993. Untersuchungen uber rassenunterschiede in der milchleistung bein schwein. I. Mitteilung: Methodik des maschinellen milchentzuges und milchmengenleistug [Investigations on breed differences in milk yield of swine. I. Commitment: Methodology of mechanical milking and milk yield]. Reprod. Domest. Anim. 28:14.

Gunzburg, W. H., B. Salmons, B. Zimmermann, M. Muller, V. Erfle, and G. Brem. 1991. A mammary-specific promoter directs expression of growth hormone not only to the mammary gland, but also to Bergman glia cells in transgenic mice. Mol. Endocrinol. 5:123.

Ha, J., and K. H. Kim. 1994. Inhibition of fatty acid synthesis by expression of an acetyl-CoA carboxylase-specific ribozyme gene. Proc. Natl. Acad. Sci. USA 91:9951.

Halter, R., J. Carnwath, G. Espanion, D. Herrmann, E. Lemme, H. Niemann, and D. Paul. 1993. Strategies to express factor VIII gene constructs in the ovine mammary gland. Theriogenology 39:137.

Hammer, R. E., V. G. Pursel, C. Rexroad, R. J. Wall, D. J. Bolt, K. M. Ebert, R. D. Palmiter, and R. L. Brinster. 1985. Production of transgenic rabbits, sheep and pigs by microinjection. Nature (Lond.) 315:680.

Hanrahan, J. P., and E. J. Eisen. 1970. A lactation curve for mice. Lab. Anim. Care 20:101.

Hansson, L., M. Edlund, A. Edlund, T. Johansson, S. L. Marklund, S. Fromm, M. Stromqvist, and J. Tornell. 1994. Expression and characterization of biologically active human extracellular superoxide dismutase in milk of transgenic mice. J. Biol. Chem. 269:5358.

Harris, S., M. McClenaghan, J. P. Simons, S. Ali, and A. J. Clark. 1990. Gene expression in the mammary gland. J. Reprod. Fertil. 88:707.

Harris, S., M. McClenaghan, J. P. Simons, S. Ali, and A. J. Clark. 1991. Developmental regulation of the sheep beta-lactoglobulin gene in the mammary gland of transgenic mice. Dev. Genet. 12:299.

Hawk, H. W., R. J. Wall, and H. H. Conley. 1989. Survival of DNA-injected cow embryos temporarily cultured in rabbit oviducts. Theriogenology 32:243.

Hill, K. G., J. Curry, F. J. DeMayo, K. Jones-Diller, J. R. Slapak, and K. R. Bondioli. 1992. Production of transgenic cattle by pronuclear injection. Theriogenology 37:222 (Abstr.).

Hochi, S., T. Ninomiya, M. Waga-Homma, J. Sagara, and A. Yuki. 1992. Secretion of bovine alpha-lactalbumin into the milk of transgenic rats. Mol. Reprod. Dev. 33:160.

Hoeschele, I. 1990. Potential gain from insertion of major genes into dairy cattle. J. Dairy Sci. 73:2601.

Houdebine, L.-M. 1994. Production of pharmaceutical proteins from transgenic animals. J. Biotechnol. 34:269.

Huang, M.T.F. 1993. Gene targeting technology for creating transgenic models of lymphopoiesis. Lab. Anim. Sci. 43:156.

James, D. C., R. B. Freedman, M. Hoare, O. W. Ogonah, B. C. Rooney, O. A. Larionov, V. N. Dobrovolsky, O. V. Lagutin, and N. Jenkins. 1995. N-Glycosylation of recombinant human interferon-gamma produced in different animal expression systems. Bio/Technology 13:592.

Kay, G. W., H. W. Hawk, R. A. Waterman, and R. J. Wall. 1988. Identification of pronuclei in in vitro fertilized cow embryos. Anim. Biotechnol. 2:45.

Krimpenfort, P., A. Rademakers, W. Eyestone, B. Van, P. Kooiman, E. Kootwijk, G. Platenburg, F. Pieper, R. Strijker, and H. A. de Boer. 1991. Generation of transgenic dairy cattle using 'in vitro' embryo production. Bio/Technology 9:844.

Krishna, M., and W. M. Generoso. 1977. Timing of sperm penetration, pronuclear formation, pronuclear DNA synthesis, and first cleavage in naturally ovulated mouse eggs. J Exp. Zool. 202:245.

Lauria, A., and F. Gandolfi. 1993. Recent advances in sperm cell mediated gene transfer. Mol. Reprod. Dev. 36:255.

Laurincik, J., P. Hyttel, and V. Kopecny. 1995. DNA synthesis and pronucleus development in pig zygotes obtained in vivo: An autoradiographic and ultrastructural study. Mol. Reprod. Dev. 40:325.

Lee, K. F., F. J. DeMayo, S. H. Atiee, and J. M. Rosen. 1988. Tissue-specific expression of the rat beta-casein gene in transgenic mice. Nucleic Acids Res. 16:1027.

Lee, S. H., and H. A. de Boer. 1994. Production of biomedical proteins in the milk of transgenic dairy cows: State of the art. J. Controlled Release 29:213.

Lin, J.-H., H. Zhao, and T.-T. Sun. 1995. A tissue-specific promoter that can drive a foreign gene to express in the suprabasal urothelial cells of transgenic mice. Proc. Natl. Acad. Sci. USA 92:679.

Lo, D., V. Pursel, P. J. Linton, E. Sandgren, R. Behringer, C. Rexroad, R. D. Palmiter, and R. L. Brinster. 1991. Expression of mouse IgA by transgenic mice, pigs and sheep. Eur. J. Immunol. 21:1001.

Logan, J. S. 1993. Transgenic animals: Beyond 'funny milk'. Curr. Opin. Biotechnol. 4:591.

Luckow, V. A., and M. D. Summers. 1988. Trends in the development of baculovirus expression vectors. Bio/Technology 6:47.

Maga, E. A., G. B. Anderson, M. C. Huang, and J. D. Murray. 1994. Expression of human lysozyme mRNA in the mammary gland of transgenic mice. Transgenic Res. 3:36.

Martin, L., M. Mora-Worms, C. Lucas, C. Reynolds, and T. A. Stewart. 1993. An evaluation of mechanisms by which tolerance to organ-specific antigens is lost using a transgenic mouse model. J. Immunol. 150:1234.

Maschio, A., P. M. Brickell, D. Kioussis, A. L. Mellor, D. Katz, and R. K. Craig. 1991. Transgenic mice carrying the guinea-pig alpha-lactalbumin gene transcribe milk protein genes in their sebaceous glands during lactation. Biochem. J. 275:459.

McFadden, T. B., R. M. Akers, and G. W. Kazmer. 1987. Alpha-lactalbumin in bovine serum: Relationships with udder development and function. J. Dairy. Sci. 70:259.

McKnight, R. A., A. Shamay, L. Sankaran, R. J. Wall, and L. Hennighausen. 1992. Matrix-attachment regions can impart position-independent regulation of a tissue-specific gene in transgenic mice. Proc. Natl. Acad. Sci. USA 89:6943.

Meade, H., L. Gates, E. Lacy, and N. Lonberg. 1990. Bovine α_{s1}-casein gene sequences direct high level expression of active human urokinase in mouse milk. Bio/Technology 8:443.

Mehtali, M., M. Munschy, D. Ali-Hadji, and M. P. Kieny. 1992. A novel transgenic mouse model for the in vivo evaluation of anti-human immunodeficiency virus type 1 drugs. AIDS Res. Hum. Retroviruses 8:1959.

Morcol, T., R. M. Akers, J. L. Johnson, B. L. Williams, F. C. Gwazdauskas, J. W. Knight, H. Lubon, R. K. Paleyanda, W. N. Drohan, and W. H. Velander. 1994. The porcine mammary gland as a bioreactor for complex proteins. Ann. N. Y. Acad. Sci. 721:218.

Ninomiya, T., M. Hirabayashi, J. Sagara, and A. Yuki. 1994. Functions of milk protein gene 5' flanking regions on human growth hormone gene. Mol. Reprod. Dev. 37:276.

Oftedal, O. T. 1984. Milk composition, milk yield and energy output at peak lactation: A comparative review. In: M. Peaker, R. G. Vernon, and C. H. Knight (Ed.) Physiological Strategies in Lactation. p 33. Academic Press, London.

Paleyanda, R., J. Young, W. Velander, and W. N. Drohan. 1991. The expression of therapeutic proteins in transgenic animals. In: L. W. Hoyer, and W. N. Drohan (Ed.) Recombinant Technology in Hemostatis and Thrombosis. p 197. Plenum Press, New York.

Pennica, D., W. J. Kohr, W.-J. Kuang, D. Glaister, B. B. Aggarwal, E. Y. Chen, and D. V. Goeddel. 1987. Identification of human uromodulin as the Tamm-Horsfall urinary glycoprotein. Science (Wash DC) 236:83.

Persuy, M. A., M. G. Stinnakre, C. Printz, M. F. Mahe, and J. C. Mercier. 1992. High expression of the caprine beta-casein gene in transgenic mice. Eur. J. Biochem. 205:887.

Petters, R. M., R. M. Shuman, B. H. Johnson, and R. V. Mettus. 1987. Gene transfer in swine embryos by injection of cells infected with retrovirus vectors. J. Exp. Zool. 242:85.

Pittius, C. W., L. Hennighausen, E. Lee, H. Westphal, E. Nicols, J. Vitale, and K. Gordon. 1988. A milk protein gene promoter directs the expression of human tissue plasminogen activator cDNA to the mammary gland in transgenic mice. Proc. Natl. Acad. Sci. USA 85:5874.

Platenburg, G. J., E. P. Kootwijk, P. M. Kooiman, S. L. Woloshuk, J. H. Nuijens, P. J. Krimpenfort, F. R. Pieper, H. A. de Boer, and R. Strijker. 1994. Expression of human lactoferrin in milk of transgenic mice. Transgenic Res. 3:99.

Pursel, V. G., and C. E. Rexroad, Jr. 1993. Status of research with transgenic farm animals. J. Anim. Sci. 71(Suppl. 3):10.

Reddy, V. B., J. A. Vitale, C. Wei, M. Montoya-Zavala, S. L. Stice, J. Balise, and J. M. Robl. 1991. Expression of human growth hormone in the milk of transgenic mice. Anim. Biotechnol. 2:15.

Rehemtulla, A., and R. J. Kaufman. 1992. Protein processing within the secretory pathway. Curr. Opin. Biotechnol. 3:560.

Richardson, T. 1985. Chemical modifications and genetic engineering of food proteins. J. Dairy Sci. 68:2753.

Riego, E., J. Limonta, A. Aguilar, A. Pérez, R. De Armas, R. Solano, B. Ramos, F. O. Castro, and J. De la Fuente. 1993. Production of transgenic mice and rabbits that carry and express the human tissue plasminogen activator cDNA under the control of a bovine alpha S1 casein promoter. Theriogenology 39:1173.

Roberts, B., P. DiTullio, J. Vitale, K. Hehir, and K. Gordon. 1992. Cloning of the goat beta-casein-encoding gene and expression in transgenic mice. Gene 121:255.

Robinson, G. W., R. A. McKnight, G. H. Smith, and L. Hennighausen. 1995. Mammary epithelial cells undergo transient differentiation in cycling virgins but require pregnancy for the establishment of terminal differentiation. Development 121:2079.

Rogers, G. E. 1990. Improvement of wool production through genetic engineering. Trends Biotechnol. 8:6.

Rosengard, A. M., N. Cary, J. Horsley, C. Belcher, G. Langford, E. Cozzi, J. Wallwork, and D.J.G. White. 1995. Endothelial expression of human decay accelerating factor in transgenic pig tissue: A potential approach for human complement inactivation in discordant xenografts. Transplant. Proc. 27:326.

Rothnagel, J. A., D. A. Greenhalgh, X.-J. Wang, K. Sellheyer, J. R. Bickenbach, A. M. Dominey, and D. R. Roop. 1993. Transgenic models of skin diseases. Arch. Dermatol. 129:1430.

Schellander, K., and J. Peli. 1992. Mammary specific secretion of human beta-interferon in transgenic mice. Theriogenology 37:291.

Schellander, K., J. Peli, F. Schmoll, and G. Brem. 1995. Artificial insemination in cattle with DNA-treated sperm. Anim. Biotechnol. 6:41.

Shamay, A., V. G. Pursel, R. J. Wall, and L. Hennighausen. 1992. Induction of lactogenesis in transgenic virgin pigs: Evidence for gene and integration site-specific hormonal regulation. Mol. Endocrinol. 6:191.

Shamay, A., S. Solinas, V. G. Pursel, R. A. McKnight, L. Alexander, C. Beattie, L. Hennighausen, and R. J. Wall. 1991. Production of the mouse whey acidic protein in transgenic pigs during lactation. J. Anim. Sci. 69:4552.

Shani, M., I. Barash, M. Nathan, G. Ricca, G. H. Searfoss, I. Dekel, A. Faerman, D. Givol, and D. R. Hurwitz. 1992. Expression of human serum albumin in the milk of transgenic mice. Transgenic Res. 1:195.

Shear, H. L., E. F. Roth, Jr., M. E. Fabry, F. D. Costantini, A. Pachnis, A. Hood, and R. L. Nagel. 1993. Transgenic mice expressing human sickle hemoglobin are partially resistant to rodent malaria. Blood 81:222.

Simons, J. P., M. McClenaghan, and A. J. Clark. 1987. Alteration of the quality of milk by expression of sheep beta-lactoglobulin in transgenic mice. Nature (Lond.) 328:530.

Simons, J. P., I. Wilmut, A. J. Clark, A. L. Archibald, and J. O. Bishop. 1988. Gene transfer into sheep. Bio/Technology 6:179.

Soulier, S., J. L. Vilotte, M. G. Stinnakrc, and J. C. Mercier. 1992. Expression analysis of ruminant alpha-lactalbumin in transgenic mice: Developmental regulation and general location of important cis-regulatory elements. FEBS Lett. 297:13.

Stinnakre, M. G., J. L. Vilotte, S. Soulier, R. L'Haridon, M. Charlier, P. Gaye, and J. C. Mercier. 1991. The bovine alpha-lactalbumin promoter directs expression of ovine trophoblast interferon in the mammary gland of transgenic mice. FEBS Lett. 284:19.

Swanson, M. E., M. J. Martin, J. K. O'Donnell, K. Hoover, W. Lago, V. Huntress, C. T. Parsons, C. A. Pinkert, S. Pilder, and J. S. Logan. 1992. Production of functional human hemoglobin in transgenic swine. Bio/Technology 10:557.

Thompson, T. C., L. D. Truong, T. L. Timme, D. Kadmon, B. K. McCune, K. C. Flanders, P. T. Scardino, and S. H. Park. 1993. Transgenic models for the study of prostate cancer. Cancer 71(Suppl. 1):1165.

Tomasetto, C., C. Wolf, M. C. Rio, M. Mehtali, M. LeMeur, P. Gerlinger, P. Chambon, and R. Lathe. 1989. Breast cancer protein PS2 synthesis in mammary gland of transgenic mice and secretion into milk. Mol. Endocrinol. 3:1579.

Velander, W. H., J. L. Johnson, R. L. Page, C. G. Russell, A. Subramanian, T. D. Wilkins, F. C. Gwazdauskas, C. Pittius, and W. N. Drohan. 1992a. High-level expression of a heterologous protein in the milk of transgenic swine using the cDNA encoding human protein C. Proc. Natl. Acad. Sci. USA 89:12003.

Velander, W. H., R. L. Page, T. Morcoel, C. G. Russell, R. Canseco, J. M. Young, W. N. Drohan, F. C. Gwazdauskas, T. D. Wilkins, and J. L. Johnson. 1992b. Production of biologically active human protein C in the milk of transgenic mice. Ann. New York Acad. Sci. 665:391.

Vilotte, J.-L., S. Soulier, M.-G. Stinnakre, M. Massoud, and J.-C. Mercier. 1989. Efficient tissue specific expression of bovine a-lactalbumin in transgenic mice. Eur. J. Biochem. 186:43.

Wagner, S. D., G. T. Williams, T. Larson, M. S. Neuberger, D. Kitamura, K. Rajewsky, J. Xian, and M. Brüggemann. 1994. Antibodies generated from human immunoglobulin miniloci in transgenic mice. Nucleic Acids Res. 22:1389.

Wall, R. J., V. G. Pursel, R. E. Hammer, and R. L. Brinster. 1985. Development of porcine ova that were centrifuged to permit visualization of pronuclei and nuclei. Biol. Reprod. 32:645.

Wall, R. J., V. G. Pursel, A. Shamay, R. A. McKnight, C. W. Pittius, and L. Hennighausen. 1991. High-level synthesis of a heterologous milk protein in the mammary glands of transgenic swine. Proc. Natl. Acad. Sci. USA 88:1696.

Wall, R. J., C. E. Rexroad, A. Powell, A. Shamay, R. McKnight, and L. Hennighausen. 1996. Synthesis and secretion of the mouse whey acidic protein in transgenic sheep. Transgenic Res. 5:67.

Ward, K. A., and C. D. Nancarrow. 1991. The genetic engineering of production traits in domestic animals. Experientia 47:913.

Weidle, U. H., H. Lenz, and G. B. Brem. 1991. Genes encoding a mouse monoclonal antibody are expressed in transgenic mice, rabbits and pigs. Gene 98:185.

Wells, K. D., and R. M. Petters. 1994. Construction of a vector for the genetic manipulation of digestion. J. Anim. Sci. 72:6.

Whitelaw, C. A., S. Harris, M. McClenaghan, J. P. Simons, and A. J. Clark. 1992. Position-independent expression of the ovine beta-lactoglobulin gene in transgenic mice. Biochem. J. 286:31.

Wilkie, T. M., R. L. Brinster, and R. D. Palmiter. 1986. Germline and somatic mosaicism in transgenic mice. Dev. Biol. 118:9.

Wilmut, I., A. L. Archibald, S. Harris, M. McClenaghan, J. P. Simons, C. B. Whitelaw, and A. J. Clark. 1990. Modification of milk composition. J Reprod. Fertil. Suppl. 41:135.

Wright, G., A. Carver, D. Cottom, D. Reeves, A. Scott, P. Simons, I. Wilmut, I. Garner, and A. Colman. 1991. High level expression of active human alpha-1-antitrypsin in the milk of transgenic sheep. Bio/Technology 9:830.

Yamamura, K.-I., F. Tashiro, S. Yi, S. Wakasugi, S. Araki, S. Maeda, and K. Shimada. 1993. Transgenic mouse model for human genetic diseases. Mol. Reprod. Dev. 36:248.

Yannoutsos, N., G. A. Langford, E. Cozzi, R. Lancaster, K. Elsome, P. Chen, and D.J.G. White. 1995. Production of pigs transgenic for human regulators of complement activation. Transplant. Proc. 27:324.

Yom, H. C., and R. D. Bremel. 1993. Genetic engineering of milk composition: Modification of milk components in lactating transgenic animals. Am. J. Clin. Nutr. 58:299S.

Domestic Animal Embryonic Stem Cells: Progress Toward Germ-Line Contribution

Steven L. Stice* and Nick S. Strelchenko†

**Advanced Cell Technology, Paige Laboratory, University of Massachusetts, Amherst, MA 01003, U.S.A. and †American Breeders Service, DeForest, WI 53232, U.S.A.*

Abstract Domestic animal embryonic stem (ES) cell technology will have a major impact on future attempts to genetically modify and improve agricultural animals. This field has made major progress in recent years. Pluripotent stem cell lines or ES-like cells have been propagated *in vitro*, while maintaining pluripotent properties. Domestic animal ES-like cells allowed to differentiate *in vitro* formed cells types representative of all three germ layers. Cell types ranged from ciliated epithelium to multinucleated striated muscle. More recent studies suggest that ES-like cells can contribute to developing fetuses and offspring. Nuclear transfer studies have tested the *in vivo* developmental capacity of cultured embryonic cells. Donor nuclei from established bovine ES-like cell lines directed fetal development through organogenesis but lacked the ability to direct normal placental development. However, cultured bovine and ovine inner cell mass cells have given rise to nuclear transfer offspring. Phenotypic evidence of porcine ES-like cell line contribution to chimeric offspring has also been observed. In the end, for ES-like cell lines to be of value, germ-line transmission must be demonstrated. Once accomplished, germ-line transmittable farm animal ES cells will be an asset to cloning and transgenic animal technologies.

Key Words: Embryonic Stem Cells, Cattle, Pig, Sheep

Potential Uses of Domestic Animal Embryonic Stems

Embryonic stem (ES) cells derived from domestic animal embryos may be the breakthrough needed in developing large-scale embryo cloning procedures and faster, more efficient means of generating valuable transgenic farm animals. The value of ES cell lines is their pluripotent characteristics. In other words, ES cells can differentiate into all cell types in a developing fetus and offspring. This includes the germ cells. Germ-line transmittable ES cells have only been produced in one species, the mouse. Recent advances in domestic ES cell research and the potential benefits of these cells, both commercial and academic, has fueled increased research efforts in this area.

One of the major barriers to widespread utilization of embryo cloning technology is the inefficiencies in the nuclear transfer procedure itself. Presently, sheep and cattle embryo cloning procedures are capable of producing at best a handful of progeny. Embryonic stem cell technology may help bypass these inefficiencies by providing a large or unlimited number of donor nuclei. Therefore, even at current cloning efficiencies, a large number of identical offspring could still be produced from an unlimited source of donor genetic material. Numerous accomplishments have been made in using embryonic cells in nuclear transfer procedures. First, cells from the ICM (sheep, Smith and Wilmut, 1989; cattle, Keefer et al., 1994; Collas and Barnes, 1994), cultured cattle ICM (Sims and First, 1993), or sheep cultured embryonic ectoderm (Campbell et al., 1995) have all been used as donor nuclei, resulting in live offspring. In addition, nuclear transfer embryos derived from bovine ES-like cell lines were able to direct development through organogenesis (Stice et al., 1994a) but failed to develop normal placentation. This led to a loss of all pregnancies prior to day 60 of gestation. However, as soon as this problem is overcome, large-scale production of cloned offspring should be possible. Already, over 1,400 genetically identical blastocyst stage ES-like cell nuclear transfer embryos have been produced (Stice, unpublished observations). Therefore, once the totipotent properties of the ICM can be maintained and the cells propagated in vitro, then it should be feasible to produce large numbers of identical offspring.

Farm animal ES cell technology could reduce the cost of producing transgenic offspring while enabling researchers to make more specific genome modifications. In cattle, millions of dollars have been spent over the last 10 years on traditional microinjection transgenic techniques. Relatively few transgenic calves have been generated from these concerted efforts. The development of an efficient ES cell system could potentially reduce the number of embryos required to generate a transgenic animal, saving time and money. Additionally, ES cells would enable researchers to place new genes in advantageous locations in an animal's genome. Alternatively, deleterious genes might be removed using ES cell technology. These types of genetic manipulations are very difficult, if not impossible, with traditional pronuclear embryo microinjection techniques. A major breakthrough would be the use of genetically modified ES cells in a cloning procedure. An unlimited number of genetically identical embryos known to be transgenic would be transferred into recipient animals. Their tremendous potential justifies the long and arduous task of pursuing farm animal ES cells.

The substantial resource and time commitments needed to conduct domestic animal ES cell studies have hampered progress in this area; however, as more uses of ES cell technology are defined, progress in this area will most likely hasten. Medical and pharmaceutical uses of ES cell technology in large animals may help stimulate more ES cell chimera and nuclear transfer studies. The ability to implement gene targeting techniques in pig ES cells might facilitate xenograft organ transplantation (for review see Dorling and Lechler, 1994). Gene targeting techniques have been used to modify, insert, or delete DNA at specific sites in the genome (for review see Melton, 1994). Gene targeting experiments in pigs, sheep, goats, and cattle will also allow more precise modifications of milk composition for pharmaceutical and agricultural purposes.

Embryonic Stem Cell Characteristics

An important and surprisingly difficult task is to define the characteristics a domestic animal embryonic cell line must possess to be considered an ES cell line. In 1981, the term embryonic stem cells or ES cells was used to denote a cell line isolated directly from mouse embryos, whereas embryonal carcinoma cells (EC) were derived from teratocarcinomas (Evans and Kaufman, 1981; Martin, 1981). Embryonic stem cells and EC cells shared many of the same characteristics. Morphologically these cells were indistinguishable from each other (Martin,1981). When mouse ES cells were subcutaneously injected into athymic mice, they formed teratocarcinomas containing derivatives of all three germ layers. In addition, mouse ES cells were capable of differentiating *in vitro* into embryoid bodies, structures resembling normal blastocyst-stage mouse embryos. Unlike EC cells, ES cells often maintained a normal karyotype after numerous passages *in vitro*. Therefore, initial descriptions of mouse ES cells relied on how they were derived, cytogenetics, *in vitro* pluripotent properties, and tumorigenicity. A number of embryonic cell lines derived from domestic animal embryos have been propagated (Table 1). These cell lines were pluripotent, and in some lines derivatives of all three germ layers were observed upon *in vitro* or *in vivo* differentiation. Many cell lines also maintain a normal karyotype. In this respect, domestic animal ES cells have already been produced.

For many researchers, the definition of what constitutes an ES cell line changed when Bradley and coworkers (1984) demonstrated that mouse ES cells formed stable germ-line chimeras. Since then, germ-line transmission of the ES cell genetics has not been reported in any species other than the mouse. Therefore, despite their pluripotent properties, domestic animal embryonic cell lines have not met this most stringent and useful characteristic of ES cells. Descriptive terms such as ES-like, pluripotent embryonic cell lines, cultured ICM cells, and cultured embryonic disc cells describe various embryonic cells in long- and short-term culture. Thus, for simplicity reasons these cells will be referred to as ES-like cells in this manuscript. Although opinions may differ on what exactly to call these cells, it is obvious that in order for them to become a useful tool in cloning and transgenic farm animals, germ-line transmission must be demonstrated.

Major inroads have been made toward demonstrating germ-line transmission, despite the many difficulties encountered in performing these experiments in domestic animals. One barrier is the excessive expense of conducting these studies. The cost of obtaining and manipulating recipient embryos or oocytes, transferring

Table 1. The potency of domestic animal embryonic stem-like cell lines as demonstrated by their ability to differentiate both *in vitro* and *in vivo*

Publication	Species	*In vitro* differentiation	*In vivo* differentiation
Handyside et al., 1987 (p)[a]	Sheep		
Piedrahita et al., 1990 (p)	Sheep	Embryoid bodies	
Notarianni et al., 1991 (p)	Sheep	Embryoid bodies	
Campbell et al., 1995	Sheep		NT lamb (3 passage)
Piedrahita et al., 1990 (p)	Pig	Embryoid bodies	
Strojek et al., 1990 (p)	Pig		
Notarianni et al., 1990 (p)	Pig	Embryoid bodies	
Talbot et al., 1993b (p)	Pig	Differing cell types	
Wheeler, 1994	Pig	Differing cell types	Chimeras (> 20 passage)
Gerfen and Wheeler, 1995 (p)	Pig	Embryoid bodies	
Saito et al., 1992 (p)	Cattle		NT embryo
Sims and First, 1993 (p)	Cattle	Embryoid bodies	NT calves (27-day culture)
Stice et al., 1994a	Cattle	Embryoid bodies	NT fetuses (> 10 passage)
Graves and Moreadith, 1993 (p)	Rabbit	Differing cell types	
Giles et al., 1993 (p)	Rabbit		Chimera (3-day culture)
Sukoyan et al., 1992 (p)	Mink	Embryoid bodies	Teratoma formation
Sukoyan et al., 1993 (p)	Mink	Embryoid bodies	Teratoma formation

[a](p) indicates a peer-reviewed article.

manipulated embryos into recipients, and maintaining the animals can be prohibitive in farm species. Because of this, it is difficult to adequately test all potential ES cell lines for germ-line transmission. Another difficulty is putting together a team of researchers with the necessary skills. Developing and testing a cell line requires researchers with differing skills, including molecular and cellular biologists, embryologists with micromanipulation skills, surgeons, and farm managers with animal husbandry skills. All of these people are required to demonstrate germ-line transmission in large animals. Finally, long gestation periods and generation intervals often equate to long-term studies. A year or more may pass before determining whether a particular ES-like cell line is germ-line transmittable. In spite of these formidable requirements and obstacles, progress continues. The following sections will review some of these accomplishments.

Attributes of Domestic Animal Embryonic Stem-Like Cells

Cell Morphology

What morphological characteristics should domestic animal ES cells maintain as they are propagated in vitro? There are basically two schools of thought. One suggests that farm animal ES cells should be morphologically similar to mouse ES cells. Undifferentiated mouse ES cells grow as multiple layer nests of cells. The colonies have highly refractile edges and the edges of individual cells within a colony cannot be determined. Individual cells have a high nuclear to cytoplasmic volume ratio. Domestic animal ES-like cell lines that maintain mouse ES-cell characteristics have been difficult to propagate in culture. Alternatively, it can be argued that domestic animal ES cells should differ morphologically from mouse ES cells because of differences in early embryonic development. The nonpolarized mouse ICM gradually gives rise to an embryonic ectoderm composed of a simple epithelium surrounding a central cavity. However, in many farm animal species the embryonic ectoderm is discoid (see review by Betteridge and Flechon, 1988).

Most published descriptions of established ES-like cell lines describe cell lines propagating as an epithelial-like sheet of cells. The morphology of the individual cells within a colony can differ considerably. Some cells maintain a high nuclear to cytoplasmic volume, characteristic of ES cells, whereas other cells seem to differentiate into larger endoderm-like cells having a low nuclear to cytoplasmic ratio. It may be useful to compare propagating ES-like cell lines to cell types found in developing embryos. By day 12 in developing cattle embryos, endodermal cells are a structurally polarized epithelial sheet of cells that contain desmosome-associated tonofilaments, and individual cells have a lens-like shape (Betteridge and Flechon 1988). Also, by day 12, the bovine ICM has differentiated into the embryonic ectoderm. The embryonic ectoderm has a polarized surface epithelium with desmosome connections. Therefore, epithelial type ES-like cells in culture may be morphologically similar to embryonic ectoderm cells in farm animal embryos. No matter what the optimal cell morphology may appear to be, the potency and eventual utility of the cells can only be determined by examining both *in vitro* and *in vivo* pluripotent properties.

In Vitro Differentiation, Karyotype, and ICM Contribution

The potential to differentiate *in vitro* is one characteristic used in determining a cell line's potency (Table 1). Methods used to examine *in vitro* differentiation vary. Often cell lines are allowed to spontaneously differentiate, as results from overcrowding in the culture dish. Embryonic stem-like cell lines will develop "domes" and eventually form embryoid bodies of various types and complexity (Figure 1). They are called embryoid bodies because, in many ways, they resemble hatched and expanding blastocyst-stage embryos. In cattle, these structures range from simple epithelial-like vesicles to more complex masses of cells (Figure 1a). More complex embryoid bodies often contain a mixture of cell types derived from different germ layers, including muscle and hemopoietic cells (Figure 1b,c) or form a syncytium (Figure 1d). In the pig, sheep, and rabbit other cell types such as fibroblast, neuronal, pigmented, and ciliated cells have been observed (Table 1). Therefore, cell types derived from all three germ layers have been observed in several species.

Karyotypic examination of ES-like cells is important since only ES cell lines that maintain a normal karyotype in culture can be used in future germ-line transmission experiments. Mouse ES cells, unlike EC lines, maintain a normal karyotype

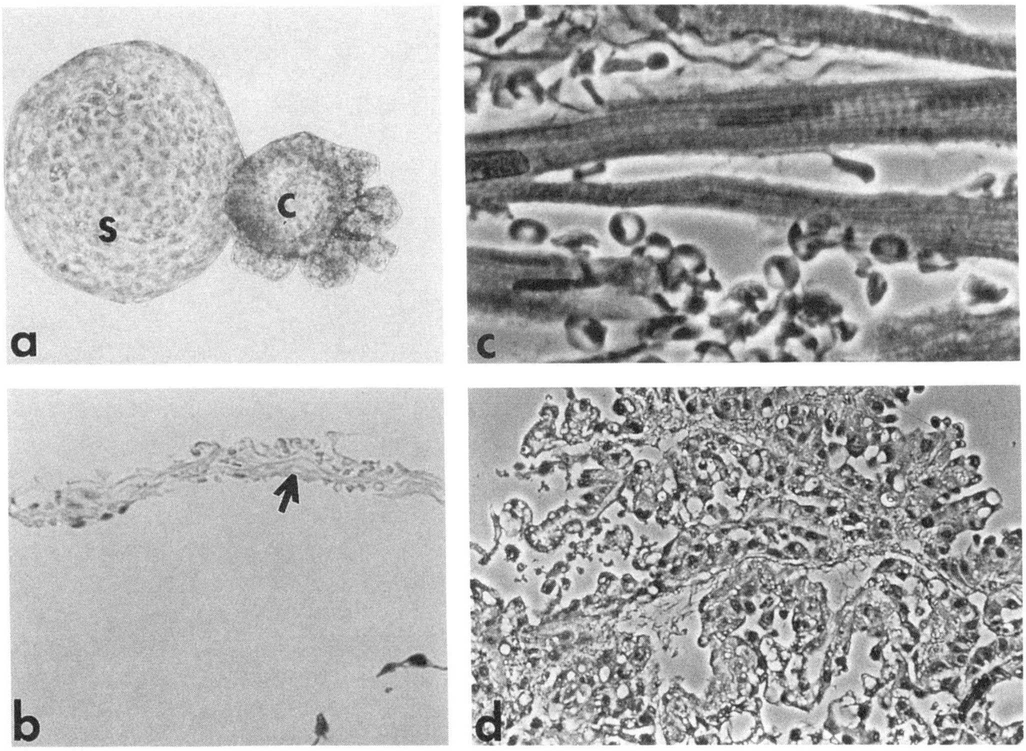

Figure 1. (a) Simple (s) and complex (c) bovine embryoid bodies form during spontaneous differentiation of bovine ES-like cells in vitro (84×); (b) Histological section of a complex embryoid body, (arrow, striated muscle and erythrocytes formation; 84×); (c) higher magnification of striated muscle and erythrocytes (336×); (d) Syncytium formation in complex embryoid bodies (84×).

(Evans and Kaufman, 1981). Although both male and female ES-like cells can maintain a normal diploid state (cattle, Saito et al., 1992; mink, Sukoyan et al., 1993; pig, Talbot et al., 1993b; Wheeler, 1994), long-term *in vitro* culture can lead to karyotype abnormalities. In particular, X inactivation (Sukoyan et al., 1992) and polyploidy (> 10% of cells, Strelchenko et al., 1995) have been observed in some domestic animal ES-like cell lines. This is not totally unexpected since multiple passage XX mouse ES cell lines commonly undergo X deletions (Robertson et al., 1983). However, XY mouse ES cell lines are often able to maintain a stable karyotype over long periods of *in vitro* culture. Thus, cell lines used in germ-line experiments have to be continually monitored for possible cytogenetic changes that could occur during *in vitro* propagation.

An obvious requirement of any ES cell line is the ability to contribute to the host embryo's ICM after being aggregated with or injected into the host embryo. In an experiment designed to test these capabilities, cattle ES-like cell membranes were labeled with the membrane-specific fluorescent marker PKH26 (Sigma, St. Louis, MO). Ten to 20 labeled, disaggregated cells (Figure 2a,b) were injected into the perivitelline space of late morula-stage host embryos. After 2 to 3 days in culture the host embryos (Figure 2c) were observed for labeled ES-like cell contribution to the ICM (Figure 2d). Embryonic stem-like cell contribution in 40 host embryos ranged

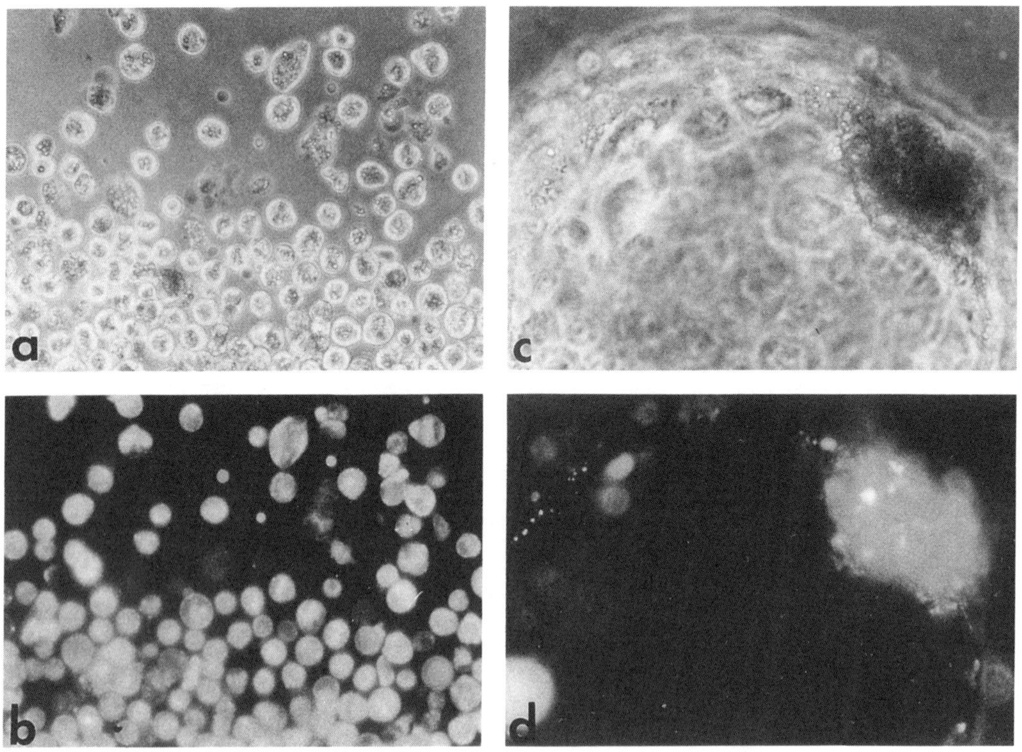

Figure 2. (a) Disaggregated bovine ES-like cells (336×); (b) Disaggregated bovine ES-like cells labeled with PKH26 (336×); (c) A host-hatched bovine blastocyst 2 days after labeled ES-like cell injection (336×); (d) Detection of PKH26-labeled ES-like cells in the ICM of the host blastocyst (336×).

from little or none to visibly high contribution. In 73% (29) of the embryos some ES-like cell contribution to the ICM could be detected. In all embryos, non-incorporated ES-like cells were also noticed (Figure 2d). As a control, membrane-labeled mouse fibroblast cells have also been injected into host cattle embryos. In this case there was no obvious contribution to the ICM of the embryo. These experiments serve as rapid initial tests, suggesting that ES-like cells can contribute to the developing ICM. This is a starting point for more informative chimera experiments examining ES-like cell contribution to the developing fetus and offspring.

Embryonic Stem Cell Specific Markers

Mouse ES cell or ICM-specific cell markers are often used in domestic ES-like cell studies. Of course, the overlying problem with using markers associated with the mouse is determining whether these markers are useful across species. Presently, markers may be useful in determining differences and similarities in ES-like cells derived from different sources of cells, passages, and among species. The presence or absence of intermediate filaments has been used to characterize domestic animal ES-like cells (Piedrahita et al., 1990; Sukoyan et al., 1993; Van Stekelenburg-Hamers et al., 1995). When probed for cytokeratin some cell lines were negative for this cytoskeletal marker. In the mouse, cytokeratin is restricted to the trophectoderm and not found in the ICM (Emerson, 1988). However, unlike the mouse ES cells, the cytokeratin negative ES-like cells have not produced chimeric offspring (Piedrahita et al., 1990; Sukoyan et al., 1993).

Numerous other cell markers have been used in ES cell studies. One marker often associated with undifferentiated mouse ES cells is alkaline phosphatase, which is a glycoprotein (Pease et al., 1990). However, mouse ES cells lose alkaline phosphatase upon differentiation in vitro. The ICM of sheep and pig embryos are positive for alkaline phosphatase but it is lost as cell lines derived from the ICM are propagated in vitro (Talbot et al., 1993a). If an ICM-like cell is desirable, then there is reason to expect that the cell line should also stain positive for alkaline phosphatase. Domestic animal ES-like cells have been screened using a number of other mouse ES cell-specific markers or ICM-specific markers (Van Steckelenburg-Hamers, 1995). An advantage to these assays is that an ES-like cell line can be easily assayed over time (numerous passages) in culture. Thus, markers can facilitate detection of differences and changes among cell lines from distinct passages within a cell line. In addition, differences between cell lines can be quantified more easily through changes in marker characteristics rather than more subjective morphological characteristics. Thus, following changes in ES-like cells using a battery of markers can be informative.

Teratocarcinomas

Originally, *in vivo* differentiating capabilities of mouse ES cells were tested by injecting these cells into athymic mice. Typically, mouse ES cells form teratocarcinomas containing derivatives of all three primary germ layers (Evans and Kaufman, 1981). Domestic animal ES-like cells have also been used in similar experiments. A number of cell types ranging from epithelial to adipocytes were produced when mink (Sukoyan et al., 1993) or pig (Wheeler, 1994) ES-like cells were injected into athymic mice. Some mink ES-like cell lines had the potential to differentiate into a wider range of cell types than other cell lines. Mink ES-like cells derived from advanced blastocyst-stage ICMs and morulae developed more diverse and highly differentiated teratomas than did early blastocyst-derived cell lines. Thus, these cells may have been more potent and better candidate ES-like cell lines for use in chimera experiments.

Chimeras

Producing chimeric domestic animal offspring is achievable when freshly isolated blastomeres (Brem et al., 1984) or ICM cells (Kasiwazaki et al., 1992) are combined with a host embryo. However, ES-like cell chimeric studies have been less successful. In the rabbit, ICM cells freshly isolated or cultured for less than 3 days contributed to chimeric offspring, whereas ICM cells cultured beyond this time did not (Giles et al., 1993). Mink ES-like cells injected into host blastocysts failed to develop into chimeric offspring (Sukoyan et al., 1992). However, the potency of these cells may have been limited since teratoma experiments using these same cell lines resulted in only fibroblastic cell types. Unfortunately, ES-like cells that produced the varied cell type teratomas were not tested in chimera experiments (Sukoyan et al., 1993).

Pig ES-like cell chimera experiments have produced mixed results. Initial attempts to produce chimeric offspring using pig ES-like cells were reported by Piedrahita and coworkers (1990) and Gerfen and Wheeler (1995). In these experiments, chimeric offspring were not produced. Similar to the rabbit, freshly isolated pig ICM cells resulted in chimeric offspring, whereas cultured ICM cells could not (Anderson et al., 1994). These studies along with those of the rabbit suggest that culture techniques can cause differentiation of ICM cells very early in culture. However, Wheeler (1994) has reported in a recent review article phenotypic evidence of chimeric pigs. The chimeras were produced by microinjecting ES-like cells into the blastocoele cavity of a host embryo. At birth, coat color was used as phenotypic evidence of chimerism. Offspring derived from these chimeras are now being generated to ascertain the incidence of germ-line transmission. Meeting the final criterion of germ-line transmission will be key to determining the utility of these cells in the production of transgenic animals.

Nuclear Transfer

Recently, a considerable amount of research has focused on using nuclear transfer instead of chimeras to test the *in vivo* potency of ES-like cells. Nuclear transfer has several advantages over chimera approaches. The nuclear transfer procedure can immediately test the totipotency of the cells in question, whereas the chimera strategy requires testing offspring derived from chimeric individuals. In cattle, testing totipotency by chimeras is a long process, since the generation interval is 2 years. Obviously, the elimination of one generation by using nuclear transfer procedures is a major advantage.

The strategy of using ES-like cell lines as donor nuclei in the nuclear transfer procedure has produced some encouraging results in cattle. Bovine nuclear transfer embryos derived from ES-like cell lines developed to the blastocyst stage and when transferred to recipient females, pregnancies were established (Stice et al., 1994a). However, ultrasound observations of over 50 initial pregnancies indicated that all the pregnancies were lost between day 30 and 55 of gestation. Examination of recovered conceptuses revealed no gross fetal abnormalities. These fetuses developed through organogenesis and were of normal size for the corresponding stage of gestation. However, the placental development in these conceptuses was abnormal (Stice et al., 1994b). The placental tissues lacked any fetal cotyledonary development and the maternal caruncles were infiltrated with erythrocytes. In addition, during a normal gestation maternal caruncles develop crypts by this stage of gestation, whereas these were absent in nuclear transfer pregnancies. An initial attempt to rescue these conceptuses used a chimera approach. Eight-cell stage blastomeres from IVF embryos were added to the nuclear transfer embryos. Pregnancies were extended at least to 85 days with this technique. DNA marker analysis of a recovered fetus at that stage revealed that only 50% of the placenta was derived from the added blastomeres. Obvious deficiencies in caruncle formation were also present in this pregnancy. New strategies that may alleviate the placental deficiencies include the following: improving nuclear reprogramming procedures, propagating less-differentiated cell lines, and chimera approaches in which the nuclear transfer embryo is not allowed to initially contribute to the developing placenta.

Mouse ES cells, when used as nuclear donors, fail to direct development of nuclear transfer embryos beyond the blastocyst stage (Tsunoda and Kato, 1993). It is possible that established ES cell lines are not totipotent when used in the nuclear transfer process. However, it should be mentioned that mouse embryo nuclear transfer experiments have been less successful than those conducted in domestic animals. To date, mouse nuclear transfer offspring have been produced only when donor nuclei from only two-, four-, and eight-cell-stage embryos were used (Cheong et al., 1993). However, when murine ES cells were combined with a tetraploid embryo, the resulting chimera gave rise to an offspring (Nagy et al., 1993). In this procedure ES cells contributed almost exclusively to the fetus, while the tetraploid cells developed into different placental tissues. Analogous to the bovine ES-like cell nuclear transfer experiments, it appears that mouse ES cells cannot form all placental components in developing conceptuses. Also, only early passage lines (second or third passage) were able to develop to live offspring in these tetraploid experiments. Therefore, mouse ES cells may also undergo subtle differentiation events early during *in vitro* propagation. Sheep nuclear transfer experiments using early passage embryonic disc cells were capable of directing development to offspring (Campbell et al., 1995). Future experiments using advanced passage cells will hopefully demonstrate that unlike the mouse, domestic animal ES-like cells used in nuclear transfer procedures result in fertile offspring.

Future Improvements in Domestic Animal Embryonic Stem-Like Cell Propagation

Although much progress has been made in propagating pluripotent ES-like cell lines, obviously improvements are still warranted. Of particular concern is the early differentiation of ES-like cell colonies in culture. Removing differentiated cells was one approach used to maintain a more homogeneous cell line (Talbot et al., 1993b). Developing procedures to propagate ES-like cells of interest via cloning may possibly decrease cell contact-induced differentiation. In addition, clonal propagation of ES cells is an important component to gene targeting procedures. Deriving colonies from single ES-like cells is difficult since individual cells tend to undergo spontaneous differentiation in culture (Stice, unpublished observations). Improvements in this area may require optimizing cell cultures using various growth factors.

To date the use of growth factors in domestic animal ES-like culture has had limited success (for review see Anderson, 1992), whereas in mouse ES and primordial germ cell culture, growth factors have played a pivotal role in propagating undifferentiated cells (Matsui et al., 1992). A number of differences may exist among species, and the use of growth factors in domestic animals may be hampered by limited cross-species reactivity. Thus, species-specific growth factors used in culture may enhance domestic animal ES-like cell propagation. Similarly, adding transgenes encoding for important growth factor receptors may prove informative.

An additional advantage to ES-like cell clonal propagation is the possibility of eliminating or reducing cell contact growth inhibition. The ES-like cells' immediate environment (cell to cell communication) may play a major role in their growth rate regulation. Pig, sheep, and cattle ES-like cell colonies' doubling time can range from 50 to 80 hours (Piedrahita et al., 1990; Stice, unpublished observations). This is considerably slower than mouse embryonic ectoderm (Snow, 1977) and cattle morula stage blastomeres (Barnes et al., 1993). Flow cytometry analysis of bovine ES-like cells indicates that close to 20% of the cells reside in S phase at any one time (Stice and Strelchenko, unpublished observations), whereas 80 to nearly 100% of undifferentiated blastomeres are in S phase (Barnes et al., 1993). This information suggests cell cycle differences between ES-like cells and developing embryos. In general, cells that proliferate slowly tend to undergo differentiation. Thus, the slow growth of ES-like cells make them more prone to differentiation. If an ES-cell culture system promotes faster proliferation, then differentiation may be diminished.

Transcription factors (e.g., Oct-3) that seem to be expressed exclusively in undifferentiated mouse ES cells may be important areas of investigation in domestic animals (Okamoto et al., 1990; Rosner et al., 1990). Genetic modifications controlling expression of endogenous Oct-3 or adding an Oct-3 transgene may help prevent spontaneous differentiation in domestic animal ES-like cells. However, these approaches can be technically difficult and not without pitfalls, since specific biological roles for transcription factors in ES cell growth are not well understood. The expression of an Oct-3 transgene may have detrimental side effects or possibly may not affect the potency of the ES-like cell lines at all.

Properly cultured primordial germ (PG) cells are another mouse cell type with germ-line transmission capabilities. In the mouse, under certain culture conditions PG cells can revert back to cells resembling ES cells (Matsui et al., 1992). Reverted PG cells have produced fertile chimeric offspring (Stewart et al., 1994), whereas PG cells not undergoing this transformation do not form chimeras (Stewart, 1982). Obviously, similar techniques may be useful in domestic animals (Cherney and Merier, 1994). However, problems associated with domestic animal ES cells may also be encountered with PG cell-derived cell cultures. In particular, germ-line transmission may be hampered by limited cross-reactivity of human or rodent growth factors with domestic animal PG cells. However, the promising results obtained with mouse PG cells will certainly encourage continued research on alternatives to culturing cells derived from domestic animal embryos.

Summary

Domestic animal ES cells may enhance both transgenic and cloning technologies. Transgenic ES cell approaches would be less expensive and faster and would permit more specific genetic modifications in farm animals. Transgenic ES cell technology may have far-reaching applications ranging from facilitating pig-to-human xenograft transplantation to altering milk protein composition in cattle. Production of large numbers of cloned cattle will be feasible when nuclear totipotent ES cells are available. Toward these goals, several recent nuclear transfer and chimera studies suggest that domestic animal ES-like cells do have *in vivo* pluripotent characteristics. Future improvements in domestic ES-like cells will certainly concentrate on propagating specific types of undifferentiated embryonic cells *in vitro* and adequately testing their potency. Given the recent progress made, the expectations of developing germ-line transmitted ES cells has heightened.

Literature Cited

Anderson, G. B. 1992. Isolation and use of embryonic stem cells from livestock species. Anim. Biotech. 3:165.

Anderson, G. B., S. J. Choi, and R. H. BonDurant. 1994. Survival of porcine inner cell masses in culture and after injection into blastocysts. Theriogenology 42:212.

Barnes, F. L., P. Collas, R. Powell, W. A. King, M. Westhusin, and D. Shepherd. 1993. Influence of recipient oocyte cell cycle stage on DNA synthesis nuclear envelope breakdown, chromosome constitution, and development in nuclear transfer bovine embryos. Mol. Reprod. Dev. 36:33.

Betteridge, K. J., and J. E. Flechon. 1988. The anatomy and physiology of pre-attachment bovine embryos. Theriogenology 29:155.

Bradley, A., M. Evans, M. H. Kaufman, and E. Robertson. 1984. Embryo-derived stem cells: A tool for elucidating the developmental genetics of the mouse. Nature (Lond.) 309:255.

Brem, G., H. Tenhumberg, and H. Krauslish. 1984. Chimerism in cattle through microsurgical aggregation of morulae. Theriogenology 22:609.

Campbell, K., J. McWhir, B. Ritchie, and I. Wilmut. 1995. Production of live lambs following nuclear transfer of cultured embryonic disc cells. Theriogenology 43:181 (Abstr.).

Collas, P., and F. L. Barnes. 1994. Nuclear transplantation by microinjection of inner cell mass and granulosa cell nuclei. Mol. Reprod. Dev. 38:264.

Cheong, H. T., Y. Takahashi, and H. Kanagawa. 1993. Birth of mice after transplantation of early cell-cycle-stage embryonic nuclei into enucleated oocytes. Biol. Reprod. 48:958.

Cherney, R. A., and J. Merier. 1994. Evidence for pluripotency of bovine primordial germ cell-derived cell lines maintained in long-term culture. Theriogenology 41:175 (Abstr.).

Dorling, A., and R. I. Lechler. 1994. Prospects for xenografting. Curr. Opin. Immunol. 6:765.

Emerson, J. A. 1988. Disruption of cytokeratin filament network in the preimplantation mouse embryo. Development (Camb.) 104:219.

Evans, M. J., and M. H. Kaufman. 1981. Establishment in culture of pluripotential cells from mouse embryos. Nature (Lond.) 292:154.

Gerfen, R. W., and M. B. Wheeler. 1995. Isolation of embryonic cell-lines from porcine blastocysts. Anim. Biotech. 6:1.

Giles, J. R., X. Yang, W. Mark, and R. H. Foote. 1993. Pluripoteny of cultured rabbit inner cell mass cells detected by isozyme analysis and eye pigmentation of fetuses following injection into blastocysts or morulae. Mol. Reprod. Dev. 36:130.

Graves, K. H., and R. W. Moreadith. 1993. Derivation and characterization of putative pluripotent embryonic stem cells from preimplantation rabbit embryos. Mol. Reprod. Dev. 36:424.

Handyside, A., M. L. Hooper, M. H. Kaufman, and I. Wilmut. 1987. Towards the isolation of embryonal stem cell lines from the sheep. Roux's Arch. Dev. Biol. 196:185.

Kasiwazaki, N., H. Nakao, S. Ohtani, and N. Nakatsuji. 1992. Production of chimeric pigs by the blastocyst injection method. Vet. Rec. 130:186.

Keefer, C. L., S. L. Stice, and D. L. Matthews. 1994. Bovine inner cell mass cells as donor nuclei in the production of nuclear transfer embryos and calves. Biol. Reprod. 50:935.

Martin, G. 1981. Isolation of a pluripotent cell line from early mouse embryos cultured in medium conditioned by teratocarcinoma stem cells. Dev. Biol. 78:7634.

Matsui, Y., K. Zsebo, and B.L.M. Hogan. 1992. Derivation of pluripotential embryonic stem cells from murine primordial germ cells in culture. Cell 70:841.

Melton, D. W. 1994. Gene targeting in the mouse. BioEssays 16:633.

Nagy, A., J. Rossant, R. Nagy, W. Abramow-Newerly, and J. C. Roder. 1993. Derivation of completely cell culture-derived mice from early-passage embryonic stem cells. Proc. Natl. Acad. Sci. USA 90:8424.

Notarianni, E., C. Galli, S. Laurie, R. M. Moor, and M. J. Evans. 1991. Derivation of pluripotent, embryonic cell lines from pig and sheep. J. Reprod. Fertil. 43(Suppl.):255.

Notarianni, E., S. Laurie, R. M. Moor, and M. J. Evans. 1990. Maintenance and differentiation in culture of pluripotential embryonic cell lines from pig blastocysts. J. Reprod. Fertil. 41(Suppl.):51.

Okamoto, K., H. Okazawa, A. Okuda, M. Sakai, M. Muramatsu, and H. Hamada. 1990. A noval octamer binding transcription factor is differentially expressed in mouse embryonic cells. Cell 60:461.

Pease, S., P. Braghetta, D. Gearing, D. Grail, and R. L. Williams. 1990. Isolation of embryonic stem (ES) cells in media supplemented with recombinant leukemia inhibitory factor (LIF). Dev. Biol. 141:344.

Piedrahita, J. A., G. B. Anderson, and R. H. BonDurant. 1990. On the isolation of embryonic stem cells: Comparative behavior of murine, porcine and ovine embryos. Theriogenology 34:879.

Robertson, E. J., M. H. Kaufman, A. Bradley, and M. J. Evans. 1983. Isolation, properties and karyotype analysis of pluripotent (EK) cell lines from normal and parthenogenetic embryos. In: L. M. Silver, G. R. Martin, and S. Strickland (Ed.) Teratocarcinoma Stem Cells. Cold Spring Harbor Conference on Cell Proliferation. p 647. Cold Spring Harbor Laboratory, Cold Spring Harbor, NY.

Rosner, M. H., M. A. Vigano, K. Ozato, P. M. Timmons, F. Poirier, P. W. Rigby, and L. M. Staudt. 1990. A POU-domain transcription factor in early stem cell and germ cells of the mammalian embryo. Nature (Lond.) 345:686.

Saito, S., N. Strelchenko, and H. Niemann. 1992. Bovine embryonic stem cell-like cell lines cultured over several passages. Roux's Arch. Dev. Biol. 201:134.

Sims, M., and N. L. First. 1993. Production of calves by nuclear transfer of nuclei from cultured inner cell mass cells. Proc. Natl. Acad. Sci. USA 91:6143.

Smith, L. C., and I. Wilmut. 1989. Influence of nuclear and cytoplasmic activity on the development in vitro of sheep embryos after nuclear transplantation. Biol. Reprod. 40:1027.

Snow, M.H.L. 1977. Gastrulation in the mouse: Growth and regionalization of the epiblast. J. Embryol. Exp. Morphol. 42:293.

Stewart, C. L. 1982. Ph.D. Thesis. University of Oxford, England.

Stewart, C. L., I. Gadi, and H. Bhatt. 1994. Stem cells from primordial germ cells can reenter the germ line. Dev. Biol. 161:626.

Stice, S. L., N. Strelchenko, J. Betthauser, B. Scott, G. Jugella, J. Jackson, V. David, C. Keefer, and L. Matthews. 1994a. Bovine pluripotent embryonic cells contribute to nuclear transfer and chimeric fetuses. Theriogenology 41:301 (Abstr.).

Stice, S. L., N. Strelchenko, C. Keefer, and L. Matthews. 1994b. Bovine nuclear transfer embryos derived from pluripotent embryonic cells results in conceptuses deficient in cotyledons. In: Symp. Mol. Cell Aspects Periimplantation Processes. Boston, MA. p 39 (Abstr.).

Strelchenko, N., M. Miltipova, and S. Stice. 1995. Further characterization of bovine pluripotent stem cells. Theriogenology 43:327 (Abstr.).

Strojek, R. M., M. A. Reed, J. L. Hoover, and T. E. Wagner. 1990. A method for cultivating morphologically undifferentiated embryonic stem cells from porcine blastocysts. Theriogenology 33:901.

Sukoyan, M. A., A. N. Golubitsa, A. I. Zhelezova, A. G. Shilov, S. Y. Vatolin, L. F. Maximovsky, L. E. Andreeva, J. McWhir, S. D. Pack, S. I. Bayborodin, A. Y. Kerkis, H. I. Kizilova, and O.

L. Serov. 1992. Isolation and cultivation of blastocyst-derived stem cell lines from American mink (Mustela vision). Mol. Reprod. Dev. 33:418.

Sukoyan, M. A., S. Y. Vatolin, A. N. Golubitsa, A. I. Zehelezova, L. A. Semenova, and O. L. Serov. 1993. Embryonic stem cells derived from morulae, inner cell mass, and blastocyst of mink: Comparisons of their pluripotencies. Mol. Reprod. Dev. 36:148.

Talbot, N. C., C. E. Rexroad, V. G. Pursel, and A. M. Powell. 1993a. Alkaline phosphatase staining of pig and sheep epiblast cells in culture. Mol. Reprod. Dev. 36:139.

Talbot, N. C., C. E. Rexroad, V. G. Pursel, A. M. Powell, and N. D. Nel. 1993b. Culturing the epiblast cells of the pig blastocyst. In Vitro Cell. Dev. Biol. 29A:543.

Tsunoda, Y. and Y. Kato. 1993. Nuclear transplantation of embryonic stem cells in mice. J. Reprod. Fertil. 98:537.

Van Stekelenburg-Hamers, A.E.P., T.A.E. Van Achterberg, H. G. Reble, J. E. Flechon, K.H.S. Cambell, S. M. Weima, and C. L. Mummery. 1995. Isolation and characterization of permanent cell lines from inner cell mass cells of bovine blastocysts. Mol. Reprod. Dev. 40:444.

Wheeler, M. B. 1994. Development and validation of swine embryonic stem cells—A review. Reprod. Fertil. Dev. 6:563.

Primordial Germ Cells for Genetic Modification of Poultry

B. Wentworth*, H. Tsai*, A. Wentworth*, E. Wong[†], J. Proudman[‡], M. El Halawani[§]

**University of Wisconsin-Madison, Madison, WI 53706, U.S.A.; †Virginia Polytechnic Institute and State University, Blacksburg, VA 24061, U.S.A.; ‡USDA-ARS, Beltsville, MD 20705, U.S.A.; and §University of Minnesota, St. Paul, MN 55108, U.S.A.*

Abstract Research was directed toward the final development in culture of five independent chicken blastodermal cell strains each of a single sex (CBCS); transfection of cultured turkey gonadal primordial germ cells (gPGCs) with foreign DNA (lacZ or antisense Prolactin DNA [asPRL]); transfer of these transfected gPGCs to the blood vascular system of stage 15–17 recipient turkey embryos (F_0); hatching some of the recipient embryos and recovery of the DNA transgenes from the gonads of F_0s and the muscle and blood tissue of the F_1 and F_2 progeny of the mated recipients. The transfection efficiency of primary cultured gPGCs with either lacZ or asPRL DNA using Lipofectamine® as the carrier was 26%. The exogenous DNA was detected in the stage 36 F_0 embryonic gonad by slot blot for lacZ using a P^{32} probe for lacZ or by PCR and/or Southern blot using primers for a combined ASVLTR-asPRL fragment or a nick-translated P^{32} probe for asPRL, respectively. This non-viral method of gene transfer in the avian species appears worthy of further investigation. Methods of transgenic enhancement by the production of sterile (lacking germ cells) recipients are also presented. The injection of embryonated chicken eggs with EMA-1 antibody and the immunization of hens with germ cell-like 500-kDa antigen resulted in sterile host embryos. Germ cells from the germinal crescent (GC) of quail embryos were able to populate the gonads of antibody-sterilized host embryos. Other methods of gene transfer in the avian species and the implications of transgenic research are discussed.

Key Words: Antisense, Primordial Germ Cells, Transgenic, Avian

Introduction

Construction of transgenic animals is founded upon inserting foreign DNA into native host DNA, having it transmitted through the germ line, and having this DNA expressed in the animal without disrupting the expression of other essential genes or gene products. The foreign gene insertion can be directed toward modulating the expression of an existing gene, phenotypic expression of a novel trait, or complete abolition of an endogenous gene or gene product. Resulting transgenic animals offer substantial utility in animal agriculture toward improved production traits and enhanced nutrient products. Additionally, they provide a powerful tool for investigating basic problems in animal development, human immune and metabolic diseases, and mechanisms involved in malignant cell proliferation. Use of transgenic animals in mass production of both pharmaceuticals and proteins is in its infancy yet is already demonstrating an enormous potential.

Despite intensive efforts by investigators who desire to manipulate the avian genome, the technology to produce transgenic poultry lags behind that in mice. Distinct features of the avian reproductive system dictate original techniques to introduce exogenous DNA into the genome of the bird (reviewed by Shuman 1990, 1991; Perry and Sang, 1993). Although methods used to produce transgenic mammals are not necessarily amenable to the avian species, one would be remiss not to look toward information being gathered from research laboratory mice and farm animals as well as other vertebrate and non-vertebrate species. Collating this information and devising novel techniques to study and construct transgenic poultry will eventually result in a product that will be accepted by society. Transgenic poultry will have unparalleled product usefulness and will also provide enormous research potential in terms of avian biological development, as well as endocrinological, cellular, and molecular mechanism of avian reproduction.

Avian Reproduction and Early Embryonic Development

Prior to discussing the various methods of gene transfer into poultry that are under investigation, a general overview of the avian reproductive system (Johnson, 1986; Etches and Verrinder Gibbins, 1993) and early embryonic development is desirable to help understand the basis for each of these technologies and some of the advantages and disadvantages encountered regarding effective gene transfer into poultry. In most avian species ovulation occurs at regularly scheduled time intervals depending on the season and the species. In the domestic chicken, the female ovulates about once every 24 hours from 15 to 30 minutes after oviposition. Fertilization follows soon afterward in the infundibula region of the oviduct. Then the second meiotic maturation division occurs in the oocyte and the second polar body is expelled, forming the female pronucleus. Polyspermy is common in the avian species. Several sperm enter the oocyte and form supernumerary male pronuclei. These are indistinguishable from the one male pronucleus destined to form the zygote. The first cell division occurs 1.5 hours after syngamy (Perry, 1987). At the time of lay (oviposition), some 20 to 24 hours later, the chicken zygote is a Stage X (Eyal-Giladi and Kochav, 1976) developed blastoderm containing from 40,000 to 60,000 cells (Spratt and Hass, 1960; Kochav et al., 1980; Watt et al., 1993). In this report, stages of chick embryonic development before oviposition are designated in Roman numerals (Eyal-Giladi and Kochav, 1976) and stages during incubation in Arabic numerals (Hamburger and Hamilton, 1951)(Table 1).

Primordial Germ Cells

Primordial germ cells (PGCs) are the first identifiable gamete precursor cells of an organism. Introducing a heritable trait by altering the genome of an animal requires access to these developmentally committed cells that will become the eggs and sperm of the next generation. They possess all the genomic information of an individual species and the unique cellular and molecular mechanisms with which to pass this information from generation to generation. Isolation of PGCs, modification of their genome, and return of these genetically altered cells to a recipient embryo is a very attractive approach toward modifying in a defined manner virtually any gene in the bird. This technology will provide the facility for incorporating specific genotypic modifications into successive generations (Wentworth et al., 1992). Analysis of the phenotypic expression of birds carrying such mutations could contribute insight into the biological role of specific genes whose functions can now only be a matter of speculation.

Origin, Identification, and Migration of Primordial Germ Cells

In all avian species studied thus far a small population of extraembryonic, extragonadal, migratory precursor germ cells, primordial germ cells (PGCs), are identified very early in embryogenesis to form the germ line lineage. In the chicken a limited number can first be recognized morphologically and histochemically or immunohistochemically, as EMA-1 positive isolated cells or cell clusters, randomly distributed in the central-most zone, the area pellucida, of the freshly oviposited Stage X chicken blastoderm (Ginsburg and Eyal-Giladi, 1987, 1989; Urven et al., 1988; Wentworth et al., 1989; Muniesa and Dominguez, 1990). They are believed to

Table 1. Stages of embryonic development of the chick referenced in this manuscript; preoviposition stages contributed by Eyal-Giladi and Kochav (1976) and incubation stages by Hamburger and Hamilton (1951)

Stage	Description
Preoviposition	
X	Twenty-hour uterine age: formation of area pellucida complete with border between it and area opaca.
XI	Koeller's sickle: most posterior aggregates coalesce to form Koeller's sickle, anterior to transparent belt.
XII	Hypoblast: a hypoblast is forming, progressing in an anterior direction. At this stage, hypoblast underlies only posterior half of the epiblast.
XIII	Full hypoblast: Koeller's sickle is still visible at posterior margin; transparent belt surrounds area occupied by hypoblast, separating it from area opaca.
XIV	Posterior bridge: at posterior side of blastoderm, cellular bridge develops, connecting hypoblast with area opaca.
Incubation	
3	Intermediate streak: (12 to 13 hours) streak extends from posterior margin to approximate center of area pellucida.
4	Definitive streak: (18 to 19 hours) the primitive streak has reached its maximal length. The primitive groove, primitive pit, and Hensen's node are present. The area pellucida has become pear-shaped.
5	Head process: (19 to 22 hours) notochord or head process is visible.
7	A somite pair: (23 to 26 hours) one pair of somites; neural folds visible.
10	Ten somites: (33 to 38 hours) first pair of somites becoming dispersed and indication of cranial flexure, and heart bent right.
12	Sixteen somites: (45 to 49 hours) head turned to left, auditory pit and heart slightly S-shaped.
15	Limb bud: (50 to 55 hours) lateral body folds, limb primordia, amnion extends to somites 7 to 14.
17	Limb and leg buds: (52 to 64 hours) both limb and leg buds lifted off blastoderm. Amnion variable, may extend to Somite 26 to closure with oval hole over Somite 28 to 36; nasal pits, allantois not formed.
20	Allantois vesicular: (70 to 72 hours) allantois vesicular, eye pigmented a faint grayish hue.
30	Feather germ: (6.5 days) feather germ two dorsal rows on either side of spinal cord; egg-tooth distinct.

originate in the epiblast of the Stage X blastoderm and translocate onto the anteriorly moving secondary hypoblast (Sutasurya et al., 1983; Ginsburg and Eyal-Giladi, 1986; Urven et al., 1988). Prior to Stage X no "precursor" PGCs have been identified, either because identifiable markers do not exist, or the cells that transcend to the hypoblast are segregating from the somatic cells to form the first cells of the germ line lineage. As of yet no experimental evidence exists to show at what stage and at what cell number the totipotency of the one-celled avian fertilized egg changes to the pluripotency of the 40,000 to 60,000 celled Stage X blastoderm at oviposition (Spratt and Hass, 1960; Kochav et al., 1980; Watt et al., 1993). Likewise, the question remains of whether the one-celled fertilized egg is in theory the Primordial Germ Cell and the somatic cells become gradually committed from this totipotent cell, or whether a limited number of cells first become committed PGCs when they segregate from the epiblast cells.

By Stage 7, from 125 to 250 Periodic Acid Schiff (PAS) positive PGCs are identified in an area anterior and anteriorlateral to the developing chick embryo at the junction of the area pellucida and area opaca known as the "germinal crescent" (Swift, 1914; Fujimoto et al., 1976b; Ginsburg and Eyal-Giladi, 1986, 1989). Early germ cell segregation and distribution are similarly described in the quail embryo (Eyal-Giladi et al., 1981; Pardanaud et al., 1987; Ginsburg et al., 1989). They soon migrate to and colonize the gonadal primordia forming the complete differentiated gonad(s) during organogenesis (Swift, 1914; Fujimoto et al., 1976b; England, 1983; Pardanaud et al., 1987; Ukeshima et al., 1987, 1991; Yoshinaga et al., 1993).

Avian PGCs are large cells (12 to 20 μm in diameter) with a conspicuously large, eccentrically placed nucleus and prominent, often fragmented, nucleoli (Fujimoto et al., 1975, 1976b). The PGCs in the chicken are most readily identified at these early stages by morphological criteria coupled with 1) the PAS reaction, which stains their intracytoplasmic glycogen granules a brilliant magenta (Meyer, 1960, 1964; Clawson and Domm, 1969; Fujimoto et al., 1975, 1976b) and by 2) a mouse monoclonal IgM antibody, EMA-1 (Hahnel and Eddy, 1986), which recognizes chicken PGCs in the epiblast of the freshly oviposited Stage X chicken embryo through to their colonization in the embryonic gonads (Urven et al., 1988; Tsai, 1995). Turkey (*Meleagris gallopavo*) PGCs stain with PAS from stage 17 through to their colonization in the gonads, and are also recognized by the EMA-1 antibody (A. Wentworth, unpublished data). Quail (*Corturnix corturnix japonica*) PGCs do not stain with PAS (Pardanaud et al., 1987); however, they do react to the Feulgen-Rossenbeck stain (Eyal-Giladi et al., 1981), the QH 1 antibody (Pardanaud et al., 1987), the lectin *Wistaria floribunda*, WFA (Nakamura et al., 1992; Yoshinaga et al., 1992), and the QCR_1 antibody (Ono and Yokoi, 1994). Ultrastructurally chick PGCs show glycogen granules, lipid droplets, "nuage" (Climent Peris et al., 1979), and cytoplasmic processes that extend to varying degrees when the cells move along a feeder cell substratum (Ando and Fujimoto, 1983; Kuwana and Fujimoto, 1984). Quail PGCs lack cytoplasmic glycogen granules but do have electron-dense and membrane-bound granules during stages of the sexually indifferent gonads. Additionally, they show a prominent nucleolus associated with condensed chromatin (heterochromatin) (Yoshinaga et al., 1993).

Unlike the PGCs in mammals, germ cells in birds and some reptiles (Gilbert, 1991) migrate to the gonads primarily by means of the embryonic bloodstream. From their location in the germinal crescent, PGCs incorporate passively or actively into the developing blood vascular system of the Stage 12–18 chicken (Singh and Meyer, 1967; Fujimoto et al., 1975, 1976a,b; Ando and Fujimoto, 1983; Kuwana and Fujimoto,

1984; Al-Thani and Simkiss, 1991; Tsai, 1995) and quail (Ginsburg et al., 1989; Yoshinaga et al., 1993) embryos. Most of the intravascular PGCs actively extravasate the blood vascular system in the region posterior to the omphalomesenteric arteries of the embryo at Stages 15 to 18 (Fujimoto et al., 1975; Ando and Fujimoto, 1983; Kuwana et al., 1987; Ukeshima et al., 1987). This area coincides with the dorsal mesentery through which the PGCs actively migrate to the developing gonads (Urven et al., 1988; Ukeshima et al., 1991). Although the germinal ridge is the sole migratory target, ectopic localizations of PGCs are recorded particularly in the head, heart, and trunk regions of the developing embryo (Nakamura et al., 1988, 1991a,b). The processes by which PGCs migrate to the gonads is still unclear. Two mutually synergistic mechanisms may direct their migration: chemotaxis (i.e., toward a concentration gradient) and contact guidance along the ECM secreted by the cells that line their migratory routes (Kuwana and Fujimoto, 1984; Fujimoto et al., 1985; Fujimoto and Yoshinaga, 1986; Kuwana et al., 1986; Urven et al., 1989; Ffrench-Constant et al., 1991; Kuwana, 1993). However, the specific mechanisms involved in avian PGC migration are not yet well defined. After colonization in the gonads the PGCs rapidly proliferate and differentiate into the spermatogonia or oogonia of the bird (Swift, 1914; van Limbrough, 1958; Meyer, 1964; England, 1983).

Methods to Produce Transgenic Poultry

Various methods are currently being developed to generate transgenic birds. These include direct microinjection of DNA into the newly fertilized egg and *in vitro* incorporation of DNA into PGCs selected from various accessible sites during their migration and then reintroducing these cells back into an appropriately staged recipient embryo. Each method has unique technical advantages and difficulties, and routine application of any one technology in birds has not yet demonstrated.

Direct Microinjection into the Fertilized Ovum

Direct microinjection of DNA into the male pronucleus of a newly fertilized avian ovum in a manner similar to that used to produce transgenic mice is hypothetically ideal but inherently problematic. By the time of oviposition the initial stages of early cleavage and zygotic formation have already occurred (Stages I through X–XI in the chicken [Eyal-Giladi and Kochav, 1976; Kochav et al., 1980] and Stages I–VII in the turkey [Gupta and Baskt, 1993]). However, these are the stages that are most critical to the effective microinjection of foreign DNA. By analogy to the work in mice (reviewed by Palmiter and Brinster, 1986; Wilmut et al., 1991) and some farm animals (Sims and First, 1994), these very early developmental stages contain "totipotent" cells with the faculty to individually or collectively develop to term in a recipient host. To experimentally manipulate a newly fertilized avian ovum at this same stage of development, one, out of necessity, has to gain access to the egg 1.5 to 4 hours after fertilization, some 20 hours prior to normal oviposition. Inducing premature ovulation (Hargrove and Ottinger, 1992; Gupta and Baskt, 1993; Watt et al., 1993), manipulating the newly fertilized ovum, and then reinserting it into a surrogate hen's oviduct for *in vivo* embryonic development and normal shell formation involves major surgery through a fistula (Wentworth, 1960; Pancer et al., 1989). Although this type of procedure is possible, it is hampered by the fragility of the large yolk mass and problems encountered during the surgery of the host hen. Recently, very early embryos, obtained from the anterior portion of the magnum and cultured *in vitro* in recipient eggshells, resulted in a 19.5% hatchability (Naito et al., 1995). Elegant techniques for injecting preoviposited ova obtained after euthanatizing the hen, identifying the male pronucleus or the fertilized ovum among the many supernumerary male pronuclei (Perry, 1987), and culturing an embryo prior to the first cleavage divisions, *ex vivo*, in surrogate eggshells after DNA microinjection (Rowlett and Simkiss, 1987; Perry, 1988; Naito et al., 1990; Ono et al., 1994a) did result in the production of a transgenic cockerel (Love et al., 1994a). This cock sired both normal and transgenic offspring. In cases of similar manipulation the introduced DNA was episomal (Sang and Perry, 1989; Naito et al., 1991a) and after a period of transient expression disappeared. The complexities of this system seem at present to prevent routine applications of these technologies.

Manipulation of Primordial Germ Cells in the Blastoderm

Evidence is accumulating that blastodermal (BD) cells derived from the Stage X embryo have begun a gradual commitment toward their cell fate. A further commitment to mesodermal, endodermal, and ectodermal cells occurs soon after incubation is initiated (Stage XIII) as a result of inductive influences among the cells within the blastoderm (Canning and Stern, 1988; Stern and Canning, 1990; Kuwana, 1993). Totipotency of any Stage X blastodermal cell has not been demonstrated; however, several researchers have unequivocally shown evidence for the presence of a limited number of identifiable primordial germ cells to be present in the freshly oviposited fertilized Stage X chicken egg (Eyal-Giladi et al., 1981; Eyal-Giladi, 1984; Ginsburg and Eyal-Giladi, 1986, 1987, 1989; Urven et al., 1988; Muniesa and Dominguez, 1990; Tsai, 1995). The data presented to date regarding blastodermal cell transfer are insufficient to answer the question of whether the somatic and occasional germ line chimerism is due to the random incorporation and differentiation of pluripotent cells or due to the selective incorporation of committed cells. Currently, the data indicate that the few committed PGCs in the Stage X-XI blastoderm are the only cells in the blastoderm capable of forming germ line chimera when transferred to host embryos. Consequently, the subsequent discussions concerning development of germ line transgenic birds will pertain to the manipulation of avian PGCs at the progressive stages along their migratory route, beginning with the PGCs in the blastoderm.

Retroviral Infection of PGCs in the Blastoderm Ten to fifteen years ago the potential for modification of the chick genome was proposed (Shuman, 1981; Shuman and Shoffner, 1982; Crittenden and Salter, 1985). One theory of introducing foreign DNA into the genome of a bird is through cells derived from the Stage X blastoderm of the freshly oviposited egg. Exogenous DNA is incorporated into the genome of the chicken by injecting a retrovirus vector containing the DNA of interest close to the blastoderm of the unincubated fertile egg (Salter et al., 1986). Since then, they and other investigators, using this technology with either replication-competent or replication-defective viral vectors, have provided evidence of transferring exogenous DNA to the progeny of infected birds (Shuman and Shoffner, 1986; Hughes et al., 1987; Salter et al., 1987; Salter and Crittenden, 1989; Bosselman et al., 1989a,b; Chen et al., 1990; Lee and Shuman, 1990). The most successful shows incorporation in 8% of the birds that hatched (Bosselman et al., 1989a,b). All four birds carried the exogenous gene in their semen and transmitted the vector to their progeny at a frequency of 2 to 8%. To date this is the most effective means to introduce foreign DNA into the avian species. Unfortunately, the use of retroviral vectors to produce commercial transgenic poultry is clouded because of the potential hazards inherent in the system. In the past, many of the injected embryos often became viremic. A sufficiently high titer for efficient transfection is often difficult to obtain. In addition, retroviruses direct a strong random integration into the host genome that conceivably can override any future attempts at homologous recombination in non-viremic birds. However, the vector system recently devised in which the env protein is supplied in *trans*, may circumvent some of these obstacles (Boerkoel et al., 1993). Without doubt, retroviral vectors will be valuable tools to study the control of gene expression and function in the avian species.

Non-Viral Manipulations of PGCs in the Blastoderm Direct microinjection of exogenous DNA into the Stage X blastoderm of a freshly oviposited egg recently demonstrated incorporation of this DNA into various somatic cells of the embryos tested after 1.5 to 5 days of incubation (Han et al., 1994c), as well as in embryonic gonads and hatched chicks (Naito et al., 1994a). The exogenous DNA, the lacZ gene under the CMV early promotor, was detected in the blood cells and sperm for more than 1 year after hatching in two of the injected chickens. The results suggested to these authors that the injected DNA ligated to form concatamers after microinjection and then was maintained in a relatively small number of cells during the early stages of development with no gross structural rearrangements. To date, this DNA has not yet been detected in the progeny of these males (Naito et al., 1994a).

The technology of transferring BD cells into similarly staged recipient embryos has been extensively exploited as a feasible approach toward the development of transgenic poultry (reviewed in Etches et al., 1993a,b; Etches and Verrinder Gibbins, 1994). Intra- and interspecies somatic and, in some cases, germ line, chimera can be obtained simply by transferring freshly isolated BD cells from one embryo to another (Marzullo, 1970; Petitte et al., 1990; Naito et al., 1991b; Shaw et al., 1992; Watanabe et al., 1992; Ono et al., 1994b). The extent of somatic chimerism is not correlated with the extent of germ line chimerism. A major obstacle yet to be overcome in BD cell transfer is the difficulty encountered in hatching embryos that have been manipulated at this stage of development. Of the studies that incubated the BD cell recipients to hatching, 1 to 22% of embryos manipulated at the BD stage (Stage X) of development hatched (Petitte et al., 1990; Carsience et al., 1993; Ono et al., 1994b; Thoraval et al., 1994; Tsai, 1995). When the injection volume was reduced to 1 μL a substantial increase in hatchability of quail-chick feather chimera was observed (40%); however, no germ line chimera resulted (Naito et al., 1991b). Apparently under these circumstances an insufficient number of determined PGCs are available in the blastodermal cell population to transfer to recipient embryos. From these data one might postulate that all BD cells derived from the Stage X embryo have begun a gradual commitment. The cells that have descended to the hypoblast are the committed PGCs, and only these cells have the facility to migrate to, colonize the gonads and differentiate into the gametes of the bird. Consequently, large numbers of embryos have to be manipulated and an even larger number of progeny have to be screened for germ line incorporation. Increased efficiency of somatic or germ line chimerism occurs when development of the recipient embryos is compromised by exposure to low doses of γ irradiation (Carsience et al., 1993; Thoraval et al., 1994). Initial studies (Brazolot et al., 1991; Fraser et al., 1993) directed toward transfer of exogenous DNA into blastodermal cells in suspension or cultured overnight showed a 30% transient transfection efficiency using lipofectin transfection procedures. When injected into host stage X blastoderms the transfected cells were observed 48 hours later to locate both in the embryo proper and in the extraembryonic tissues. Recently, relatively pure populations of early migratory murine PGCs have been isolated by using a murine PGC-specific antibody in combination with immunoaffinity adhesion (De Felici and Pesce, 1995). Advances such as this will pave the way for more successful *in vitro* culture, genetic manipulation, and transfer to recipient embryos of pure PGC populations.

Culture of BD Cells and Development of Chicken Embryonic Blastodermal Cell Strains Targeted mutations in specific genes by homologous recombination in selected, transfected embryonic stem (ES) cells of the laboratory mouse produce transgenic mouse lines (Smithies et al., 1985; Thomas and Capecchi, 1987; Capecchi, 1989,a,b; Mansour et al., 1990; Valancius and Smithies, 1991; Deng and Capecchi, 1992; Thomas et al., 1992; Deng et al., 1993). However, ES cell lines that pass through the germ line of mice consistently and with relatively high frequency (20 to 30%) are primarily limited to ES cell lines established from the mouse strain 129 and are markedly dependent on the host strain (reviewed in Smith, 1992a,b). This technique is not yet as effective in other mammals and to date no avian ES cell lines have been reported. Cultures of BD cells for up to 48 hours (Etches et al., 1993b) or for up to 7 days (Tsai and Wentworth, 1992) were inserted into recipient stage X chicken embryos. However, they produced only a limited number of somatic and fewer germ line chimera. If BD cells could be expanded in culture over several weeks, selection and development of blastodermal PGCs bearing specific genetic changes through homologous recombination would become feasible. Transferring these targeted cultured cells into recipient embryos could produce germ line chimera. Toward this end, Tsai (unpublished data) developed five chicken embryonic blastodermal cell strains from BD cell cultures of individual blastoderms. Since each strain is from a single blastoderm, it is of one sex. Karyotype analysis shows the strains to be either ZZ or ZW. Initially the primary cultures increased in BD cell and in PGC cell number to day 5 of culture (Figure 1). The cultures then remained "obligatory quiescent" for 3 months, at which time they began to proliferate, doubling every 72 hours. They can be cryopreserved at –70°C for 3 months (80% viability) and will resume the same growth pattern. Specific characterization of these strains has yet to be demonstrated; however, they seem to show a diversity of cell morphology as well as a gradual spontaneous differentiation. Some cells both in early and late passages produce melanin, demonstrating neural crest derivation. Other cells (<1%)in each of these strains react to the PGC-specific immunoreactive marker EMA-1 and could conceivably contribute to the germ line of recipient embryos. Selection and expansion of these cells in culture may provide the important link to the production of transgenic poultry. When 10^5 cells are placed into a 50-μL hanging drop culture they quickly aggregate (24 hours) into as-yet-undefined colonies. When these individual colonies are placed into a flask *in toto*, they plate down and then demonstrate a rapid proliferation of cell outgrowth in 24 hours. Medium from passages 7 and 25–33 and medium from cultures of BRL cells in an 18% reducing SDS-PAGE show a similar pattern of migrating proteins (Figure 2). This is the first reported instance of a continuous culture (currently

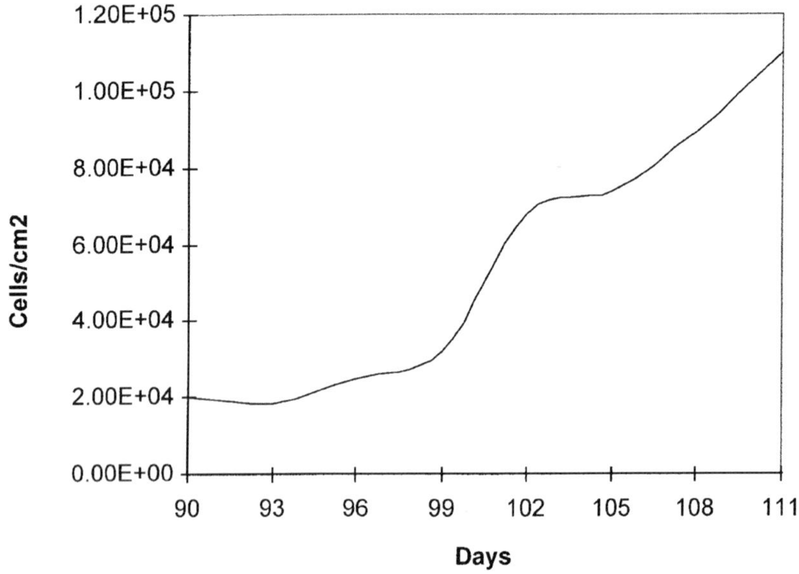

Figure 1. Cumulative growth curve of blastodermal cells in culture. After 3 months of initial delayed growth, cells show doubling every 72 hours for 50 passages (data shown are for 23 passages).

to 50 passages) in the absence of feeder cells of any avian blastodermal cell strain. This seminal advance in chicken blastodermal cell culture should be extremely useful in studies of the control of avian cellular determination and differentiation.

Production of Germ Line Chimera by Transfer of PGCs from the Germinal Crescent and Blood

An increasing number of intra- and interspecies transfers of premigratory/migratory PGCs, sequestered from the germinal crescent (GC) of Stage 7 embryos, or from blood of Stage 15 to 17 embryos, are resulting in germ cell chimeric gonads (Simkiss et al., 1989; Guido et al., 1991; Nakamura et al., 1991b, 1992; Yasuda et al., 1992; Ono and Yokoi, 1994) or hatchlings (Gonzales, 1989; Wentworth et al., 1989; Simkiss and Li, 1992; Tajima et al., 1993; Vick et al., 1993b; Naito et al., 1994b,c). The original studies established methods to enrich for PGCs in cell suspensions using a delayed plate-down procedure in an overnight culture (Shuman, 1981). Subsequently, suspensions of GC cells from a strain of chickens carrying the homozygous chromosome marker [MNt(1;4)(q22;q23)] were enriched for PGCs at a Percoll density of 1.032 to 1.047 (Gonzales, 1989). These cells were injected into the vitelline vein of Stage 15–17 host embryos. Viable chicks were raised to sexual maturity and the males were mated with normal hens. Out of six potential germ line chimeric males assayed, two were documented to have transferred the chromosomal marker, one of which gave a 1.5% incidence of transfer to its progeny. A significant advantage in the use of PGCs derived from the GC over that of the BD cells is that a larger number of PGCs, 150 to 250, can be obtained per embryo than in the Stage X BD (4 to 27 per embryo; Muniesa and Dominguez, 1990).

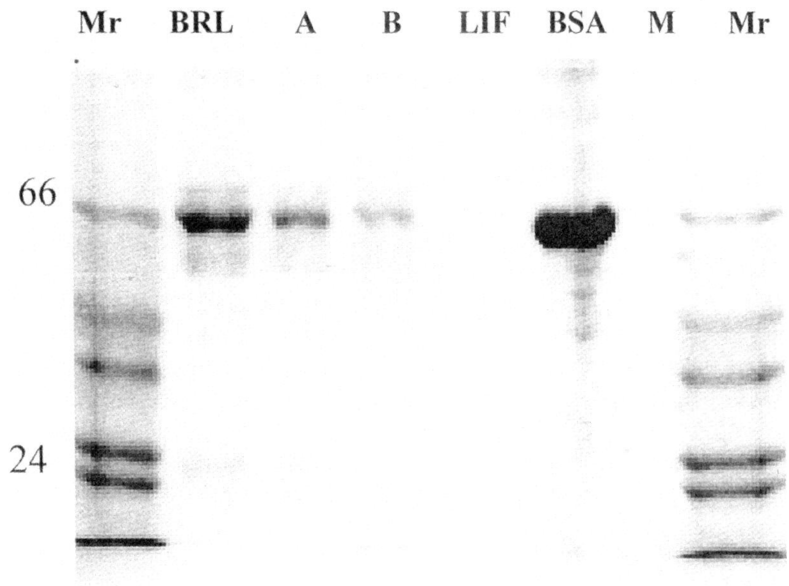

Figure 2. SDS-PAGE (18% reducing) of media from two individual chicken blastodermal cell strains (CBCS) compared with media from BRL cell line. Coomassie Blue stain. Lanes 1 and 8 = M_r protein standards; Lane 2 = Buffalo Rat Liver Cell Line (BRL); Lane 3 = Strain A; Lane 4 = Strain B; Lane 5 = Media + rLIF (quantity loaded below limits of detection); Lane 6 = Media + BSA (size marker); Lane 7 = Media alone. Pattern of protein migration for CBCS A and B is similar to that of BRL cells.

Production of Germ Line Chimera by Transfer of PGCs from the GC or Blood Carrying Exogenous DNA

Using a PGC separation procedure by lack of attachment in culture, PGC-enriched germinal crescent cells were transfected in cell suspension with Lipofectin® and injected into GC of host embryos (Han et al., 1994a,b). One embryo out of a total of 59 embryos and hatchlings that survived tested positive for the lacZ gene in the gonad(s). Other tissues from these same embryos were not tested. More recently, ballistic transfection of GC cells with the LacZ reporter gene were injected into the blood vascular system of recipient embryos (Li et al., 1995). When these embryos hatched they successfully produced cockerels carrying the foreign DNA in their sperm. Interestingly, offspring (F_1) that were initially positive lost the foreign DNA as they matured.

Although fewer PGCs per embryo can be obtained from the blood of a Stage 15 to 17 embryo, approximately 10 to 40 PGCs/10 µL (Al-Thani and Simkiss, 1991), a pure population of PGCs in the migratory phase, can be isolated using Ficoll separation procedures (Chang et al., 1992; Yasuda et al., 1992; Tajima et al., 1993). This provides an attractive avenue to obtain a relatively pure cell population of PGCs for producing germ line chimera. To acquire a sufficiently large number of PGCs from the blood of appropriately staged embryos and inject them into an appropriately staged recipient is not trivial (Petitte et al., 1991). However, using Ficoll separation of PGCs from the blood of stage 13–15 chicken embryos, and then selecting individual PGCs for injection into similarly staged host embryos, chimeric chickens were produced that transferred the donor-derived germ cells to their progeny at a frequency of up to 12% (Tajima et al., 1993). A much higher rate of transmission (23% to 95%) was obtained when blood containing endogenous PGCs was removed from the host embryo prior to the transfer of donor PGCs (Naito et al., 1994c). To genetically manipulate these PGCs, some studies used retroviral vectors to introduce the lacZ gene into cell suspensions obtained from either the germinal crescent region (Savva et al., 1991; Simkiss et al., 1992) or blood (Vick et al., 1993a). They then injected these cells into the blood vascular system of Stage 17 recipient embryos. Incorporation of the transfected cells into the recipients' gonad(s) was detected after 18 days of incubation by PCR amplification of a fragment of the lacZ gene with subsequent hybridization to a ^{32}P probe specific for the amplified fragment using Dot or Southern blots. Cell suspensions containing PGCs from GC demonstrate a higher efficiency of incorporation, 23%, than the PGC suspensions from blood samples, 3%. Of five hatchlings, two cocks produced sperm containing vector DNA. One male occasionally transmitted the foreign DNA to its progeny, as demonstrated by phenotypic color expression and DNA analysis (Vick et al., 1993a). Finally, intravascular injection of a liposome:DNA mixture to Stage 10–15 chick embryos enabled introduction of exogenous DNA (lacZ) into PGCs of these embryos. A limited number of PGCs in 9% to 41% of the gonads from the injected embryos demonstrated β-gal expression at Stage 24–27 (Watanabe et al., 1994).

The sex effect of donor-derived cells on the phenotypic and functional sex of the recipient embryo as well as the reciprocal relationship was suggested to affect blastodermal cell transfers (Shaw et al., 1992). The genetic sex of germ cells seems to have no role in the phenotypic sexual differentiation of the gonadal rudiments, although entrance into the gonads by the germ cells may be required to initiate their differentiation (Haffen, 1975; McCarrey and Abbott, 1978, 1979, 1982; Hajji et al., 1988). Both sexes of quail PGCs were reported to colonize the gonad(s) of both sexes of recipient chicken embryos (Ono and Yokoi, 1994), yet the donor PGCs cannot differentiate into functional gametes in the recipient gonads if the sexes of the donor PGCs and recipient embryos are different (Naito et al., 1994c). These data are of valuable significance to selectively produce progeny that is of predominantly one sex.

Manipulation and Transfer of PGCs from the Embryonic Gonads to Produce Transgenic Birds

Another potential source of germ cells for genetic manipulation and transfer into recipient embryos are the PGCs that have entered the gonadal primordia, gPGCs. Gonadal PGCs offer several advantages over the other three sources, BD, GC, and blood. For example, the indifferent gonads provide an easily accessible, abundant supply of undifferentiated gonadal germ "stem" cells for manipulation, 1,000 to 2,000 per gonad at Stages 26–30 (Wentworth and Wentworth, 1991). The germ cell:somatic cell proportion (1:10) is considerably higher than in either BD (approx. 1:2000), GC (125 to 250 per embryo), or blood (10 to 40 PGCs per embryo sampling)(Simkiss and Li, 1992). Gonadal PGCs can be expanded in culture when co-cultured along with a gonadally derived somatic cell feeder layer (Wentworth and

Wentworth, 1991), or with the addition of SCF, LIF, and bFGF (Allioli et al., 1993, 1994). Foreign DNA can be transfected into these cultured cells, which can then be transferred into a host embryo (Wentworth et al., 1992; Allioli et al., 1993, 1994). These properties make gPGCs attractive candidates as vehicles for genetic manipulation and germ line transmission of foreign DNA in the avian species. Pertinent unanswered questions regarding this approach to the production of transgenic poultry involve the inherent biology of the PGC. Will gPGCs maintained in culture for *in vitro* incorporation of exogenous DNA migrate to the recipient's gonadal primordia after injection into the blood vascular system of an appropriately staged host embryo? Data from studies on the mouse indicate that during migration in the dorsal mesentery, PGCs extend long processes with which they link up to each other and form extensive networks. They subsequently aggregate into clusters and lose their processes as well as their motility. When cultured, these cells perform the same sequences (Gomperts et al., 1994). *In vitro* cultures of germ cells isolated from gonads of 12.5 to 13.5 dpc murine embryos resume locomotory and invasive phenotype (De Felici et al., 1989). Similarly, gPGCs cultured from gonads of Stage 28–30 quail, chicken, and turkey initially exhibit random motility; however, after several days in culture, they extend processes to form a network of non-motile proliferating cells that seem to be associated with a layer of fibronectin-secreting gonadal somatic cells (A. Wentworth, unpublished data). These often appear as clusters of gPGCs. When subcultured, they initially regain motility (A. Wentworth, unpublished data); however, subculturing quail, chicken, or turkey gPGCs co-cultured with gonadal somatic cells is not successful beyond one or two passages. In these cases, a dramatic loss of gPGCs occurs concomitant with an increase in fibroblast cell proliferation. In sparsely seeded cell cultures gPGCs do not proliferate, although they may retain a chemokinetic (random) mobility for a longer period of time. Occasionally they are observed to move along the axis of the ECM-secreting somatic cells. Success of gPGC primary cell culture proliferation is unequivocally dependent on an initial seeding with a high concentration of gPGCs (A. Wentworth, unpublished data). In cultures of dispersed overtly male gonads (Stage 32), the authors did not observe histogenetic reaggregation of dissociated cells, as has been observed in cultures of mouse fetal gonads (Pesce et al., 1994). Neither has morphologically distinct Sertoli cell differentiation been observed. The expression of proto-oncogenes c-kit, c-myc, c-fos, and c-jun, which has been implicated in mediating proliferation and initial differentiation of murine PGCs (Coucouvanis and Jones, 1993; Urven et al., 1993; Papadopoulos and Dym, 1994), has not yet been tested in these avian PGC cultures.

A second unanswered question is whether gPGCs have a finite capacity to divide and propagate. This is of particular significance if they have already been proliferating for a time in culture. Apoptosis may play a useful biological role in controlling gonadal PGC proliferation *in vivo* (Pesce et al., 1993; Pesce and De Felici, 1994), especially in the female of the species. It may also be involved in eliminating ectopically placed PGCs (Shibuya et al., 1993). Data from mice indicate that prolonged cultures of murine gPGCs are capable of contributing to chimeric embryos (Matsui et al., 1992) and produce normal chimera with normal functional gametes derived from the injected cells (Labosky et al., 1994; Stewart et al., 1994). This possibility could exist in the avian species, also.

Formation of lacZ Chimeric Gonads in Recipient Turkey Embryos A preliminary study was designed to test the ability of gPGCs to transfer to a recipient host embryo and colonize in the gonads of that host. The general paradigm, injection, transmission,

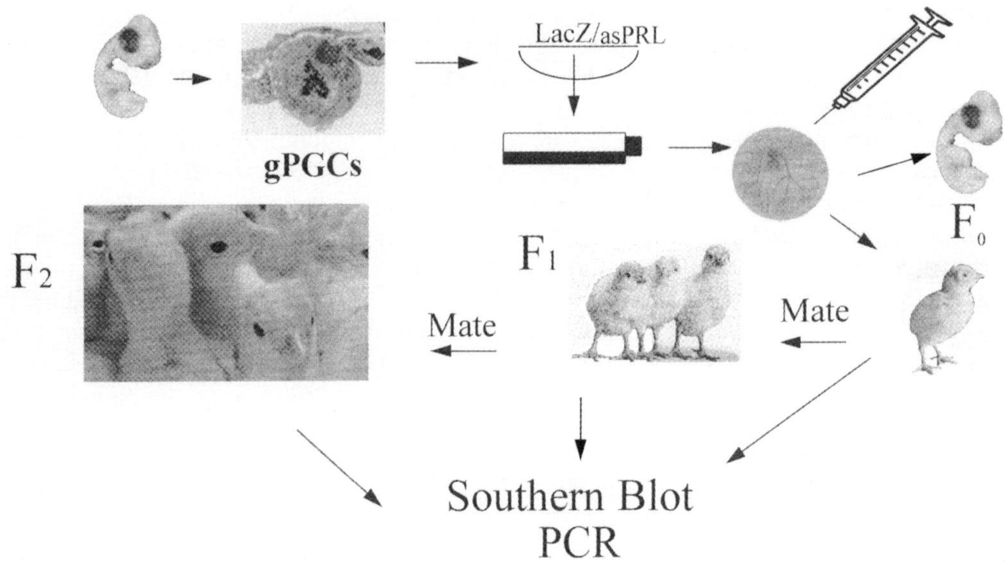

Figure 3. Paradigm for producing transgenic poultry. Gonadal PGCs from indifferent gonads of stage 28-30 turkey embryos are cultured and transfected with exogenous DNA (pSV40 β-gal or ASVLTR-asPRL). These cells are injected into the blood vascular system of Stage 15–17 recipient turkey embryos. Gonads and muscle are removed from these embryos at Stage 34 for testing for the presence of the foreign DNA in these tissues. Others are allowed to hatch. Hatched poults, putative gonadal chimera, are designated the founder line (F_0). These are mated. Progeny of these matings, F_1, are tested for the presence of the foreign DNA. Individuals testing positive for the asPRL gene using PCR/Southern blots are mated. Progeny of these matings (F_2) are similarly tested for the asPRL gene.

Table 2. Results from slot blot of pooled tissue samples from turkey embryos injected with lacZ transfected cultured gPGCs. F_0 and F_1 turkeys probed for the lacZ gene. F_0 is designated as the founder (injected) line; F_1 is the line derived from mating F_0 putative gonadal PGC chimera

Tissue	No. of samples	No. of pooled samples[a]	No. positive
F_0 Gonads	38	18	4
Controls	30	4	0
F_1 Gonads	650	65	6
F_1 Muscle	130	15	1
Controls	50	8	0
F_1 Semen		6	4

[a]Samples pooled by date.

and results of these procedures are outlined in Figure 3. Primary cultures of gPGCs from indifferent gonads of the stage 28–30 quail or turkey embryo were transfected with pSV-40 β-gal during their log phase of growth in culture using Lipofectin®-mediated procedures. The gPGCs were co-cultured with undifferentiated gonadal somatic cells and were transfected after 4 to 8 days of culture. Following 3 to 4 days of recovery, 500 cells in 5 to 7 μL of PBS were injected into the vitelline vein of Stage 15-17 recipient quail (52-hour incubation) or turkey (90- to 94-hour incubation) embryos. After 12 days (quail) or 17 days (turkey) of incubation samples were removed from some of the embryos for gonad, muscle, and liver analysis, and others were allowed to hatch. The injected embryos were designated as the founder line, F_0. The hatchlings were raised to maturity and mated. Progeny of these birds are designated as F_1. Tissue was removed from the F_1 to test for the presence of the lacZ sequence. To construct the probe, the β-gal coding region was cut out of the expression vector and subcloned into pBSSK+ (Strategene). This was linearized with *Hin*dIII and employed for preparation of the riboprobe using α ^{32}P UTP and T_3 polymerase. The probe detected the lacZ gene by slot blot analysis in 4 of the 18 pooled F_0 samples from gonads, one from the pooled muscle samples of the F_0 injected embryos and in seven of the pooled tissue samples from F_1 progeny (Table 2).

Introducing Antisense Prolactin DNA into the Genome of the Turkey A compelling practical application of transgenic technology in poultry is the concept of modulating the expression of the prolactin gene such that incubation behavior is prevented in the turkey multiplier hens. A strategy that can be used to down-regulate the expression of an endogenous gene is to transfer into the genome the antisense DNA, which interferes with the function of the normal gene. With respect to the prolactin gene in the turkey, this approach is ideal since complete abolition of normal prolactin gene expression is not the desired end point. Prevention of the sudden 10- to 100-fold increase in prolactin mRNA and protein (Wong et al., 1991) that is correlated with incubation behavior is the desired end point. If the transgene is driven by the promoter sequences of its endogenous counterpart, then expression is limited to those sites in which it can reduce the amount of targeted gene product. In addition to the commercially important aspects of this technology, prolactin antisense gene transfer can substantially increase our understanding of key biological events that regulate prolactin gene expression in the turkey.

Several critical steps are of particular significance to perfecting the technology for gene transfer, genomic incorporation, and expression of the prolactin antisense gene. An antisense prolactin DNA construct with the prolactin gene promotor/enhancer sequences is necessary such that tissue-specific expression of the antisense prolactin gene is endogenously regulated at the same time as its counterpart, the normal prolactin gene, is regulated. A mechanism is needed to incorporate this construct into the germ line of the turkey so that it will be passed onto the progeny. A valid assay is warranted to determine the percentage of transmission of the antisense gene from the germ line chimeric turkeys to their progeny. Finally, the effective expression of the normal and the antisense genes in the "transgenic" turkeys has to be evaluated in relation to the fecundity and overall well-being of these birds. Toward this end the following published and preliminary unpublished data are discussed.

Wong et al. (1991) cloned the turkey prolactin gene from which was constructed the antisense prolactin sequence. The construct, ASVLTR-asPRL, expresses antisense turkey prolactin RNA under the control of the avian sarcoma virus long terminal repeat (ASV-LTR) promoter/enhancer. This plasmid (ASVLTR-asPRL) was designed to direct constitutive synthesis of an antisense RNA that is complementary to 28 nucleotides of the 5'-untranslated sequence and the nucleotides that encode the first 12 amino acids of the signal peptide in the sense strand.

Effect of asPRL in Primary Anterior Pituitary Cell Cultures Primary anterior pituitary cells were isolated from turkey hens and transfected with plasmid ASVLTR-asPRL or a linear "antisense" oligodeoxynucleotide (ODN) using Lipofectamine®. The ODN was 15 nucleotides long and was complementary to three nucleotides of the 5'-untranslated sequence, the ATG start codon, and the next four codons. A control "sense" ODN covering the same region was also synthesized. When challenged with vasoactive intestinal peptide (VIP), a hypothalamic stimulator of prolactin synthesis/secretion (El Halawani et al., 1992, 1995; El Halawani and Rozenboim, 1993; Pitts et al., 1994), prolactin secretion was significantly ($P < .01$) reduced in both the antisense ODN and the ASVLTR-asPRL transfected cells compared with the controls (Sun et al., 1995). Since prolactin antisense ODNs and expression vectors transiently repress prolactin synthesis/secretion in VIP-challenged primary turkey pituitary cells, one might concur that a prolactin "antisense" strategy could effectively modulate the expression of the prolactin gene in the turkey.

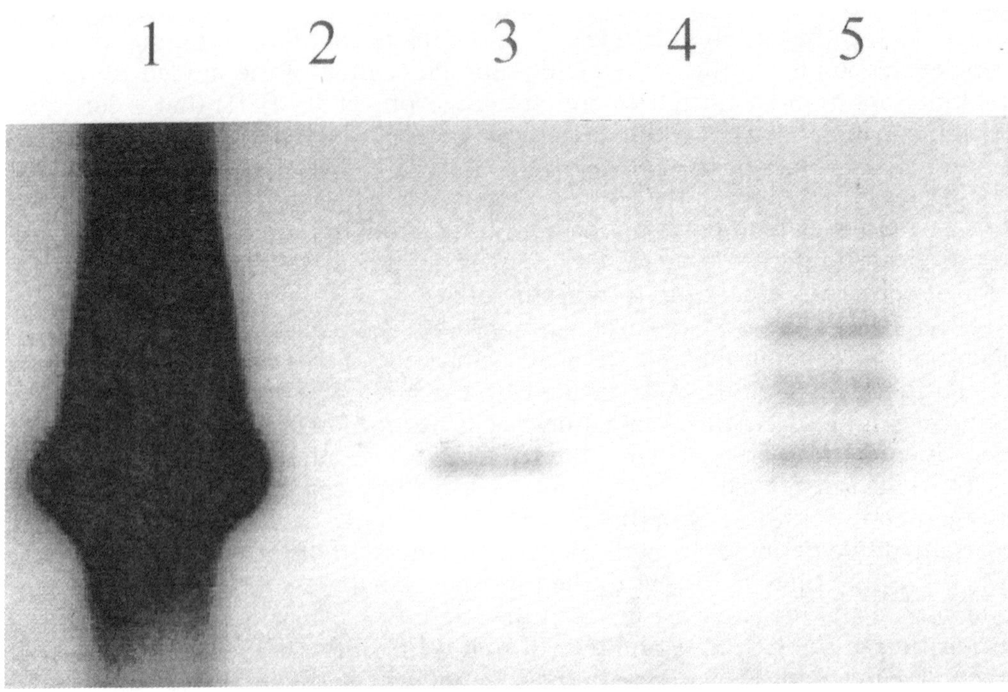

Figure 4. Southern blot of cultured gPGCs transfected with asPRL. Lane 1 = asPRL plasmid; Lane 2 = plasmid without asPRL insert; Lane 3 = transfected gPGCs; Lane 4 = untransfected gPGCs; Lane 5 = transfected gPGCs (*Eco*RI) digest.

Incorporation of asPRL into the Genome of the Turkey The asPRL construct was transfected into gPGCs enriched primary gonadal cell cultures. Southern analysis of the transfected cells in culture indicated that the asPRL construct had incorporated into the cultured cells (Figure 4). The founder line, F_0 injected embryos and hatchlings, should be gonadal chimera (i.e., contain a mix of transfected and non-transfected germ cells in the gonad[s]). Gonads, muscle, and liver were removed from some of the injected F_0 birds, and others were allowed to hatch. Hatchlings were raised to maturity and mated. Muscle tissues were removed from some of the F_1 embryos for analysis and the remaining embryos were allowed to hatch. Tissues collected were pooled in random groups of 3 to 12 samples. Blood was collected from the F_1 hatchlings for analysis. DNA was extracted from the pooled muscle and individual blood samples and analyzed for the presence of the ASVLTR-asPRL construct by polymerase chain reaction (PCR). The primers were complementary to adjoining sequences of the ASV-LTR and the asPRL gene construct and should generate a fragment of 190 bp. Verification of the PCR product was determined by Southern blot. In Trial I (1993), three of the pools of F_0 gonadal tissue tested positive for the ASVLTR-asPRL transgene by PCR. Because the PGCs target primarily to the developing gonads in F_0 birds, the transgene should be detected mainly in gonadal tissue. Examination of pools of liver and muscle tissue from the same F_0 birds revealed a positive signal in 2 of these 18 pools. Approximately 5% of the pooled muscle tissues from embryonic F_1s were positive for the transgene in Trial I (1993) and 50% in Trial II (1994). If the gonadal tissues of F_0 birds are chimeric (i.e., consist of both transgenic and non-transgenic germ cells), then only a small percentage of the F_1 birds would contain the transgene. The F_1 birds that are positive should carry the antisense transgene in every cell. The analyses of individual blood samples from adult living F_1 birds from Trial I indicated a 14.5% incidence for the presence of the transgene (Table 3). Particularly intriguing in the data summarized from Trial II (1994) is the variable incidence of the transgene in the muscle of the F_1 embryos with time (Table 4). The authors hypothesize that this is an effect of the presence of host endogenous primordial germ cells.

Table 3. Results from Trial I (1993) and Trial II (1994) of tissue analysis for asPRL in F_0 (injected with cultured and asPRL transfected gPGCs) and F_1 (progeny of mated F_0) turkeys. F_0 is designated as the founder line; F_1 is the line derived from mating putative F_0 gonadal PGC turkey chimera

Tissue	No. of samples	No. of pooled samples	No. positive
	1993 (Trial I)		
F_0 Gonads	98	15	3 (20%)
F_0 Muscle	98	18	2 (11.1%)
F_1 Muscle	233	80	4 (5%)
F_1 Blood	200		29 (14.5%)
	1994 (Trial II)		
F_1 Muscle	292	72	36 (50%)

Table 4. Antisense prolactin DNA (Trial II) in embryonic muscle samples from F_1 progeny of matings with pooled semen from putative F_0 gonadal PGC chimera. Samples pooled by dam weekly (A-I = 9 weeks). Subjective evaluations of autoradiograms from asPRL PCR analyses

	Percentage positive (subjective evaluation)				
Set ID (weekly)	Strong Positive +++	++	+	+/–	Negative –
A		25.0	37.5	37.5	0
B		37.5	37.5	12.5	12.5
C		12.5	50.0	12.5	25.0
D	12.5	12.5	37.5	37.5	0
F			25.0	12.5	62.5
G			12.5		87.5
H					100
I			25.0	25.0	50.0

Table 5. Summary of embryo injections with gPGCs from various donor embryos. Percentage of hatch, somatic chimera, germ line chimera, and transmission to progeny are reported when applicable

Author	No. of embryos injected	No. of cells injected	No. survive (age)	Tissues positive (embryo age)	Hatch (% of remaining)	Trans-mission to progeny
Wentworth 1992 (lacZ)	36 (T)	500 mix culture T gPGC (lacZ)	28 (17 d)	Yes	15 (42%)	Yes
Wentworth 1992 (lacZ)	10 (Q)	500 mix culture Q gPGC (lacZ)	8 (10 d)	Yes	4 (40%)	Yes
Wentworth asPRL Trial I 1993	168 LWT (+330)	500 mix culture LWTgPGC (asPRL)	139 (17 d) 83%	59 (17 d) pooled 3/18 GM	59 (73.8%)	Yes (F_1) 14.5% Individual
Wentworth Trial II 1994 asPRL	82 LWT (+402)	500–800 mix culture Bronze gPGCs (asPRL)	NA	NA	16/33 (48% of transferred)	Yes F_1, 50% (Pooled embryonic muscle samples)

[a]Q = quail; C = chicken; LWT = large white turkey; Bronze = bronze turkey; M = muscle; G = gonad; LWTgPGC = cultured gonadal PGC from embryonic gonads of large white turkeys; bronze gPGC = cultured gonadal PGC from embryonic gonads of bronze turkeys; lacZ = cultured gPGCs transfected with lacZ gene; asPRL cultured gPGCs transfected with antisense prolactin gene; NA = not analyzed.

The detection of the antisense transgene in gonadal tissue and semen of F_0 birds, and in muscle and blood tissue of F_1 birds, indicates that the transgene can be passed on to subsequent generations. The finding of asPRL in some of the muscle and liver tissues of the founder line, F_0, embryos indicates that 1) transfer of some of the co-cultured, transfected somatic cells could have resided in these tissues; 2) the transferred gPGCs migrated to ectopic sites, as occurs with endogenous and exogenous PGCs circulating in the blood (Nakamura et al., 1988, 1991a; Fry, 1992; Yasuda et al., 1992; Chaganti et al., 1994); or 3) cultured gPGCs could become totipotent and contribute to somatic cell lineages, as well as reentering the germ line (Stewart et al., 1994). This warrants further investigation.

Approximately 40 to 60% hatch rate was obtained from the founder, F_0, injected birds (Table 5). Taken together, these results indicate that transfection of gonadal primordial germ cells and subsequent injection of these cells into stage 15–17 turkey recipients seem to be worthy of further investigation as a non-viral method for gene transfer in avian species.

Transgenic Enhancement Through Production of Sterile Hosts

Early investigators conjectured that the efficiency of incorporating exogenously introduced germ cells into a recipient host gonad could be dramatically enhanced if one were to use a sterile host (i.e., a host lacking germ cells). The ideal sterile host is one produced without the use of invasive, potentially harmful chemicals. The authors produced sterile birds, "quickens," by crossing a rooster with a female quail (Wentworth et al., 1989). Successful invasion of the quicken embryonated egg with PGCs could substantially increase the efficacy of producing a transgenic model. To date, 37 embryonic quickens have been injected with quail gPGCs; however, none have survived to hatch. Other early methods to produce sterile hosts were founded upon irradiation (Reynaud, 1976) or ablation experiments in which germinal crescents were surgically excised (McCarrey and Abbott, 1978, 1982). These procedures often resulted in either a limited number of sterile hybrids, embryonic mortality, or teratomic embryos. The use of a chemosterilant, busulfan, which is cytotoxic to migrating germ cells, results in effective sterilization of avian embryos when injected into the albumen of unincubated quail eggs (Hallett and Wentworth, 1991). Busulfan administered directly onto the embryo or into the yolk prior to or during PGC migration from the GC region is also effective in sterilizing the host embryos (Aige-Gil and Simkiss, 1991). The observed abnormal embryonic formation seemed to be dose-related. By optimizing the dose of busulfan and site of injection, sterilizing embryos with busulfan prior to transferring exogenous PGCs increased the efficiency of gonadal germ cell chimerism from 4% to 14% or more (Vick et al., 1993b; Bresler et al., 1994).

Two other approaches are providing intriguing data to effectively sterilize embryos for subsequent use as recipients for exogenous germ cells. The EMA-1 antibody, which reacts specifically to chick PGCs (Urven et al., 1988), can be used to sterilize embryos (Tsai, 1995). Direct injection of EMA-1 antibody into the unincubated fertile egg results in complete absence of germ cells in gonads of these embryos after 6 days of incubation. Quail PGCs from quail germinal crescents can be injected into the blood vascular system of EMA-1 antibody-sterilized embryos and colonize the chick gonads (Tsai, 1995). The quail cells residing in the sterile chick gonads are identified by the quail characteristic heterochromatin as determined by the Feulgen-Rossenbeck reaction. Thus, quail PGCs from the germinal crescent are capable of repopulating EMA-1 antibody-sterilized gonads of the chick. However, this procedure results in a high mortality rate of the injected embryos. A second innovative procedure that is also successful in producing a sterile embryo is by immunizing hens with a 500 kDa protein isolated from Nulli SCC1 cells derived from a mouse teratocarcinoma (Tsai, 1995). These cells contain a PGC surface marker epitope that can be identified by the EMA-1 antibody (Hahnel and Eddy, 1986; Urven et al., 1988). Circulating antibodies were present after the second immunization. Fertile eggs were collected from these hens and incubated. Embryonic gonads removed from 7-day-old embryos show no immunohistochemically detectable PGCs.

Cryopreservation of the Genome

Cryopreservation of chicken semen is a procedure that will provide some fertility and is one route for species preservation. However, techniques for semen preservation in other avian species are not as well established. Additionally, preservation of the female germ line has not yet been possible. Initial reports demonstrate that cryopreservation of BD cells (Naito et al., 1992) and subsequent transfer to BD of recipient embryos result in somatic but not germ line chimera (Naito et al., 1992; Petitte et al., 1993). However, cryopreserved PGCs isolated from the blood of donor chick embryos can be transferred to the blood vascular system of recipient chick embryos, F_0, and give rise to viable offspring, F_1 (Naito et al., 1994b). This is the first indication of the ability to preserve the female of a chicken line. Cryopreservation of BD cell strains (Tsai, 1995) may likewise provide a means for genome preservation. The technology for preservation of genetic material in the avian species is now tangible.

Summary

The reported preliminary data taken together with data compiled from other laboratories demonstrate attractive strategies toward the production of transgenic poultry and preservation of the avian genome. Avian primordial germ cells provide a vehicle for innovative approaches to genomic manipulation. Useful sources of PGCs can be obtained from the Stage X blastoderm at oviposition, Stage 7–10 germinal crescent, Stage 15 embryonic blood, and Stage 28 embryonic gonads. These sources of PGCs can be transfected with exogenous DNA using retroviruses or other facilitating agents such as ballistic insertion or Lipofectin®/Lipofectamine®. Reports of somatic and in some cases genomic chimera result when using PGCs from the embryonic blastoderm, germinal crescent, blood, and gonads as vectors for new DNA insertion. The embryonic gonadal primordial germ cells (gPGCs) are shown in a lacZ model and prolactin antisense (asPRL) application to incorporate antisense prolactin DNA into the genome of the turkey, thereby potentiating the control of incubating behavior of turkeys. The asPRL transgene is transmitted at a rate of 14.5% to the F_1 poults. Future work with a PGC vector can be enhanced by the use of sterile hosts.

Cryopreservation of PGCs obtained from embryonic blood or blastodermal cell lines opens important avenues of investigation for genome preservation. The early vision of mechanisms to develop the various technologies necessary for these events is becoming a reality. A stable incorporation of foreign DNA into PGCs is crucial to the production of transgenic poultry. As the ability to screen for stably transfected PGCs to transfer to recipient embryos is increased, the efficiency of generating germ line chimera will also be manifested. Producing transgenics is only the first step in a complex series of events that will culminate in understanding the consequences of imposing novel genetic material on an intact, highly integrated living system. The level of expression of transgenes can be very dependent on their site of integration into the genome. In addition, their site of integration can markedly affect the expression of essential endogenous genes. Practical and routine use of transgenic animals will depend substantially on time, effort, and money spent in investigating and validating the phenotypic consequences of gene transfer.

Acknowledgements

The authors wish to thank collaborators Mohammed El Halawani, John Proudman, Eric Wong for their contributions as part of a Regional Project S-213 "Reproductive Efficiency of Turkeys." Research from the authors' laboratory was supported by USDA-ARS Cooperative Agreement No. 58-3K47-0-009 and College of Agricultural and Life Sciences, Univ. of Wisconsin-Madison. The authors express sincere thanks to E. M. Eddy, NIH, Research Triangle, NC for graciously providing the EMA-1 antibody used in these studies.

Literature Cited

Aige-Gil, V., and K. Simkiss. 1991. Sterilization of avian embryo with busulphan. Res. Vet. Sci. 50:139.

Al-Thani, R., and K. Simkiss. 1991. Effects of an acute in vivo application of concanavalin A on the migration of avian primordial germ cells. Protoplasma 161:52.

Allioli, N., C. Legras, J. L. Thomas, Y. Chebloune, V. M. Nigon, and G. Verrider. 1993. Use of gonadal primordial germ cells (PGCs) as tool for gene transfer in chicken. In: Proc. 31st NIBB Conference, Japan (Abstr.).

Allioli, N., J. L. Thomas, Y. Chebloune, V. M. Nigon, G. Verdier, and C. Legras. 1994. Use of retroviral vectors to introduce and express the beta-galactosidase marker gene in cultured chicken primordial germ cells. Dev. Biol. 165:30.

Ando, Y., and T. Fujimoto. 1983. Ultrastructural evidence that chick primordial germ cells leave the blood-vascular system prior to migrating to the gonadal anlagen. Dev. Growth Differ. 25(4):345.

Boerkoel, C. F., M. J. Federspiel, D. W. Salter, W. Payne, L. B. Crittenden, H. J. Kung, and S. H. Hughes. 1993. A new defective retroviral vector system based on the Bryan strain of Rous sarcoma virus. Virology 195:669.

Bosselman, R. A., R. Y. Hsu, T. Boggs, S. Hu, J. Bruszewski, S. Ou, L. Kozar, F. Martin, M. Nicolson, J. A. Schultz, K. M. Semon, and R. G. Stewart. 1989a. Germline transmission of exogenous genes in the chicken. Science (Wash DC) 243:533.

Bosselman, R. A., R. Y. Hsu, T. Boggs, S. Hu, J. Bruszewski, S. Ou, L. Souza, L. Kozar, F. Martin, F. Green, F. W. Jacobsen, M. Nicholson, J. A. Schultz, K. M. Semon, W. Richell, and R. G. Stewart. 1989b. Replication-defective vectors or reticuloendotheliosis virus transduce exogenous genes into somatic stem cells of the unincubated chicken embryo. J. Virol. 63:2680.

Brazolot, C. L., J. N. Petitte, R. J. Etches, and A. Verrinder Gibbins. 1991. Efficient transfection of chicken cells by Lipofectin, and introduction of transfected blastodermal cells into the embryo. Mol. Rep. Dev. 30:304.

Bresler, M., J. Benham, G. Luke, and K. Simkiss. 1994. Manipulations of germ-cell populations in the gonad of the fowl. Br. Poult. Sci. 35:241.

Canning, D. R., and C. D. Stern. 1988. Changes in the expression of the carbohydrate epitope HNK-1 associated with mesodermal induction in the chick embryo. Development 104:643.

Capecchi, M. R. 1989a. Altering the genome by homologous recombination. Science, (Wash DC) 244:1288.

Capecchi, M. R. 1989b. The new mouse genetics: altering the genome by gene targeting. Trends in Genetics 5:70.

Carsience, R. S., M. E. Clark, A.M.V. Gibbins, and R. J. Etches. 1993. Germline chimeric chickens from dispersed donor blastodermal cells and compromised recipient embryos. Development 117:669.

Chaganti, R.S.K., E. Rodriguez, and S. Mathew. 1994. Origin of adult male mediastinal germ-cell tumours. Lancet 343:1130.

Chang, I. K., A. Tajima, Y. Yasuda, T. Chikamune, and T. Ohno. 1992. Simple method for isolation of primordial germ cells from chick embryos. Cell Biol. Int. Rep. 16:853.

Chen, H. Y., E. A. Garber, E. Mills, J. Smith, J. J. Kopchick, A. G. DiLella, and R. G. Smith. 1990. Vectors, promoters and expression of genes in chick embryos. In: W. Hansel and B. J. Weir (Ed.) Genetic Engineering in Animals. J. Reprod. Fertil. (Suppl.) 41:173.

Clawson, R. C., and L. V. Domm. 1969. Origin and early migration of primordial germ cells in the chick embryo: A study of the stages definitive streak through eight somites. Am. J. Anat. 145:87.

Climent Peris, S., V. J. Gotzens, and L. Dominguez. 1979. "Nuage" in the primordial germ cells of the chick embryo. Zentrabl. Veterinaermed. Reihe C. Anat. Histol. Embryol. 8:200.

Coucouvanis, E. C., and P. P. Jones. 1993. Changes in protooncogene expression correlated with general and sex-specific differentiation in murine primordial germ cells. Mech. Dev. 42:49.

Crittenden, L. B., and D. W. Salter. 1985. Genetic engineering to improve resistance to viral diseases of poultry: A model for application to livestock improvement. Can. J. Anim. Sci. 65:553.

De Felici, M., S. Dolci, and G. Siracusa. 1989. Fetal germ cells establish cell coupling with follicle cells in vitro. Cell Differ. Dev. 28:65.

De Felici, M., and M. Pesce. 1995. Immunoaffinity purification of migratory mouse primordial germ cells. Exp. Cell Res. 216:277.

Deng, C., and M. R. Capecchi. 1992. Reexamination of gene targeting frequency as a function of the extent of homology between the targeting vector and the target locus. Mol. and Cell. Biol. 12(8):3365.

Deng, C., L. R. Thomas, and M. R. Capecchi. 1993. Location of crossovers during gene targeting with insertion and replacement vectors. Mol. Cell. Biol. 13(4):2134.

El Halawani, M. E., and I. Rozenboim. 1993. The ontogeny and control of incubation behavior in turkeys. Poult. Sci. 72:906.

El Halawani, M. E., J. L. Silsby, T. I. Koike, and B. Robinzon. 1992. Evidence of a role for the turkey posterior pituitary in prolactin release. Gen. Comp. Endocrinol. 87(3):436.

El Halawani, M. E., J. L. Silsby, I. Rozenboim, and G. R. Pitts. 1995. Increased egg production by active immunization against vasoactive intestinal peptide in the turkey (Meleagris gallopavo). Biol. Reprod. 52:179.

England, M. 1983. The migration of primordial germ cells in avian embryos. In: A.C.C. McLaren (Ed.) Current Problems in Germ Cell Differentiation. p 91. Cambridge University Press, Cambridge, U.K.

Etches, R. J., R. S. Carsience, M. E. Clark, R. A. Fraser, A. Toner, D. L. Shaw, and A. Verrinder Gibbins. 1993a. Novel approaches to studies of avian endocrinology using transgenic chickens produced via chimeric intermediates. In: P. J. Sharp (Ed.) Avian Endocrinology. p 381. Burgess Science Press, Bristol, U.K.

Etches, R. J., R. S. Carsience, M. E. Clark, R. A. Fraser, A. Toner, and A. Verrinder Gibbins. 1993b. Chimeric chickens and their use in manipulation of the chicken genome. Poult. Sci. 72:882.

Etches, R. J., and A. Verrinder Gibbins. 1993. The anatomy of reproduction in birds, with emphasis on poultry. In: Manipulation of the Avian Genome. p 15. CRC Press, Boca Raton, FL.

Etches, R. J., and A. Verrinder Gibbins. 1994. Gene transfer: Overcoming the avian problems. Proc. 5th World Cong. Genet. Appl. Livest. Prod. 20:97.

Eyal-Giladi, H. 1984. The gradual establishment of cell commitments during the early stages of the chick development. Cell Differ. 14:245.

Eyal-Giladi, H., M. Ginsburg, and A. Farbarov. 1981. Avian primordial germ cells are of epiblastic origin. J. Embryol. Exp. Morphol. 65:139.

Eyal-Giladi, H., and S. Kochav. 1976. From cleavage to primitive streak formation: A complementary normal table and a new look at the first stages of the development of the chick. Dev. Biol. 49:321.

Ffrench-Constant, C., A. Hollingsworth, J. Heasman, and C. Wylie. 1991. Response to fibronectin of mouse primordial germ cells before, during and after migration. Development 113(4):1365.

Fraser, R. A., R. S. Carsience, M. E. Clark, R. J. Etches, and A.M.V. Gibbins. 1993. Efficient incorporation of transfected blastodermal cells into chimeric chicken embryos. Int. J. Dev. Biol. 37:381.

Fry, R. C. 1992. The effect of leukaemia inhibitory factor (LIF) on embryogenesis. Reprod. Fertil. Dev. 4:449.

Fujimoto, T., T. Menomiya, and A. Ukeshima. 1975. Light and electron microscopic studies on origin and migration of the primordial germ cells in the chick. Acta Anat. Nippon. 50:22.

Fujimoto, T., T. Ninomiya, and A. Ukeshima. 1976a. Observations of the primordial germ cells in blood samples from the chick embryo. Dev. Biol. 49:278.

Fujimoto, T., A. Ukeshima, and R. Kiyofujii. 1976b. The origin, migration, and morphology of the primordial germ cells in the chick embryo. Anat. Rec. 185:139.

Fujimoto, T., and K. Yoshinaga. 1986. The role of fibronectin in the interstitial migration of primordial germ cells in amniotes. Congen. Anom. 26:187.

Fujimoto, T., K. Yoshinaga, and I. Kono. 1985. Distribution of fibronectin on the migratory pathway of primordial germ cells in mice. Anat. Rec. 211:271.

Gilbert, S. F. 1991. Developmental Biology (3rd Ed.). Sinauer Associates, Sunderland, U.K.

Ginsburg, M., and H. Eyal-Giladi. 1989. Primordial germ cell development in culture of dispersed central disks of stage X chick blastoderms. Gamete Res. 23:421.

Ginsburg, M., and H. Eyal-Giladi. 1987. Primordial germ cells of the young chick blastoderm originate from the central zone of the area pellucida irrespective of the embryo-forming process. Development 101:209.

Ginsburg, M., and H. Eyal-Giladi. 1986. Temporal and spatial aspects of the gradual migration of primordial germ cells from the epiblast into the germinal crescent in the avian embryo. J. Embryol. Exp. Morphol. 95:53.

Ginsburg, M., J. Hochman, and H. Eyal-Giladi. 1989. Immunohistochemical analysis of the segregation process of the quail germ cell lineage. Int. J. Dev. Biol. 33:389.

Gomperts, M., M. Garcia Castro, C. Wylie, and J. Heasman. 1994. Interactions between primordial germ cells play a role in their migration in mouse embryos. Development 120:135.

Gonzales, D. S. 1989. Gonadal germ cell chimeras produced by microinjecting primordial germ cells into the vasculature of early chicken embryos. M.S. Thesis. Univ. of Wisconsin, Madison.

Guido, T. C., U. K. Abbott, and J. R. McCarrey. 1991. The interspecific transfer of avian primordial germ cells. Manipulation of the Avian Genome, Symposium. Cl 113:204 (Abstr.).

Gupta, S. K., and M. R. Baskt. 1993. Turkey embryo staging from cleavage through hypoblast formation. J. Morphol. 217:313.

Haffen, K. 1975. Sex differentiation of avian gonads in vitro. Am. Zool. 15:257.

Hahnel, A. C., and E. M. Eddy. 1986. Cell surface markers of mouse primordial germ cells defined by two monoclonal antibodies. Gamete Res. 15:25.

Hajji, K., C. Martin, A. Perramon, and F. Dieterlen-Lievre. 1988. Sexual phenotype of avian chimeric gonads with germinal and stromal cells of opposite genetic sexes. Biol. Struct. Morphogen. 1(3):107.

Hallett, J. S., and B. C. Wentworth. 1991. The effect of busulfan on gonadal differentiation in Japanese quail (*Corturnix corturnix japonica*). Poult. Sci. 70:1619.

Hamburger, V., and H. L. Hamilton. 1951. A series of normal stages in the development of the chick embryo. J. Morphol. 88:49.

Han, J. Y., D. S. Seo, H. Y. Hong, R. N. Shoffner, and K. S. Guise. 1994a. Expression of transfected primordial germ cells in the chicken gonad and germline chimeric chicken. In: Proc. 31st Nibb Conf. p 108 (Abstr.).

Han, J. Y., R. N. Shoffner, and K. S. Guise. 1994b. Gene transfer by manipulation of primordial germ cells in the chicken. Asian-Australasian J. Anim. Sci. 7:427.

Han, J. Y., R. N. Shoffner, and K. S. Guise. 1994c. Microinjection and expression of marker gene in the early chicken embryo. Korean J. Anim. Sci. 36:244.

Hargrove, T. L., and M. A. Ottinger. 1992. Induced oviposition of precalcified eggs following prostaglandin administration. Poult. Sci. 71:548.

Hughes, S. H., J. J. Greenhouse, C. J. Petropoulos, and P. Sutrave. 1987. Adaptor plasmids simplify the insertion of foreign DNA into helper-independent retroviral vectors. J. Virol. 61:3004.

Johnson, A. L. 1986. Reproduction in the Female. In: P. D. Sturkie (Ed.) Avian Physiology. Vol. 18. p 403. Springer-Verlag, New York.

Kochav, S., M. Ginsburg, and H. Eyal-Giladi. 1980. From cleavage to primitive streak formation: A complementary normal table and a new look at the first stages of development of the chick. II. Microscopic anatomy and cell population dynamics. Dev. Biol. 79:296.

Kuwana, T. 1993. Migration of avian primordial germ cells toward the gonadal anlage. Dev. Growth Differ. 35:237.

Kuwana, T., and T. Fujimoto. 1984. Locomotion and scanning electron microscopic observations of primordial germ cells from the embryonic chick blood *in vitro*. Anat. Rec. 209:337.

Kuwana, T., H. Maeda-Suga, and T. Fujimoto. 1986. Attraction of chick primordial germ cells by gonadal anlage in vitro. Anat. Rec. 215:403.

Kuwana, T., Y. Miyamama, Y. Kajiwara, and T. Fujimoto. 1987. Behavior of chick primordial germ cells moving toward gonadal primordium in vitro: Scanning electron microscopic study. Anat. Rec. 219:164.

Labosky, P. A., D. P. Barlow, and B.L.M. Hogan. 1994. Mouse embryonic germ (EG) cell lines: Transmission through the germline and differences in the methylation imprint of insulin-like growth factor 2 receptor (Igf2r) gene compared with embryonic stem (ES) cell lines. Development 120:3197.

Lee, M. R., and R. M. Shuman. 1990. Transgenic quail produced by retrovirus vector infection transmit and express a foreign marker gene. Proc. 4th World Congr. Genet. Appl. Livest. Prod. 16:107.

Li, Y., J. Behnam, and K. Simkiss. 1995. Ballistic transfection of avian primordial germ cell in ovo. Transgen. Res. 4:26.

Love, J., C. Gribbin, C. Mather, and H. Sang. 1994. Transgenic birds by DNA microinjection. Bio/Technology 12:2.

Mansour, S. L., K. R. Thomas, C. Deng, and M. R. Capecchi. 1990. Introduction of a lac Z reporter gene into the mouse *int-2* locus by homologous recombination. Proc. Natl. Acad. Sci. USA 87:7688.

Marzullo, G. 1970. Production of chick chimaeras. Nature (Lond.) 225:72.

Matsui, Y., K. M. Zsebo, and B.L.M. Hogan. 1992. Derivation of pluripotential embryonic stem cells from murine primordial germ cells in culture. Cell 70:841.

McCarrey, J. R., and U. K. Abbott. 1978. Chick gonad differentiation following excision of primordial germ cells. Dev. Biol. 66:256.

McCarrey, J. R., and U. K. Abbott. 1982. Functional differentiation of chick gonads following depletion of primordial germ cells. J. Embryol. Exp. Morphol. 68:161.

McCarrey, J. R., and U. K. Abbott. 1979. Mechanisms of genetic sex determination, gonadal sex differentiation, and germ cell development in animals. Adv. Genet. 20:217.

Meyer, D. B. 1960. Application of the periodic acid-schiff technique to whole chick embryos. Stain Technol. 35:83.

Meyer, D. B. 1964. The migration of primordial gem cells in the chick embryo. Dev. Biol. 10:154.

Muniesa, P., and L. Dominguez. 1990. A morphological study of primordial germ cells at pregastrular stages in the chick embryo. Cell Differ. Dev. 31:105.

Naito, M., K. Agata, K. Otsuka, K. Kino, M. Ohta, K. Hirose, M. M. Perry, and G. Eguchi. 1991a. Embryonic expression of beta-actin-lacZ hybrid gene injected into the fertilised ovum of the domestic fowl. Int. J. Dev. Biol. 35:69.

Naito, M., K. Nirasawa, and T. Oishi. 1995. An in vitro culture method for chick embryos obtained from the anterior portion of the magnum of oviduct. Br. Poult. Sci. 36:161.

Naito, M., K. Nirasawa, and T. Oishi. 1990. Development in culture of the chick embryo from fertilised ovum to hatching. J. Exp. Zool. 254:322.

Naito, M., K. Nirasawa, and T. Oishi. 1992. Preservation of quail blastoderm cells in liquid nitrogen. Br. Poult. Sci. 33:449.

Naito, M., E. Sasaki, M. Ohtaki, and M. Sakurai. 1994a. Introduction of exogenous DNA into somatic and germ cells of chickens by microinjection into the germinal disc of fertilized ova. Mol. Reprod. Dev. 37:167.

Naito, M., A. Tajima, T. Tagami, Y. Yasuda, and T. Kuwana. 1994b. Preservation of chick primordial germ cells in liquid nitrogen and subsequent production of viable offspring. J. Reprod. Fertil. 102:321.

Naito, M., A. Tajima, Y. Yasuda, and T. Kuwana. 1994c. Production of germline chimaeric chickens with high transmission rate of donor-derived gametes, produced by transfer of primordial germ cells. Mol. Reprod. Dev. 39:153.

Naito, M., D. Watanabe, M. Kinutani, K. Nirasawa, and T. Oishi. 1991b. Production of quail chick chimaeras by blastoderm cell transfer. Br. Poult. Sci. 32:79.

Nakamura, M., T. Kuwana, Y. Miyamama, and T. Fujimoto. 1988. Extragonadal distribution of primordial germ cells in the early chick embryo. Anat. Rec. 222:90.

Nakamura, M., T. Kuwana, Y. Miyayama, K. Yoshinaga, and T. Fujimoto. 1991a. Ectopic colonization of primordial germ cells in the chick embryo lacking the gonads. Anat. Rec. 229:109.

Nakamura, M., H. Maeda, and T. Fujimoto. 1991b. Behavior of chick primordial germ cells injected into the blood stream of quail embryos. Okajimas Folia Anat. Jpn. 67:473.

Nakamura, M., K. Yoshinaga, and T. Fujimoto. 1992. Histochemical identification and behavior of quail primordial germ cells injected into chick embryos by the intravascular route. J. Exp. Zool. 261(4):479.

Ono, T., T. Murakami, M. Mochii, L. Agata, K. Kino, K. Otsuka, M. Ohta, M. Mizitani, M. Yoshida, and G. Eguchi. 1994a. A complete culture system for avian transgenesis supporting quail embryos from the single-cell stage to hatching. Dev. Biol. 161:126.

Ono, T., S. I. Muto, M. Mizutani, K. Agata, M. Mochii, K. Kino, K. Otsuka, M. Ohta, M. Yoshida, and G. Eguchi. 1994b. Production of quail chimera by transfer of early blastodermal cells and its use for transgenesis. Jpn. Poult. Sci. 31:119.

Ono, T., and R. Yokoi. 1994. Transfer of male and female primordial germ cells of quail into chick embryonic gonad of the same and opposite sexes. Proc. 31st Nibb Conf. p 97 (Abstr.).

Palmiter, R. D., and R. L. Brinster. 1986. Germline transformation of mice. Annu. Rev. Genet. 20:465.

Pancer, Z., N. Snapir, B. Robinzon, A. Friedman, and H. Eyal-Giladi. 1989. Recovery of ova and their re-insertion into the hen's oviduct through a fistula. Br. Poult. Sci. 30:953.

Papadopoulos, V., and M. Dym. 1994. Sertoli cell differentiation on basement membrane is mediated by the c-fos protooncogene. Proc. Natl. Acad. Sci. USA 91:7027.

Pardanaud, L., C. Buck, and F. Dieterlen-Lievre. 1987. Early germ cell segregation and distribution in the quail blastodisc. Cell Differ. 22:47.

Perry, M. M. 1987. Nuclear events from fertilisation to the early cleavage stages in the domestic fowl. J. Anat. 150:99.

Perry, M. M. 1988. A complete culture system for the chick embryo. Nature (Lond.) 331:70.

Perry, M. M., and H. M. Sang. 1993. Transgenesis in chickens. Transgen. Res. 2:125.

Pesce, M., and M. De Felici. 1994. Apoptosis in mouse primordial germ cells: A study by transmission and scanning electron microscope. Anat. Embryol. 189:435.

Pesce, M., M. G. Farrace, M. Piacentini, S. Dolci, and M. De Felici. 1993. Stem cell factor and leukemia inhibitory factor promote primordial germ cell survival by suppressing programmed cell death (apoptosis). Development 118:1089.

Pesce, M., G. Siracusa, Q. Giustiniani, and M. De Felici. 1994. Histotypic in vitro reorganization of dissociated cells from mouse fetal gonads. Differentiation 56:137.

Petitte, J. N., C. L. Brazolot, M. E. Clark, G. Liu, A. Verrinder Gibbins, and R. J. Etches. 1993. Assessing the genome of the chicken using germline chimeras. In: R. J. Etches and A. Verrinder Gibbins (Ed.) Manipulation of the Avian Genome. p 81. CRC Press, Boca Raton, FL.

Petitte, J. N., M. E. Clark, and R. J. Etches. 1991. Assessment of functional gametes in chickens after transfer of primordial germ cells. J. Reprod. Fertil. 92:225.

Petitte, J. N., M. E. Clark, G. Liu, A. Verrinder Gibbins, and R. J. Etches. 1990. Production of somatic and germ line chimeras in the chicken by transfer of early blastodermal cells. Development 108:185.

Pitts, G. R., O. M. Youngren, J. L. Silsby, I. Rozenboim, Y. Chaiseha, R. E. Phillips, D. N. Foster, and M. E. El Halawani. 1994. Role of vasoactive intestinal peptide in the control of prolactin-induced turkey incubation behavior: II. Chronic infusion of vasoactive intestinal peptide. Biol. Reprod. 50(6):1350.

Reynaud, G. 1976. Etude de la localisation des cellules germinales primordiales chez l'embryon de Caille japonaise au moyen d'une technique d'irradiation aux ultra-violets. C. R. Acad. Sci. (Paris) 282:1195.

Rowlett, K., and K. Simkiss. 1987. Explanted embryo culture: *in vitro* and *in ovo* techniques for the domestic fowl. Br. Poult. Sci. 28:91.

Salter, D. W., and L. B. Crittenden. 1989. Artificial insertion of a dominant gene for resistance to avian leukosis virus into the germline of the chicken. Theor. Appl. Genet. 77:457.

Salter, D. W., E. J. Smith, S. H. Hughes, S. E. Wright, and L. B. Crittenden. 1987. Transgenic chickens: Insertion of retroviral genes into the chicken germ line. Virology 157:236.

Salter, D. W., E. J. Smith, S. H. Hughes, S. E. Wright, A. M. Fadly, R. L. Witter, and L. B. Crittenden. 1986. Gene insertion into the chicken germline by retroviruses. Poult. Sci. 65:1445.

Sang, H. M., and M. M. Perry. 1989. Episomal replication of cloned DNA injected into the fertilised ovum of the hen. Mol. Reprod. Dev. 1:96.

Savva, D., N. Page, L. Vick, and K. Simkiss. 1991. Detection of foreign DNA in transgenic chicken embryos using the polymerase chain reaction. Res. Vet. Sci. 50:131.

Shaw, D. L., R. S. Carsience, R. J. Etches, and A. Verrinder Gibbins. 1992. The fate of female donor blastodermal cells in male chimeric chickens. Biochem. Cell Biol. 70:1218.

Shibuya, T., T. Murota, N. Horiya, H. Matsuda, and T. Hara. 1993. The induction of recessive mutations in mouse primordial germ cells with N-ethyl-N-nitrosourea. Mut. Res. 290:273.

Shuman, R. M. 1981. Primordial germ cell transfer in the chicken, *Gallus domesticus*. M.S. Thesis. University of Minnesota, St. Paul.

Shuman, R. M. 1990. Genetic Engineering. In: R. D. Crawford (Ed.) Poultry Breeding and Genetics. Vol. 14. p 585. Elsevier, New York.

Shuman, R. M. 1991. Production of transgenic birds. Experientia 47:897.

Shuman, R. M., and R. N. Shoffner. 1982. Potential genetic modification in the chicken, *Gallus domesticus*. Proc. 2nd World Cong. Genet. Appl. Livest. Prod. 6:157.

Shuman, R. M., and R. N. Shoffner. 1986. Gene transfer by avian retroviruses. Poult. Sci. 65:1437.

Simkiss, K., and Y. Li. 1992. The production of transgenic birds from PGC manipulated embryos. V Int. Symp. Avian Endocrinol. V:24.

Simkiss, K., K. Rowlett, N. Bumstead, and B. M. Freeman. 1989. Transfer of primordial germ cell DNA between embryos. Protoplasma 151:164.

Simkiss, K., L. Vick, G. Luck, N. Page, and D. Savva. 1990. Infection of primordial germ cells with defective retrovirus and their transfer to developing embryo. Proc. 4th World Cong. Genet. Appl. Livest. Prod. 16:111.

Sims, M., and N. L. First. 1994. Production of calves by transfer of nuclei from cultured inner cell mass cells. Proc. Natl. Acad. Sci. USA 91:6143.

Singh, R. P., and D. B. Meyer. 1967. Primordial germ cells in blood smears from chick embryos. Science (Wash DC) 156:1503.

Smith, A. G. 1992a. Mammalian stem cell systems. Semin. Cell Biol. 3:383.

Smith, A. G. 1992b. Mouse embryo stem cells; their identification, propagation and manipulation. Sem. Cell Biol. 3:385.

Smithies, O., R. G. Gregg, S. S. Boggs, M. A. Koralewski, and R. S. Kucherlapati. 1985. Insertion of DNA sequences into the human chromosomal β-globin locus by homologous recombination. Nature (Lond.) 317:230.

Spratt, N. T., and H. Hass. 1960. Integrative mechanisms in development of the early chick blastoderm II. Role of morphogenetic movements and regenerative growth in synthetic and topographically disarranged blastoderms. J. Exp. Zool. 144:139.

Stern, C. D., and D. R. Canning. 1990. Origin of cells giving rise to mesoderm and endoderm in chick embryo. Nature (Lond.) 343:273.

Stewart, C., I. Gadi, and H. Bhatt. 1994. Stem cells from primordial germ cells can reenter the germ line. Dev. Biol. 161:626.

Sun, S., E. A. Wong, J. A. Proudman, B. C. Wentworth, and M. E. El Halawani. 1995. Prolactin antisense DNA and RNA blocks VIP-stimulated prolactin secretion from cultured turkey pituitary cells. Poult. Sci. 74 (Suppl. 1):157.

Sutasurya, L. A., S. Yasugi, and T. Mizuno. 1983. Appearance of primordial germ cells in young chick blastoderms cultured *in vitro*. Dev. Growth Differ. 25:517.

Swift, C. H. 1914. Origin and early history of the primordial germ cells in the chick. Am. J. Anat. 15:483.

Tajima, A., M. Naito, Y. Yasuda, and T. Kuwana. 1993. Production of germ line chimera by transfer of primordial germ cells in the domestic chicken (Gallus domesticus). Theriogenology 40:509.

Thomas, K. R., and M. R. Capecchi. 1987. Site-directed mutagenesis by gene tarageting in mouse embryo-derived stem cells. Cell 51:503.

Thomas, K. R., C. Deng, and M. R. Capecchi. 1992. High-fidelity gene targeting in embryonic stem cells by using sequence replacement vectors. Mol. Cell. Biol. 12:2919.

Thoraval, P., F. Lasserre, F. Coudert, and G. Dambrine. 1994. Somatic and germline chicken chimeras obtained from brown and white leghorns by transfer of early blastodermal cells. Poult. Sci. 73:1897.

Tsai, H. 1995. Stem cells—The way to the avian germline. Ph.D. Dissertation. University of Wisconsin, Madison.

Tsai, H., and B. C. Wentworth. 1992. Long-term culture of chick primordial germ cells isolated from the blastodisc and germinal crescent. Poultry Sci. 71:72 (Abstr.).

Ukeshima, A., M. Kudo, and T. Fujimoto. 1987. Relationship between genital ridge formation and settlement site of primordial germ cells in chick embryos. Anat. Rec. 219:311.

Ukeshima, A., K. Yoshinaga, and T. Fujimoto. 1991. Scanning and transmission electron microscopic observations of chick primordial germ cells with special reference to the extravasation in their migratory course. J. Elect. Microsc. 40:124.

Urven, L. E., U. K. Abbott, and C. A. Erickson. 1989. Distribution of extracellular matrix in the migratory pathway of avian primordial germ cells. Anat. Rec. 224:14.

Urven, L. E., C. A. Erickson, U. K. Abbott, and J. R. McCarrey. 1988. Analysis of germ line development in the chick embryo using an antimouse EC cell antibody. Development 103:299.

Urven, L. E., D. E. Weng, A. L. Schumaker, J. D. Gearhart, and J. R. McCarrey. 1993. Differential gene expression in fetal mouse germ cells. Biol. Reprod. 48:564.

van Limbrough, J. 1958. Number and distribution of the primary germ cells in duck embryos in the 28- to 36-somite stages. Acta Morphol. Neeri. Scand. 2:119.

Valancius, V., and O. Smithies. 1991. Testing an "in-out" targeting procedure for making subtle genomic modifications in mouse embryonic stem cells. Mol. Cell. Biol. 11:1402.

Vick, L., Y. Li, and K. Simkiss. 1993a. Transgenic birds from transformed primordial germ cells. Proc. R. Soc. Lond. Ser. B Biol. Sci. 251:179.

Vick, L., G. Luke, and K. Simkiss. 1993b. Germ-line chimaeras can produce both strains of fowl with high efficiency after partial sterilization. J. Reprod. Fertil. 98:637.

Watanabe, M., M. Kinutani, M. Naito, O. Ochio, and Y. Takashima. 1992. Distribution analysis of transferred donor cells in avian blastodermal chimeras. Development 114:331.

Watanabe, M., M. Naito, E. Sasaki, M. Sakurai, T. Kuwana, and T. Oishi. 1994. Liposome-mediated DNA transfer into chicken primordial germ cells in vivo. Mol. Reprod. Dev. 38(3):268.

Watt, J. M., J. N. Petitte, and R. J. Etches. 1993. Early development of the chick embryo. J. Morphol. 215:165.

Wentworth, A. L., A. Frank, and B. C. Wentworth. 1992. Transfection of Japanese quail primordial germ cells with DNA expressing lacZ gene product. Poult. Sci. 73(Suppl. 1):71 (Abstr.).

Wentworth, B. C. 1960. Fistulation of the hen's oviduct. Poult. Sci. 39:782.

Wentworth, B. C., H. Tsai, J. H. Hallett, D. S. Gonzales, and S. G. Rajcic. 1989. Manipulation of avian primordial germ cells and gonadal differentiation. Poult. Sci. 68:999.

Wentworth, B. C., and A. L. Wentworth. 1991. Primordial germ cell characteristics in the Japanese quail embryo. Poult. Sci. 70(Suppl. 1):129 (Abstr.).

Wentworth, B. C., A. L. Wentworth, H. Tsai, T. Lee, E. A. Wong, and J. A. Proudman. 1992. Avian primordial germ cells as a vehicle for gene transfer. V Int. Symp. Avian Endocrinol. V:24 (Abstr.).

Wilmut, I., M. L. Hooper, and J. P. Simons. 1991. Genetic manipulation of mammals and its application on reproductive biology. J. Reprod. Fertil. 92:245.

Wong, E. A., N. H. Ferrin, J. L. Silsby, and M. E. El Halawani. 1991. Cloning of a turkey prolactin cDNA: expression of prolactin mRNA throughout the reproductive cycle of the domestic turkey (Meleagris gallopavo). Gen. Comp. Endocrinol. 83:18.

Yasuda, Y., A. Tajima, T. Fujimoto, and T. Kuwana. 1992. A method to obtain avian germ-line chimaeras using isolated primordial germ cells. J. Reprod. Fertil. 96:521.

Yoshinaga, K., T. Fujimoto, M. Nakamura, and H. Terakura. 1992. Selective lectin-binding sites of primordial germ cells in chick and quail embryos. Anat. Rec. 233:625.

Yoshinaga, K., M. Nakamura, and A. Ukeshima. 1993. Ultrastructural characteristics of primordial germ cells in the quail embryo. Anat. Rec. 236:547.

Session IV

Acceptance of Biotechnology: Political and Social Issues

Food and Drug Administration Concerns Regarding Genetic Improvement of Farm Animals

Steven R. Bauer

Molecular Immunology Laboratory, FDA, Center for Biologics Evaluation and Research, Division of Cellular and Gene Therapies, Bethesda, MD 20892, U.S.A.

Abstract The biotechnology explosion that began in the 1970s has led to the development of techniques enabling molecular biologists to directly manipulate the germline of a variety of animals through the introduction of new genetic material. The newly introduced DNA becomes part of the genome of the animal and can be inherited by subsequent generations. The techniques of transgenesis allow both removal of undesired traits and the introduction of new traits. An enormous number of applications of this technique can be envisioned and are being developed. These applications include use of transgenic animals as producers of biological therapeutics or as human disease models for testing of potential pharmaceutical agents. Depending on the application, several federal agencies can be involved in regulation of the use of transgenic animals. The Food and Drug Administration regulates the use of transgenic animals through scientific assessment of the risks and benefits associated with the use of this technique. Issues beyond the scope of this scientific risk/benefit assessment are considered in conjunction with outside advisory bodies including the NIH Recombinant DNA Advisory Committee and FDA Advisory Committees convened to give advice on specific issues arising from the use of biotechnology.

Key Words: Transgenic Animals, Biotechnology, FDA, Regulation

Risk/Benefit Analysis in Regulation

The regulatory approach currently in use by reviewers in the Center for Biologics Evaluation and Research (CBER) is to scientifically assess the risks and benefits inherent in the use of a new therapeutic product or novel biotechnology-based approaches to develop therapeutic products. Genetic alteration of livestock animals is a case falling into the latter category.

The consideration of risks associated with a new therapeutic product or technology to develop therapeutic products includes factors of a wide scope. The potential risks could affect individuals, family members, health care workers, pets and other animals, and even the environment. Some risks appear to be clear and predictable. However, new technologies often involve new, unclear, and unpredictable risks. In contrast, potential benefits of a new biotechnology or product to individuals and to society are often well defined and predictable.

The balancing of these considerations depends on the willingness to accept the possibility of some risk while doing everything possible to ensure the safety of a new approach. Thus, preclinical studies should be designed not only to examine efficacy but also to investigate risks so that a realistic assessment can be made.

Policy Development

As new technologies develop, CBER scientists recognize the possibility that these developments could be used to generate novel therapeutic approaches. As CBER's scientific staff gains expertise relevant to the new technology, working groups are formed to formulate some guidance for manufacturers who may utilize the new technique to generate a therapeutic product. The CBER then informs the biotechnology industry of its scientific concerns through the "Points to Consider" series of documents. Due to rapid technological advancement in the field of biotechnology, Points to Consider documents, although published in the *Federal Register*, are presented as drafts in revision. These documents present scientific issues which are associated with a particular product class or manufacturing approach and which should be addressed by manufacturers intending to test new therapeutic products.

FDA Concerns Regarding Transgenic Animals

The recent development of techniques to alter the genome of animals has created opportunity for the use of livestock animals as producers of therapeutic products. With this consideration in mind, a task force of FDA staff from several Centers formulated a new Points to Consider Document that addresses scientific issues that affect the evaluation of Investigational New Drug (IND) submissions based on the use of animals to manufacture therapeutic products (CBER, 1995). Availability of this document has recently been announced in the *Federal Register* (1995). It is entitled "Points to Consider in the Manufacture and Testing of Therapeutic Products Intended for Human Use Derived from Transgenic Animals." Several previous Points to Consider Documents address overlapping issues stemming from the use of recombinant DNA to manufacture novel therapeutics. The new document primarily addresses issues that are associated with the use of transgenic animals. Seven major concerns that are unique to this approach are summarized below.

1. Transgene Construct Information describing the transgene and vector should be detailed and allow a complete understanding of the intended gene product, its transcriptional control, its intended tissue distribution, and other salient data.

2. Generation and Characterization of Transgenic Animals The methods used to introduce DNA into the animal and the techniques used to identify and characterize transgenic animals should be described in detail.

3. Stability of the Transgene Transgenic animal strains should be characterized to ensure that the transgene is stable in terms of integration, structural integrity, and expression. Single sites of integration are desirable so that a transgenic strain contains a stable number of transgene copies since copy number often influences expression levels. Stable structural integrity implies that no mutation, amplification, deletion, or other aberration has deleteriously affected the transgene during or subsequent to integration into the animal genome. When possible, stability of expression should be examined within generations, between different generations, and using the genetic backgrounds most likely to be used for production animals.

4. Long-Term Availability of a Useful Strain Biotechnology production methods traditionally have relied on cell culture in which a Cell Bank system has been developed to ensure that highly characterized clones of valuable cells are cryopreserved so that a desirable biological can be available for an extended period of time. Animals, particularly livestock, have no equivalent method of cloning and cryopreservation. Outbred genetic backgrounds can affect production of transgene products. Therefore, manufacturers should give careful consideration to a mechanism that would preserve a transgenic strain over long periods and ensure that descendants of that strain will be capable of production at a useful and acceptable level.

5. Breeding Scheme The intended scheme of breeding of transgenic animals should be presented in detail. This should include information on the breeding partners and whether or not the transgenic animals will be maintained as a heterozygous or homozygous strain.

6. Production Scheme The manner in which the product is to be harvested, pooled, characterized, and tested should be carefully planned. Production runs should meet acceptable criteria. For example, harvesting of a therapeutic product produced in milk should be described so that it is clear how batches are defined, how milk is pooled and stored, and how production proceeds after harvest.

7. Adventitious Agents The potential use of a wide variety of new animal hosts for production of therapeutics presents the possibility of exposure of humans to new pathogens. Therapeutic products intended for parenteral use, possibly in immunosuppressed patients, need to be free of adventitious agents. Testing or process validation for adventitious agent removal should be fully described. Veterinary advice regarding possible zoonotic agents should be obtained. Animals from geographical locations known to harbor transmissible spongiform encephalopathy agents should be avoided as production hosts.

A wide variety of publications, Points to Consider documents and other information can be obtained from CBER by mail, FAX or e-mail. The CBER FAX Information System can be reached at (301)-594-1939. Points to Consider in the Manufacture and Testing of Therapeutic Products Intended for Human Use Derived from Transgenic Animals can be obtained by this mechanism.

Mechanisms for Public Input Regarding the FDA Regulatory Process

Several mechanisms exist whereby members of the public can interact with the FDA. Less formal mechanisms include audience questions and responses to talks given on particular topics by CBER representatives at scientific or biotechnology industry conferences. Opportunities for public commentary are presented on a more formal basis as well. When Points to Consider Documents are issued, an appeal for comment from interested parties is included in the Federal Register Notice of Availability and in the document itself. Another mechanism is provided through FDA Advisory Committees. These committees contain panels of experts from outside government who are asked to advise the Agency on scientific issues that affect the use of new therapeutics. These meetings include public sessions during which members of public can comment. The FDA also participates in NIH Technology Assessment Conferences, which also have sessions for public comment. Finally, the Recombinant DNA Advisory Committee (RAC) holds public meetings to consider gene therapy protocols that will be regulated by the FDA.

Literature Cited

CBER. 1995. Points to Consider in the Manufacture and Testing of Therapeutic Products for Human Use Derived From Transgenic Animals. Docket No. 95D-0131. Center for Biologics Evaluation and Research, FDA, Rockville, MD.

Federal Register. 1995. Notice of Availability: Points to Consider in the Manufacture and Testing of Therapeutic Products for Human Use Derived from Transgenic Animals. Fed. Reg. 60(164):44036.

Acceptance of Biotechnology: Food Safety and Inspection Service Concerns

P. Basu

Science & Technology, Food Safety and Inspection Service, USDA, Washington, DC 20250, U.S.A.

Abstract Recent progress in the field of biotechnology and the production of transgenic livestock has raised the need for guidance on the disposition at slaughter of these animals. National surveys in the United States and in Great Britain have shown that although the public knowledge of biotechnology has increased, the use of this technology is more acceptable in human medicine and in production of plants than in the production of transgenic food animals. Therefore, there is a need to regulate food animals resulting from transgenic animal research. It is anticipated that several governmental agencies will be involved in regulatory issues pertaining to these animals. The Food Safety and Inspection Service (FSIS) of the United States Department of Agriculture will ultimately be responsible for ensuring that transgenic animals intended for human consumption are wholesome, unadulterated, and properly labeled. Scientists in FSIS, in conjunction with other governmental agencies and the Agricultural Biotechnology Research Advisory Committee, have developed guidelines for the slaughter of animals resulting from transgenic research.

Key Words: Biotechnology, Transgenic Animals, Regulation, Food Safety, Research Animals

Introduction

Over the last half of this century, modern techniques have been introduced to improve animal lines. Artificial insemination has already had an enormous impact on the dairy industry. Techniques such as *in vitro* fertilization, embryo cloning, nuclear transplantation, and transgenesis are reaching maturity. While supporters predict a number of important benefits from the use of biotechnology, opponents express concerns about the safety, government oversight, and the ethics of biotechnology. "What is needed to balance our technological prowess is a renewed engagement in the debate over biotechnology policy—not just the ethics of genetic engineering, but the entire relationship between biotechnology and our future," said Vice President Al Gore in 1991.

Biotechnology has the potential to improve and enhance animal production and offers a number of new tools to improve the health of food animals. However, this potential will be realized only if consumers accept the use of these new technologies for food production, and for this to happen, consumers must have confidence that the products of agricultural biotechnology are safe and effective and that government is exercising proper oversight.

Recent progress in the field of animal biotechnology and the production of transgenic livestock has raised questions as to the need for governmental oversight of this process and for guidance on the disposition at slaughter of these animals, to ensure that animals produced through biotechnology are safe for human consumption. National surveys conducted in the United States by Drs. Hoban and Kendall

(1993) and in Great Britain by the Committee on Ethics of Genetic Modification and Food Use (1993) chaired by Reverend Dr. J. C. Polkinghorne have shown that public knowledge of biotechnology has increased, but the use of this technology is more acceptable in human medicine and in the production of plants than in the production of transgenic food animals. Therefore, there is a need to assure the public that these products are safe by regulating food animals resulting from transgenesis.

The Food Safety and Inspection Service (FSIS) of the United States Department of Agriculture (USDA) administers the Federal Meat Inspection Act (FMIA) and the Poultry Products Inspection Act (PPIA). Under these acts, FSIS inspects food-producing establishments to ensure that meat and poultry products sold in interstate and foreign commerce for human consumption are wholesome, unadulterated, and accurately labeled. Species that are inspected under these acts include cattle, sheep, swine, goats, horses, mules, or other equines, and any domesticated birds. The FSIS conducts inspection with a work force of nearly 8,000 inspectors and veterinarians in 6,700 slaughter and processing plants nationwide. In 1992, FSIS inspected approximately 127 million red meat animals, 7 billion birds, and 161 billion pounds of processed meat and poultry products. The Agency is mandated to inspect each individual livestock and bird presented for slaughter, including experimental animals that result from research. USDA's coordinated framework for the regulation of biotechnology (Federal Register notice 1986), titled, "Final Policy Statement for Research and Regulation of Biotechnology Processes and Products," states the Department's intention to regulate foods produced by new methods, such as recombinant DNA techniques, within the existing regulations. Therefore, at the present time, there are no new regulations, and there should not be any new regulation proposed for the slaughter of transgenic food animals.

Guidelines for Nontransgenic Research Animals

In March of 1989, FSIS received a petition from a biotechnology corporation in the southwestern United States asking to allow the slaughter of experimental animals that were "failures" of transgenic animal research, or in other words, those animals in which an attempt to insert a foreign gene into an embryo was unsuccessful. This petition prompted FSIS to realize the need for regulatory oversight to ensure the food derived from these animals is safe for human consumption.

In responding to this request, FSIS sought additional information, including description of the research protocol; the gene constructs; methods used to determine whether or not the animals were transgenic; copies of the results of southern hybridization and polymerase chain reaction (PCR); and the number, age, sex, and identification of involved animals.

It was during this phase that the company indicated that they anticipated large numbers of these nontransgenic animals resulting from biotechnology research due to the extreme inefficiency of producing transgenic livestock. The efficiency of producing transgenic livestock is very low (< 2%). As a result of extreme inefficiency, there is a large number of these "no-takes," or nontransgenic animals. Due to the large number of nontransgenic animals that result from transgenic animal experiments, there may also be a high level of interest from other corporations and researchers in slaughtering these animals for food.

The FSIS is regulating the slaughter of animals derived from biotechnology experiments under the experimental animal regulations set forth in 9 CFR 309.17 and 9 CFR 381.75. The term "experimental animal" is defined by 9 CFR 301.2 (Aa) as *animals used in any research investigation involving the feeding or other administration of, or subjection to, an experimental biological product, drug, or chemical.* Transgenic animals can be included under this definition of experimental animals because the technique for producing the first generation of transgenic animals would involve the introduction of new genetic material considered to be either an experimental biological product, drug, or chemical. Experimental status would remain with future generations of these animals because the genetic material remains in edible tissues through subsequent generations.

Over the course of the next 13 months, a policy was formulated and in June 1990 FSIS presented a proposal for the handling of these animals to USDA's Agricultural Biotechnology Research Advisory Committee (ABRAC). ABRAC includes representatives from government, industry, academia, and consumer groups and advises USDA with respect to policies, programs, operations, and activities associated with the conduct of agricultural biotechnology research. The FSIS proposal was unanimously endorsed by ABRAC, whose meetings were open to the public.

On December 27, 1991, FSIS published a notice (Federal Register 67054-67055, vol. 56, No. 249) titled "Livestock and Poultry Connected with Biotechnology Research." This notice announced the availability of the document "Decision Criteria for the Evaluation of Nontransgenic Animals from Transgenic Research."

The Decision Criteria document was reviewed by the Animal and Plant Health Inspection Service (APHIS), the Food and Drug Administration (FDA), and was unanimously endorsed by ABRAC. The document outlines the FSIS regulation for the slaughter of nontransgenic animals resulting from transgenic animal research experiments. The following criteria will be used to determine that an animal is nontransgenic: 1) failure to detect the presence of the transgene by genetic methods, 2) absence of a measurable gene product, 3) absence of transgene-associated traits, and 4) physical condition and appearance of the animal prior to slaughter.

In January 1992, a team of FSIS scientists visited the site of the biotechnology corporation that had requested the disposition of nontransgenic animals and reviewed the data from each individual animal that was included in the request for slaughter. The FSIS reviewed the results on DNA isolation from two tissues: blood and ear notches. This review included southern hybridization and PCR data submitted by the company. The company provided sufficient data for FSIS to conclude that animals found negative for the attempted gene insertion were not changed in such a manner as to be made unfit for human consumption by virtue of this attempted gene insertion. Sixty-six animals were released for slaughter since both tests were negative for the foreign gene. Thus, permission for slaughter of these animals for food was granted according to 9 CFR 309.17.

Guidelines for Transgenic Research Animals

In January 1993, FSIS petitioned ABRAC to address the scientific questions associated with the food safety of human products prepared from transgenic animals. In response, ABRAC formed a transgenic animal working group composed of specialists in the field of animal biotechnology from the industry, academia, and government.

A public meeting of the working group was held on April 8, 1993, in Washington, DC. Minutes of the working group meeting were published. The findings and recommendations of the working group were presented to the main ABRAC body in a public meeting held at the North Carolina Biotechnology Center, Research Triangle Park, North Carolina, on June 29-30, 1993. In this meeting, the recommendations from the working group were discussed and further refined by ABRAC. The recommendations from ABRAC to FSIS are noted below.

ABRAC Recommendations

The ABRAC supported the concept of addressing human food safety of food derived from transgenic animals and proposed the following definition of transgenic animals: *transgenic animals are animals and their progeny whose genetic composition has been changed by introducing specific genes (e.g. recombinant DNA). The scope of this definition is intended to also include animals developed by antisense, amplification, and deletion technologies.* The ABRAC also concluded the following: that the presence/detection of the gene is the starting point, rather than gene expression *per se,* that animals will be declared to be "not transgenic" based on the criteria in the FSIS guidelines published earlier, and that animals will be declared to be "transgenic" based on at least one of the following: 1) detection of the transgene by southern hybridization or PCR; 2) presence of measurable gene product; and/or 3) presence of transgene-associated traits. ABRAC suggested several steps for determining the human food safety of 1) gene additions, gene modifications, mosaics, marker genes, and somatic cell therapy, and 2) viral (and other) vectors and stated that "if there are no effects of human food safety concern, the transgenic animals and their progeny may enter the human food chain." Copies of the ABRAC document are available for public review.

Based on the ABRAC recommendation, on March 17, 1994, the FSIS published a notice (Federal Register Vol. 59, No. 52) titled "Update on Livestock and Poultry Connected with Biotechnology Research." This notice informs the public of the availability of a document, "Points to Consider in the Slaughter of Transgenic Animals from Transgenic Animal Research." For researchers, the document should provide guidance on the type of information that could be used to demonstrate that the transgenic change(s) will not result in adulterated products from such livestock or poultry. For consumers, this document should provide assurance that the regulatory oversight in ensuring food safety of meat and poultry derived from transgenic research animals is adequate.

Conclusion

The process of developing regulatory oversight for the slaughter of nontransgenic and transgenic animals resulting from transgenic research required over 4 years to complete. The FSIS feels confident that by being very cautious, the Agency has provided guidelines that ensure that these products are safe for the consuming public. These guidelines in turn will help in the acceptance of biotechnology by the public.

Literature Cited

Hoban, J. H., and P. A. Kendall. 1993. Consumer Attitudes About Food Biotechnology. Project Report.

Report of the Committee on the Ethics of Genetic Modification and Food Use, Ministry of Agriculture, Fisheries and Food. 1993. London, England.

Federal Register Notice Vol. 51, No. 123, 23336, June 26, 1986. Final Policy Statement for Research and Regulation of Biotechnology Processes and Products, USDA, Washington, DC.

Federal Register Notice Vol. 56, No. 249, December 1991. Livestock and Poultry Connected with Biotechnology Research. FSIS, USDA, Washington, DC.

Federal Register Notice Vol. 59, No. 52, March 1994. Update on Livestock and Poultry Connected with Biotechnology Research. FSIS, USDA, Washington, DC.

Public Perceptions of Animal Biotechnology

Robin Yeaton Woo, Ann M. Holbrow, and Gerald E. Gaull

The Center for Food and Nutrition Policy and The Ceres Forum™ of the Graduate Program in Public Policy, Georgetown University, Washington, D.C. 20007, U.S.A.

Abstract Biotechnology has evolved from the genetic manipulation of microbes to the development of transgenic animals in just a quarter of a century. Although the American public readily has accepted human insulin produced by genetically altered microbes and tacitly has accepted chymosin for use in the manufacture of cheese, in Germany both have engendered outright antagonism. The modification of animals through genetic engineering already has attracted intense public interest and opposition in The Netherlands and may also in the United States. Opinion surveys in various countries show that the publics are more opposed to genetic engineering in animals than to such applications in plants or microbes. While the potential uses of genetic engineering in animals could significantly improve human health and the global food supply, consumer acceptance is needed for continued advancement. Efforts must be made to understand public perception of the issues unique to genetic engineering in animals, as well as the general issues pertaining to the technology. Some of the unique issues include ethical concern for preserving genetic integrity, concern for the welfare of animals as sentient beings, and fear of some implications of genetic engineering in humans. By recognizing both unique and general issues in public perception, corporations, scientists, and regulators will be better able to assess acceptable uses of animal biotechnology, consider ethical issues in designing projects, and address consumer concerns early in the process.

Key Words: Biotechnology, Public Policy, Public Perceptions, Animal Genetic Engineering

Introduction

Genetic engineering has changed biology from a largely descriptive to a manipulative science. With the new tools of biotechnology, man now can control aspects of biological structure and/or function. With this control comes the potential for beneficial applications affecting human life in diverse ways. Application to animals is the latest frontier of biotechnology where benefits are beginning to be realized. Animals have been engineered to display human genetic or chronic health disorders in order to serve as more effective models for research on possible treatments and cures. Genetic engineering of animals may result in more efficient agricultural production, more nutritious food products, and improved agricultural products. Bioactive substances (e.g., lactoferrin) may soon be economically produced by genetically engineered animals for more effective use by humans. Animals may also be genetically engineered to produce pharmaceuticals as well as more suitable sources of organs for transplant. But there is also the realization that this new manipulative ability implies a potential for harmful effects.

There is greater public acceptance of genetic engineering in microbes and plants than in animals (Zechendorf, 1994). In the United States, people are more likely to accept genetic engineering of plants (70%) than of animals (42%) (Figure 1). In the European Union, effects of biotechnology are more positively perceived in applications to plant-based foods (38%) than in applications to animals (33%). In Japan, respondents were more likely to accept genetic engineering of plants and microbes (81%) than genetic engineering of animals (54%). The lesser acceptance of animal biotechnology is based on social and ethical concerns unique to it that are likely to add an additional barrier to public acceptance.

Differences in consumer acceptance in microbial, plant, and animal biotechnology between the United States and Europe have translated into more stringent regulatory and political controls of biotechnology products in Europe. In March 1995, the European Parliament rejected a proposed regulation providing for patenting of life forms. Although the European Patent Office has approved animal patents (e.g., the Harvard oncomouse), rejection of the proposal has sent legislators back to the drawing board after 6 years of negotiation and has hindered establishment of consistent regulation throughout the European Union. In the United States, patents have been issued for life forms (from microbes to animals) since the Supreme Court case of *Diamond vs. Chakrabarty* in 1986. Another example of the stringency of European controls is recombinant bovine somatotropin (rBST), which was approved for use in the United States in 1994, while the European Union issued a moratorium on approval for commercial use until 1999, with a recent exception for preliminary efficacy testing. In Germany, operation of a human insulin facility was delayed for years due to public concerns and permit difficulties. In The Netherlands, the disposition of cows carrying the gene for human lactoferrin has yet to be determined.

Animal Biotechnology Issues

Debates regarding the risks, benefits, regulation, and ethics of biotechnology have occurred since the initial manipulations of genes in microbes. Concerns regarding

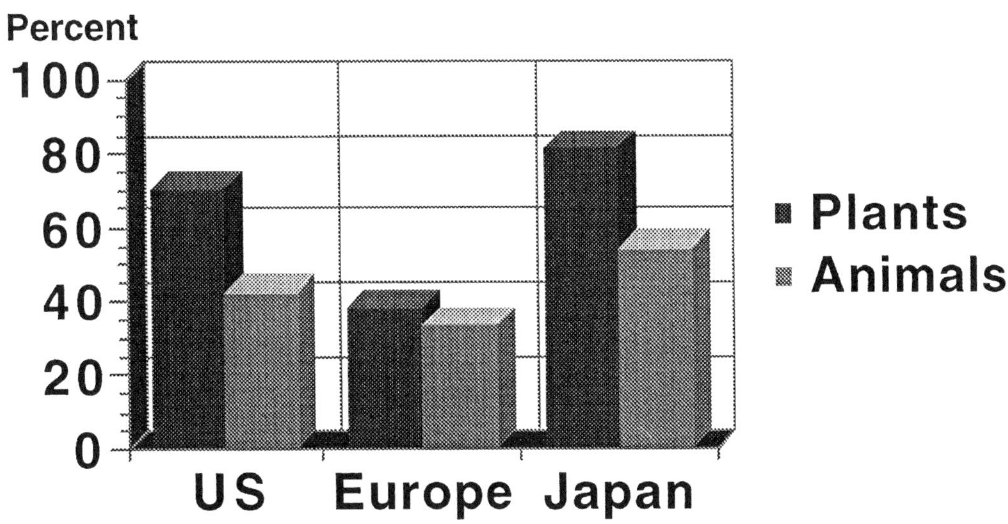

Figure 1. Consumer acceptance of genetic engineering in the United States, Europe, and Japan, grouped by type of species (based on data taken from Zechendorf (1994).

putative reduction of genetic diversity, risks of unintentional gene transfer, and development of new uncontrollable species are central to biotechnology debates in a variety of applications, including animal biotechnology. New dimensions added by animal biotechnology to the debate include ethical concerns about manipulating the natural order and destroying the genetic integrity of animal species. Others fear that genetic engineering in animals will culminate in genetic engineering in humans, haunted by images of Nazi medical experiments and eugenics. In popular culture, cartoons, and literature, human emotions and feelings are often attributed to animals, adding further credibility to the passionate response from animal rights activists and consumers against potential pain and suffering by animals.

In order for the full potential of animal biotechnology to be realized, scientists, regulators, consumer activists, and industry will have to understand these issues, achieve concurrence, and communicate that concurrence to the public. In this section, we examine the issues that have generated the greatest debate.

Issue 1. Changes in Species Diversity May Result from Applications of the New Biotechnology

Some proponents of sustained genetic diversity consider genetic engineering a threat to their interests. Genetic engineering, like conventional breeding, tends to result in consistent genetic traits, eliminating other traits or species that may be necessary for survival under future selective pressures. Conventional plant breeding has resulted in fewer than 20 plant species that produce 90 percent of the food supply (Buffet, 1994). Species selection using animal breeding techniques, such as artificial insemination and embryo transfer, have resulted in genetically consistent offspring with consistent strengths (e.g., size) and weaknesses (e.g., susceptibility to disease).

Extinction of species is estimated at a rate 400 times that recorded through recent geologic time (Wilson, 1985), but quantitative changes in diversity cannot be specified because funding has been inadequate for the immense effort necessary to catalog all species on the planet. Since we don't know how many species there are, we cannot ascertain whether the rate of creation of new species has kept up with the rate of disappearance. Nonetheless, the efforts being made to conserve species and reduce the rate of extinction are prudent. For example, the United Nations Economic, Social, and Cultural Organization (UNESCO), the United States Agency for International Development (USAID) and other organizations are sponsoring genetic conservation programs, particularly in developing countries where species diversity is greatest (Wilson, 1985).

Issue 2. Genetic Traits May Be Transferred to Unintended Species with Adverse Results

Transfer of genetic material between transgenic animals is generally less likely to occur than transfer in plants and microbes because of the greater human control of breeding of domesticated and laboratory animals. However, the risk of uncontrolled transfer of genes may be significant in transgenic fish. Aquaculture presents a special case as species can be produced on fish farms where they may be released or escape to lakes, rivers, and streams where spawning with non-transgenic fish cannot be controlled. The concern is that all new fish eventually will retain the transgenic gene and associated phenotypic trait, which might have adverse effects on the ecosystem. For instance, larger fish may require more food, resulting in an imbalance in the ecosystem, or fish with a cold-tolerance gene may be able to migrate to new ecosystems that they could disrupt.

However, innovative approaches to fish production are necessary to help sustain the predicted growth of the human population. World aquaculture, faced with overexploited fisheries, will need to increase production sevenfold to feed the world's population by 2025 (Colwell, 1995). An example of research efforts to achieve increased fish production is the growth-hormone-enhanced salmon, which is up to 11 times heavier than normal salmon (Devlin et al., 1994). Rather than automatically overlooking potentially vital benefits simply because the large salmon are a product of genetic engineering, it is reasonable to develop this technology while simultaneously developing systems to minimize the potential risks (e.g., captive management). The USDA's science advisory committee for biotechnology (ABRAC) has developed such guidelines for aquaculture.

Issue 3. Aggressive Species May Be Created by Gene Transfer Technology

The products of genetic engineering are associated in laypersons' minds with the problem of exotic species: organisms transplanted to a new environment absent their natural predators that reproduce and become dominant (provided that adequate space and nutrients are available). The adverse effects of an exotic species are tangible to the public from well-publicized examples, such as damages caused by the Africanized ("killer") bees, the kudzu vine, and rabbits in Australia (see for example Fox, 1992).

By associating the negative effects of such exotic species with the products of biotechnology, opponents can launch an effective, albeit specious, argument to the public. However, they ignore the fact that new genes must be integrated into the total genome or the organism will be less likely to survive (Davis, 1986). Our ability to manipulate genes that might produce a dominant species is limited, since such fitness traits are typically polygenic, and some new traits may lessen an animal's ability to survive without human assistance. Additionally, a dominant exotic species arises when a species with a genome that evolved in a balanced manner to its present environment is introduced into a new environment where it is more competitive. In the case of plants, laboratory testing of species with a single or few modified traits and gradual scale-up of field testing in its natural habitat can significantly reduce the potential for the accidental creation of a dominant, exotic species.

Issue 4. Species Boundaries Are Defined by Nature and Should Not Be Manipulated

According to some, genetic engineering disrupts the natural order of species boundaries, particularly between species that could not transfer genetic material through natural reproduction (e.g., between a bird and a cow). The new biotechnology breaks these reproductive barriers by successfully transferring genetic information from one species to another. Whereas microbes of differing species are known to interchange DNA fragments, animals are generally unable to transfer DNA between species that do not reproduce naturally. From the ethical perspective of some, transgenic modifications of animals extend beyond the laws imposed by Nature.

At the heart of this issue is the definition of species as independent, unrelated organisms with a unique telos. In the system developed by Linnaeus, species are defined by phenotypic traits to establish some organization to and cataloging of the life forms on earth. A basic tenet of this system is that different species cannot interbreed and produce fertile offspring. Upon further study, however, we are learning that behavioral and physical plasticity are common, and species boundaries are not as distinct as heretofore thought. Species are no longer considered fixed, static categories representing some unique essence, but rather as dynamic, fluid relationships. For instance, Bullock's and Baltimore orioles have distinct plumages and were initially classified as separate species. When mating between these species was observed 22 years ago, they were reclassified as one species. Recently, these oriole species were again separated because each exhibits a preference for its own type when mating (Daniel, 1995). While orioles are closely related species that are not as distinct as a bird and a cow, this example illustrates the arbitrary nature of species classifications. "These (interspecies) crosses have played a significant role in evolution" (Berkowitz, 1990).

Transfer of single genes (naturally or through biotechnology) does not change the nature of the species, but only the manifestation of specific phenotypic traits. Many of these traits could be selected through conventional breeding programs, although recombinant DNA (rDNA) techniques significantly enhance specificity of the results and reduce the time required for traits to be altered. Such unpredictable mixtures of phenotypic traits may have a greater effect on species integrity than the transfer of a single gene. Interestingly, the use of conventional breeding techniques to improve or change species has not been subjected to the same criticisms as genetic engineering, likely due to their familiarity after centuries of use.

Issue 5. Animal Biotechnology Intensifies the Debate over the Rights of Animals as Sentient Beings

Animal biotechnology extends the ongoing debate regarding animal welfare, an issue not relevant to plants or microbes. Adoption of the Federal Animal Welfare Act in 1966, related amendments, and state legislation have instituted standards of care for animals used in laboratory research. Although the Act was established to reduce animal suffering, the use of animals in research was not banned because the benefits to humans were generally considered to justify the use of animals in laboratories. Perceptions of animal rights and potential suffering, cited by activists and publicized by the media, will most certainly play a role in consumer acceptance of animal biotechnology.

Biotechnological research in animals should be based on the same premise as all other types of animal research. For example, the human growth hormone gene was transferred to pigs in an attempt to produce more efficient growth and leaner meat. Although the gene was successfully transferred in mice and rats, transgenic pigs displayed arthritis, lethargy, defective vision, high mortality, and reduced immunity. Opponents of animal biotechnology would suggest that "predictions and assurances about the safety and humaneness of genetic engineering cannot be generalized from one animal species to another" (Fox, 1992), with the possible adverse result of inhumane conditions for the animals. Scientists would argue that outcomes of genetic experiments cannot be known with certainty, but are conducted under controlled circumstances using the best information available. Progress in science, which is based on trial and error, would be halted entirely if *all* uncertainty regarding the outcomes were required to be known in advance (zero risk).

Genetic engineering in animals may actually benefit animal welfare by optimizing animal use in research to reduce the numbers of animals needed for research and to develop model systems for disease using lower-order mammals (mice, rats) as effective replacements for higher-order mammals (chimpanzees) (Holtzman, 1989). Transgenic animal models of human diseases with no known corollary in the animal kingdom also now can be studied. For example, a mouse model for phenylketonuria has been developed to facilitate the study of the pathogenesis and treatment of that disorder (Shedlovsky et al., 1993). Ironically, the inbred health problems of purebred dog species indicate some acceptance of the unintended creation of dysfunctional animals through traditional breeding activities of seemingly less significant benefit to humans (Holtzman, 1989).

Issue 6. Applications of Biotechnology Are Viewed as the Next Step in the Genetic Manipulation of Humans

Because animals are perceived as closer to humans than plants or microbes, extension of genetic engineering to animals is a more highly charged issue. The ultimate concern, the prospect of widespread human gene therapy, was placed within the realm of possibility in 1983 when the genetic code for the enzyme defective in patients with phenylketonuria was sequenced (Woo et al., 1983), paving the way for identification of carriers of PKU and prenatal screening. Since that time, clinical trials of gene therapy in humans have occurred on a limited basis for diseases such as severe combined immunodeficiency disease (SCID), which affected the well-publicized "boy in the plastic bubble."

Opponents of animal biotechnology equate human gene therapy with animal genetic engineering, thereby confusing the issues for the general public and reducing acceptance of animal biotechnology. Genetic engineering in animals is portrayed as inevitably leading to the engineering of a perfect human race and elimination of those who are less than perfect. By suggesting that the evolution from one application to the other is inevitable, opponents further bolster public opposition to animal biotechnology.

However, the issues in each application are distinct: one deals with the many issues discussed herein and the other with acceptability and management of risks associated with obvious potential benefits to human health and welfare. While some of the concerns regarding human gene therapy may have validity if applied to germ cells, they are distinct from genetic engineering in animals. Ironically, no geneticist known to us has ever espoused the use of germ line therapy in humans. Public acceptance of genetic engineering in animals depends on an accurate portrayal of the distinction between these applications to limit public debate to only the relevant issues.

Development of Consumer Acceptance of Animal Biotechnology

Successful applications of animal biotechnology will depend on consideration of consumer needs, consistent regulatory oversight, consumer knowledge and understanding of the process and products, and acknowledgment of moral concerns. Many of these lessons can be learned from other products of biotechnology that have made it to market through diligent effort. It is important for scientists to understand the concerns of the public and answer them in their own terms—without condescension and without recourse to scientific jargon.

Lesson 1. Products Must Meet Consumer Needs and Appear Desirable to Consumers

Chymosin and rBST provide contrasting cases for considering consumer needs and reactions prior to product introduction. Chymosin is an enzyme produced by recombinant bacteria that abets the coagulation process in cheese making. Recombinant chymosin was developed to replace rennet, usually isolated from the stomachs of slaughtered calves at great expense. Chymosin has been successfully marketed to cheese manufacturers (primary consumers) in the United States without threatened secondary consumer boycotts or demands for labeling. It is now used to manufacture approximately 70% of cheese in the United States, demonstrating at least tacit consumer acceptance.

In contrast, the introduction of rBST resulted in significant media attention, threatened boycotts, legislation in individual states, and demands for federal labeling requirements. Extra doses of this natural hormone, obtained from genetically engineered bacteria, increase milk production in cows by 10 to 20%, thereby increasing efficiency in dairy production. Early reports of 40% increases in milk production sent a shock wave through the dairy farming community, foreboding the demise of the family farm. Consumers, activists, and lobbying groups for dairy farmers questioned the necessity of increasing production of a commodity that currently required price supports to offset surplus production. Efficiency gains that would obtain larger yields with smaller herds were not considered useful by the general public, and this technology was not perceived as scale-neutral by small dairy farmers. Regulatory delays, Congressional debate, and negative representations by the media were encouraged by the reaction of both primary and secondary consumers to a product that was perceived by the farmers as either bad or unnecessary and by the public as of no appreciable advantage.

Now that potential products are more numerous, developers of animal biotechnology should try to select products that consumers would consider useful rather than just because they are technically easier to develop. In the case of rBST, agbiotech companies made the strategic decision to develop what was feasible without foreseeing the issues of public concern. Products of biotechnology with pharmaceutical applications, in contrast, have been positively received by the American public because of their ability to directly improve human health. Such products will increase public awareness of potential benefits of biotechnology in general and may facilitate the subsequent introduction of other products with benefits less immediately obvious to the public.

Lesson 2. Products Must Be Seen by the Public as Having Received Regulatory Approval and Support for Safety

The Flavr Savr tomato and rBST provide examples of the importance of regulatory support in achieving consumer acceptance. Because Flavr Savr tomatoes were considered substantially similar to regular tomatoes, Calgene was not under regulatory obligation to obtain pre-market approval from the FDA. In anticipation of public concern, Calgene voluntarily submitted its product to FDA review so that consumers would have the imprimatur of FDA in addition to the company's safety assurances when the product was marketed. The review process was fraught with public debate, media attention, and negative portrayals by some consumer advocates (i.e., the Foundation on Economic Trends and the Pure Food Campaign). Nevertheless, Calgene considered this step necessary in order to provide the guarantee of safety from a public agency that was not perceived as profit-motivated. While submitting the tomato to FDA review ostensibly delayed marketing, final FDA approval encouraged consumer acceptance. Flavr Savr tomatoes are currently available in limited markets; Calgene is expanding production to meet demand (Genetic Engineering News, 1995).

In attempts to prevent marketing of rBST, doubts were raised about the safety of the hormone to cows and the safety of milk from rBST-treated cows for human consumption. Monsanto, the product's prime developer, had several respected, nonpartisan health-and-safety-related organizations review the test results. Groups such as the American Medical Association and the American Dietetic Association provided independent product safety endorsements. While significant controversy was generated regarding health issues and the spread of misinformation continued, endorsements by respected organizations helped vitiate misinformation and contributed to acceptance of the product. Final approval from FDA allowed product marketing to begin in 1994. While considerable opposition continues in some dairy farming states, 7% of dairy farmers are using rBST (Schneider, 1994). Interestingly, total milk consumption increased 4% in 1994, and milk labeled "rBST-free" captured less than 5% of the market.

Regulatory support for Monsanto's and Calgene's products was integral to their acceptance by the public. In a national consumer survey, non-government health associations, professors, and farmers or farm groups had the highest levels of consumer trust (categories indicating "some" or "a lot" of trust), followed closely by regulatory agencies (Hoban and Kendall, 1993). The high quality and quantity of technical review by both governmental and health-related organizations increased consumers' comfort levels with these new products, crucial in controverting arguments made by opponents.

Lesson 3. Consumer Education and Participation Will Encourage Acceptance of Biotechnology

As with any new technology, consumer understanding is critical to acceptance. An informed public is less likely to be swayed by misinformation and is more prone to make rational decisions regarding products. Regulators, scientists, educators, industry and professional organizations can help in this long-term project. Monsanto continues to gather post-approval information regarding the use, efficacy, and adverse effects of rBST for the FDA. During the sixth meeting of the National

Agricultural Biotechnology Council (NABC), recommendations included attendance at the National Association of Biology Teachers (NABT) and the National Science Teachers Association (NSTA) to encourage the inclusion of agricultural biotechnology in K through 12 science curricula (MacDonald, 1994). We think innovative programs for young children will be the most likely to succeed in the long run.

Lesson 4. Moral Objections by Consumers Must Be Acknowledged, but Are Difficult to Change

Over one-half of consumers surveyed by Hoban and Kendall believed that the use of biotechnology to change animals is morally wrong (1993). Moral objections are often based on religious beliefs or ideals. "Religious ideals are without doubt the most potent source of claims against animal biotechnology.... People holding religiously based views have successfully opposed the teaching of evolution, Greek philosophy, and even stress management in public schools" (Thompson, 1992). Moral or religious views are not changed through economic or scientific arguments because they are intrinsic personal beliefs that are strongly held. Particularly in areas where moral objectors constitute the majority, significant efforts must be made to acknowledge and understand the objections and respond where possible with equal fervor.

Consumer activists convincingly influence public opinion with passionate extrapolation of ethical arguments. Their extreme stances capture the attention of the media, are amplified by them, and effectively motivate opposition to products of biotechnology. Although their arguments are not always based on current science, their effectiveness cannot be underestimated, particularly in light of the proportion of people who have ethical concerns regarding biotechnology.

Scientists, regulators, and industry leaders must enter into ethical dialogues with opponents of biotechnology to address and resolve ethical conflicts (Sandoe, 1994).

Conclusions

Products of newly introduced technologies have always been met with public apprehension due to uncertainties regarding health and safety and potential changes to the economic or social *status quo*. Opponents of the introduction of the steam locomotive argued that "speed was harmful for the human body, cows in pastures would be so frightened that milk production would cease, and that God did not intend to let inanimate objects move by their own force" (van Vugt and Brom, 1994). Opponents become more passionate when the technologies affect something as basic to life as food and nutrition. While "damage to the small family farm" and "interference with nature's perfect food" are arguments made by critics of rBST, similar arguments were voiced against pasteurization for over 50 years after its development (Satin, 1992). Activists' protests even recently reopened, at considerable cost to taxpayers (United States Department of Health and Human Services, 1991), the issue of safety of fluoridated drinking water which was instituted shortly after World War II (Crain et al., 1969).

History has shown that all parties involved, scientists, regulators, industry, consumer advocates, and consumers, should carefully consider the advisability of new technologies and strive to understand fully their long-range effects *a priori*. The goal of scientists, regulators, and industry should be to respond to concerns and increase public awareness of benefits without over promising. Recognition by scientists of the primary issues in the animal biotechnology debate and the rationale for historical success and failure in other applications of biotechnology can significantly improve the likelihood that they will be able to communicate honestly and effectively with consumers.

Literature Cited

Berkowitz, D. B. 1990. The food safety of transgenic animals. Bio/Technology 8:819.

Buffet, H. G. 1994. Nothing to fear from biodiversity. Washington Post, September 4, Sec. C, p 7.

Colwell, R. 1995. Technology Trends and Applications: Biotechnology. National Academy Press, Washington, DC.

Crain, R. L., E. Katz, and D. B. Rosenthal. 1969. The Politics of Community Conflict: The Fluoridation Decision. Bobbs-Merrill, Indianapolis, IN.

Daniel, M. 1995. Birds do it, but not with any old oriole. U.S. News & World Report. March 20, p 19.

Davis, B. D. 1986. Review of Sociobiology: The New Synthesis. Storm over Biology. Prometheus Books, Buffalo, NY.

Devlin, R. H., T. Y. Yesaki, C. A. Biagi, and E. M. Donaldson. 1994. Extraordinary salmon growth. Nature (Lond.) 371:209.

Fox, M. W. 1992. Superpigs and Wondercorn: The Brave New World of Biotechnology and Where It All May Lead. Lyons & Burford, New York.

Genetic Engineering News. 1995. Calgene expands network for tomato packing and distribution. January 15.

Hoban, T. J., and P. A. Kendall. 1993. Consumer Attitudes about Food Biotechnology: Project Report. Extension Serv., USDA, North Carolina State Univ. and Colorado State Univ., Raleigh, NC.

Holtzman, S. 1989. Testimony of Steven Holtzman before the Subcommittee on Courts, Intellectual Property and the Administration of Justice, House Judiciary Committee. September 13.

MacDonald, J. F. 1994. Agricultural Biotechnology and the Public Good (NABC Rep. 6). National Agricultural Biotechnology Council, Ithaca, NY.

Sandoe, P. 1994. Public perception: The need for ethical dialogue. An International Workshop on Animal Biotechnology Issues. April 6-9, 1994, Davis, CA.

Satin, M. 1992. Original objections to pasteurization. Conf. on Free and Informed Consumer Choices, October 27, Orlando, FL.

Schneider, K. 1994. Dairymen increasing use of a growth hormone in cows. New York Times, October 30, p 28.

Shedlovsky, A., J. D. McDonald, D. Symula, and W. F. Dove. 1993. Mouse models of human phenylketonuria. Genetics 134:1205.

Thompson. P. B. 1992. Science and ethics of new developments in animal reproduction. J. Dairy Sci. 75:2294.

United States Department of Health and Human Services. 1991. Review of fluoride benefits and risks. Ad Hoc Subcommittee on Fluoride of the Committee to Coordinate Environmental Health and Related Programs, Public Health Service.

van Vugt, F., and F.W.A. Brom. 1994. Animal biotechnology and society: What are the issues? Proc. of the International Workshop on Animal Biotechnology Issues. University of California, Davis.

Wilson, E. O. 1985. The biological diversity crisis. Bioscience 35:700.

Woo, S. L., A. S. Lidsky, F. Guttler, T. Chandra, and K.J.H. Robson. 1983. Cloned human phenylalanine hydroxylase gene allows prenatal diagnosis and carrier detection of classical phenylketonuria. Nature (Lond.) 359:151.

Zechendorf, B. 1994. What the public thinks about biotechnology. Bio/Technology 12:870.

Changing Paradigms in Animal Agriculture: Addressing Animal Issues in Washington by a Professional Society[1]

Robert G. Zimbelman*, David K. Waggoner†, and Daniel B. Waggoner‡

**American Society of Animal Science, Bethesda, MD 20814, U.S.A.; †Consultant, National Association of State Universities and Land Grant Colleges, Washington, DC 20003, U.S.A.; and ‡Former Staff Director, Subcommittee on Livestock, Dairy, and Poultry, United States House of Representatives, Washington, DC 20003, U.S.A.*

Abstract This article discusses key points that we believe are important as we consider how a professional society in Washington could address animal issues with a changing paradigm as expressed by the public. In summary, our professional perspectives need to change as much as or more than society does. We must understand the advocacy groups and where they are coming from. We have to provide relevant messages to the diverse audiences that play an important role in the public policy process. We have to be clear about our goals and objectives as professionals so as to be more effective as we participate with others in trying to achieve stated goals. We need to understand the broad view of the role of science that exists today, particularly focusing on accountability and relevance of our work to assure the public that their funds are well spent. We need to ensure that all the questions are addressed so that society will maintain trust and respect for the role of science in society.

Key Words: Public, Professional Society, Accountability, Paradigm, Congress

Introduction

We believe the theme of this symposium, changing paradigms in animal agriculture, is a very important one. We had the benefit of reading the manuscript by Darwin Braund (1995) and are going to refer, on occasion, to some of his very insightful comments.

First off, let us say there is good news and bad news. The good news is that we are changing very rapidly. The bad news is that we seem to be changing more slowly than the rest of society. "We" refers to the American Society of Animal Science. Changes made by recent boards of directors, including the new Strategic Plan, the establishment of an office in Washington to communicate with other groups, and other changes, are proceeding in the right direction. However, as mentioned above, questions may arise as to whether ASAS is progressing rapidly enough and(or) far enough. This paper will offer a rationale that focuses on these questions: What are the challenges?, Who are some of the groups that are providing these challenges?,

---

[1]This article originally appeared in the *Journal of Animal Science* (J. Anim. Sci. 73:3178) and is reprinted with permission.

What is the nature of the audiences to which we need to respond?, What should be the goal of a professional society?, and How can we effectively participate? This will be followed by a general perspective regarding the current role of science in today's society and its ability to enhance animal agriculture.

Today's Challenges Defined

As noted elsewhere, "paradigm" implies a shared set of assumptions. We are going to be vulnerable, then, to challenges by those who were not involved in our old paradigm but are willing to evaluate a new paradigm. Some would describe the old paradigm as "more food at the cheapest cost possible and my discipline is the way to achieve that." Following that paradigm, we each, as individual professionals within the society, laid out some goals and often pursued research because it was possible and "vogue" to carry out. As you will see by later comments, such a paradigm is being challenged today by various parties. Two points are of particular importance about the current challenges we face. First, current research efforts should focus on impending issues facing our total society. We will use "society" to mean the citizens of the United States and when we refer to the American Society of Animal Science we will focus on "our profession." Second, we need also to understand that different discussions need to occur at different levels. In other words, to properly discuss the importance and relevance of our work to a Congressperson, a consumer advocacy group, to our Dean, to our family, or to a fellow scientist we must use different language to achieve understanding and a justification of our effort. Scientists are generally accustomed to communicating with colleagues, but at times we find it difficult to present our work to various audiences. The individuals who contest our current paradigms have achieved greater success in communicating to a host of audiences regarding agriculture's acclaimed inability to serve society well.

Nature of Advocacy Groups

Public policy advocacy groups can be described in a spectrum from those who possess sincere commitment toward serious change to those who wish to use an "extreme" push for change leading to restrictions upon, and eventual eradication of, modern animal agriculture. Some groups, particularly in the latter category, are best characterized by the cunning phrase "often in error, rarely in doubt." Capitol Hill has a tendency to accommodate situations whereby instant expertise is granted without meritorious challenge. Why does this happen? The number and complexity of issues that a Member of Congress must strive to comprehend range from a superconducting supercollider to tax issues to health care to agricultural research to food safety and many others. It is difficult, with a typically small staff, to have experts on each and every issue.

Many Congressional Members arrive in Washington from the legal profession and understand little about the scientific underpinning of many critical issues. As a result, members of their staff have to become "instant experts" on a broad range of issues. Generally, such expertise is obtained through an inquiry to the Congressional

Research Service, a division of the Library of Congress that employs professionals to deal with the literature and who provide either pre-written documents on the subject or new pieces from literature searches, or some other knowledgeable source. However, it may also be gleaned from such non-scientific sources as reading the newspaper or viewing a television show on the topic and assuming that the information received is credible and accurate. The information and expertise may also be provided by a lobbyist whom they can trust and who will provide useful information. They may also encourage input from lobbyists with differing points of view and then judge the most reasonable-sounding story.

Even though scientific associations or the Council for Agricultural Science and Technology (CAST) may get called upon, the people or sources that get called upon most frequently are those who are easily accessible or have established a "presence." We have not yet developed an equivocal presence in Washington decision-making circles, although our Congressional Science Fellow is a provider of expertise and advice to key committees on an ongoing basis.

Another area in which special interest groups have excelled is in their ability to shape the issue and establish the framework for debates. For instance, the issues of animal well-being or the environment or food safety or consumption of fat quickly come to mind. All too often our society finds itself "behind the eight ball" lacking such ability because we deal almost exclusively with today's problems. Credit has to be given to those advocacy groups that possess a sense of where the next public policy debate lies and their ability to stay ahead of the curve in picking their next battle.

Nature of the Audiences

As I implied earlier about communication, the audiences for our message are diverse and therefore they require different responses. Seldom will a thorough, well-grounded scientific report make its way to a Congressman's collection of briefing papers. The greatest likelihood of such an occurence is if you as an individual scientist visit directly with the Representative or Senator from your district or state. For a professional society, with a Washington office, greater success can be achieved when the scientist or professional group communicates with an industry or trade association lobbyist to ensure that their information is correct. The lobbyist will then translate the supplied information into appropriate terms for Congressional use. Granted, one can be skeptical of such an approach, but we don't see much possibility that it will change markedly. Obviously, the process is dependent on the trust among the parties involved. Therefore, the lobbyist or special interest group must trust the scientist. In turn, the Congressional staff has to trust the lobbyist and the Congressperson has to trust his or her staff member.

How should we, as scientists, approach the message that the Member of Congress ultimately receives? We would ask questions such as: How do you know that?, What degree of confidence do you possess in making the case?, and other similar questions within the nature of the scientific process. Importantly, a different set of questions may capture the attention of the legislator. For example, on the issue of inadequate selenium in animal diets, scientific input has verified the need for selenium to prevent a variety of diseases in certain circumstances. Congressional staffers were informed by industry representatives that many pigs and calves faced increased chances of death if producers are not allowed to supplement .3 ppm to diets. Neither the staffers nor Members of Congress required more precision or justification on the matter. Such verifiable information was important enough for them to develop an effective strategy to ensure that producers are able to adequately supplement animal diets to prevent some diseases and potential animal deaths. Following this issue, one could not help but observe that there are great leaps of faith, based on trust of the person who was speaking. The issue moved quickly from what is believed to be the scientific merits of an issue to a justification for a change in law.

The Goal of Professional Societies

So what should be our primary goal as a professional scientific society that provides day-to-day representation in Washington? One of the first goals that comes to mind is education of the public, including public policy decision makers. It is frequently argued that the general public and decision makers lack a general understanding of where food comes from, and that issue has to be raised very carefully. As some observers pose it, that approach suggests that scientists are arrogant or paternalistic when responding, "But you don't understand, let me tell you how it really is!" Others argue that is an insult to the intelligence of the person raising an issue.

It is pertinent also to raise the question of whether the scientists are the ones who can best deliver the educational message. Can we put the message in the language of the intended recipient or is someone else more skilled and more appropriate for such endeavors? We could conclude that a joint effort is a worthy approach. Other persons and entities who communicate more freely with the general public can more closely assess the pressing questions that the public has and will know how to effectively address them in terms that will be meaningful and foster greater understanding.

So what is needed is a joint effort, including commodity groups or specialized public relations groups who have recognized greater expertise on how to deliver the message. Media training at our meetings, although a very necessary and worthy effort, is not a cure-all to make us the final word on public education initiatives.

Effective Participation by a Professional Society

So how can we participate effectively in this whole process with its many components? One of the most successful mechanisms in participating on the Washington scene is through the formation of coalitions. When a coalition is formed with commodity groups who do not necessarily have the technical background but who know what questions the public is asking, we can bolster our effectiveness. Coalition members need to trust that our profession will maintain its own standards of objectivity and truthfulness and help provide a message that is positive about their industry to the extent possible.

Congressional staff members who have visited with the candidates for the Congressional Science Fellow Program have given them advice that is particularly poignant at this point. One piece of advice is "Leave your degree at the door." What are they saying in that message? The message is a caution against the expectation that an advanced degree is a license to understand everything best. It also says "Be a true colleague, not a divine source of inspired information." Other members of the coalitions will know their area as well or better than you know the technical details. They will be able to answer these questions: What is important in this issue?, Who's interested in this issue and what message is necessary?, What are the options for obtaining assistance and(or) support for the matter at hand? One also should look for "teachable moments." There are occasions when it is timely to raise an issue and times when one's focus and resources should be targeted to other matters. Other persons in the coalition, rather than the one from a professional society, may have better judgment in that regard. Consequently, it is extremely important to be a true team player to be effective in the coalition. Those who officially represent our profession need to be very objective. It can be damaging to have an individual agenda completely separate from the coalition in which one is participating. For example, it would be easy for us to be a proponent of basic research regardless of how it is perceived by the animal industry commodity groups. There may be teachable moments at which time coalition members can be made to understand the value of basic research, but if that were a primary agenda of professional society representatives it would detract from helping develop strategies on a broad range of issues facing animal agriculture.

Science in Society Today

The two words that frequently surface regarding the role of science within the Washington scene are "accountability" and "relevance." The commonly espoused defense of basic research as a means to expand the knowledge base in a particular area is no longer tenable, it must be relevant to some societal need. With increasingly tight budgetary constraints of the last few years, congressional decision makers are saying they cannot afford everything, they can't afford any luxuries, they can only afford what is deemed relevant to answering societal problems.

A few years ago, the National Research Initiative (NRI) Competitive Grants Program and its predecessor program put their primary focus on basic research. To describe an ultimate practical goal or make it relevant was not necessary, in fact, in some ways it may have been counter-productive.

Today, describing the ultimate practical benefit is an essential component of a proposal to the NRI. Where do you expect this to be relevant? What societal problems or issues do you expect to benefit most from this research proposal? Citing examples of basic research that achieved something other than the initial intent is not germane to this discussion. Funding entities such as NRI are not asking for a promise that your basic research project "fit" only where you expect it to, but they are suggesting that one needs to have some vision of where it will fit into helping society.

Accountability seems to depend on accomplishments with past funding. As a scientific community we seem to be more adept at expounding why we think our research is needed than we are at explaining the accomplishments of recent research. To say that we have isolated the gene for X enzyme or something at that level is not necessarily meaningful to the voting public. In today's budget cutting environment, it has to be related to a societal need or benefit.

Experience and involvement with several coalition groups in Washington reveals that we have a long way to go to properly express the accomplishments of our research. The pertinent questions related to accountability are: What did you do with your last funds? Why should we give you more until you can show specifically how prior funds have been utilized? What have you accomplished with previous funds?

Another issue that affects both accountability and relevance relates to the interdisciplinary aspect of research—and that goes across the board. In terms of interdisciplinary research progress and teams working together, industry probably has made greater strides in this regard than academia. A reward system at many universities needs to be developed to encourage and foster interdisciplinary activities. We believe that this is one area in which the universities really need to take the lead. Professional societies may be able to help by communicating this need to the universities and by their publication policies. This point reminds me of the movie entitled Field of Dreams, a theme on the order of, "If we build it, they will come," referring to a baseball field. Some would say the field of dreams for scientists has been "If we do it, they will want it.". That may even have been true at one time, but it is no longer the case. When research funding was more plentiful than it is today, grantors could allow more flexibility. The present environment brings forth a growing philosophy that can be interpreted as, "Society will tell you what we want, then you go do it."

Importantly, society seems to be raising a further question that we haven't typically asked. For instance, animal scientists are inclined to take some development in basic biology or biomedicine and see if it works in our research subjects. An example of this would be the work that started in the 1930s on human growth hormone that eventually led to discoveries regarding the biology of growth hormone, its structure, and the gene regulating its production. Using all the generated basic biology and biomedical knowledge it seemed appropriate to then ask, "What will it do in cows and pigs?" Typical scientific questions are: How does it happen?, What is the mechanism of action of a given phenomenon? Are you studying the mechanism of action or are you just describing what happens? Have we really achieved biological impact? The whole area of statistical level of confidence as to whether there is a significant effect or not is a question scientists are comfortable with. Scientists are trained to include such questions in their decision making process, but those questions are not sufficient in today's public policy environment.

A growing number within the general population are indicating that they have one additional question: Should it be done? The natural instinct of researchers is that if science has made it possible to explore a new phenomenon, then we have an obligation to follow up and see that the applicable science is applied to the food animal industry. The public is now saying that whether or not it is pursued should be a question, not an automatic reflex. Why would someone assume it should not be done?

If you are going to be successful in the future you have to ask the questions that might be negative also. For example, reproductive physiologists have devoted tremendous amounts of attention and resources to increase the litter size of pigs. We have even imported Chinese pigs because of their prolific tendencies and are pursuing efforts to incorporate those genes into our more typical American breeds.

Maybe there is a connection between the great amount of back fat and the ability to nurse a larger litter. One needs to consider potentially negative questions, such as,

If we increase the litter size of our leaner pigs in this country, are there increased stress-related problems? Is the well-being of the animal and its behavior negatively affected by this larger litter? Is there a need for a change in the management of larger litters and is that management compatible with our contemporary livestock operations? Is it more compatible with smaller scale operations or with larger commercial enterprises being developed in the swine industry? Is there an impact on the lifespan of the animals?

Society is demanding that we look at all possible ramifications. We need to integrate the aforementioned questions and provide convincing evidence that we as scientists are addressing such questions. If society comes forward with unanticipated and unaddressed questions, they lose confidence in science and that can generate mistrust about our scientific pursuits.

Implications

Understanding what the public expects of scientists is important to continued funding. Reacting positively to such a changing paradigm is essential to support of our scientific efforts and the ability to do research without unreasonable regulation.

Literature Cited

Braund, D. G. 1995. Changing paradigms in animal agriculture: The role of academia and industry in technology transfer. J. Anim. Sci. 73:3173.

Session V

Marker Application

Introduction to QTL Detection and Marker-Assisted Selection[1]

J. I. Weller

Agricultural Research Organization, The Volcani Center, Bet Dagan, Israel

Abstract Expression of most economically important traits in livestock is determined by the joint effects of many environmental and genetic factors. Traditional trait-based breeding methodologies utilize only a fraction of the genetic variation present in outbred populations and are limited by expression of many important traits only in females. Via marker-assisted selection (MAS), it should be possible to increase the accuracy of genetic evaluations, increase selection intensity, decrease generation intervals, and better utilize non-additive genetic variation. The advantage of MAS will be greatest in situations in which traditional index selection is inefficient due to long generation interval, low heritability, or index selection for negatively correlated traits. Segregation of individual loci affecting quantitative traits (QTL) can be detected with the aid of genetic markers via genetic linkage. Quantitative trait loci in outbred populations have been detected by analysis of large half-sib or full-sib families. Power of detection per genotype scored can be increased by selective genotyping, replicate progeny, sequential sampling, and pooling of genotypes. New statistical methods to combine marker data on many individuals for several traits are required.

Key Words: Quantitative Trait Loci, Marker-Assisted Selection

Introduction

Expression of nearly all traits of economic importance in farm animals is determined by many genetic and environmental factors. Traditional methods of genetic selection are based on phenotypic expression of these traits and do not consider the individual segregating genes that affect these traits. It was first demonstrated by Sax (1923) that the effects of individual genes could be identified with the aid of segregating genetic markers. However, until recently, selection based on individual segregating loci was not considered a viable alternative to trait-based selection, chiefly due to the lack of known segregating genetic markers in important agricultural species. Since 1990, this problem was solved by the development of DNA-level genetic markers, especially DNA-microsatellites. Highly saturated genetic maps have been developed for most important agricultural species (e.g., Bishop et al., 1994; Rohrer et al., 1994).

Many reviews on detection and effect estimation of quantitative trait loci (QTL) and marker-assisted selection (MAS) have recently been published (Soller, 1990, 1991, 1994; Weller and Fernando, 1991; Weller, 1992; Weller and Ron, 1994). However, this field is so dynamic that frequent review is justified. Although many similar

[1]This research was supported by the United States-Israel Binational Agricultural Research and Development Fund (BARD), project no. IS-1939-91R, and the Israel Milk Marketing Board.

questions are involved for both plants and animals, problems and solutions specific to the main agricultural animals will be emphasized. I will first present a critique of traditional trait-based selection, emphasizing areas where MAS potentially can contribute. The experimental designs that have been suggested for QTL detection and methods for parameter estimation and optimization of experimental design will be reviewed. Finally, I will briefly discuss combination of information from genetic markers with phenotypic data and methods for implementation of MAS.

Trait-Based Selection: Accomplishments and Limitations

Scientific breeding programs based on the principles of quantitative genetics began in the 1950s. The most advanced programs have been for dairy cattle, poultry, and swine, for which breeding objectives are well-defined and economically important traits measured with relative ease. Numerous advances have increased the efficiency of trait-based selection programs. Trait recording has been greatly improved, due chiefly to the introduction of new technology. For example, until 1980, analysis of milk components other than fat percentage was not economically feasible. Since then, equipment has been developed for the automatic analysis and recording of many samples per day. Methods of statistical analysis have kept pace with advances in computing capabilities; first, the introduction of best linear unbiased prediction (BLUP) methodology in the 1960s (Henderson, 1984), followed by adoption of the individual animal model within the last decade (Wiggans et al., 1988).

The example of dairy cattle will be considered in detail. A number of studies have shown that rates of genetic gain in the range of .1 phenotypic SD units/year are well within the capability of advanced dairy cattle progeny testing programs (Nicholas and Smith, 1983; Van Vleck, 1986). The phenotypic SD for milk, which in the past was the main selection objective, is between 1,000 and 1,400 kg under conditions of intensive management. In general, realized genetic trends have been less than theoretical expectations for several reasons. The objective of commercial breeders is to maximize their own profit, which does not necessarily result in the maximum possible genetic gain (Weller, 1994); selection on secondary traits reduces the gain possible on the primary selection objective; in many cases, there is an antagonistic relationship between elements of natural selection and the breeding objective.

Lee et al. (1985) found that genetic trends in the U.S. population from 1960 to 1969 were only 1.55 and 2.55 kg/year, based on cow and bull evaluations, respectively. During the period 1969 to 1978, however, genetic trends were 52.55 and 83.73 kg/year for the cow and bull evaluations. Although the genetic trends during the later period were much higher, they were still less than the expected value of about 100 kg/year, which is still a conservative estimate. Weller and Ron (1989) estimated annual genetic trends for the Israeli dairy cattle population for production traits as 147 kg milk, 2.45 kg fat and –.025% fat during the decade 1978 to 1988. Between 1981 and 1990 genetic trends for this population were 128 kg milk, 2.4 kg fat, and 3 kg protein, as estimated by the individual animal model (Israel, 1993). Finally, annual genetic trends for Canadian Holsteins between 1983 and 1992, also computed by the individual animal model, were about 100 kg milk, 3 kg fat, and 3 kg protein (Robinson et al., 1994).

Despite considerable accomplishments, trait-based selection has several intrinsic limitations. It is less efficient for traits with low heritability, or for multi-trait breeding objectives with negative correlations among economic traits. Many traits of interest can be scored only on certain individuals, and these are not necessarily the optimum candidates for selection. The most salient example is sex-limited traits, such as milk production and number of offspring, which can only be scored on females. Thus, selection among dairy sires is based on the milk production of their female relatives. Certain traits cannot be scored on live animals, such as carcass composition, and infectious disease traits can only be scored on animals exposed to the disease challenge. Based on BLUP methodology, incorporating Henderson's algorithm to invert the complete numerator relationship matrix (Henderson, 1984), it is possible to obtain genetic evaluations for any individual, including those that do not express the traits, based on the records of their relatives. However, selection based on the performance of relatives will be inherently less efficient. Trait-based selection utilizes only additive genetic variance, and ignores dominance and epistasis. Finally, trait-based selection is not very useful in crossbreeding where the main objective is to utilize heterosis.

Experimental Designs for QTL Detection with the Aid of Genetic Markers

Nearly all studies on QTL detection with the aid of genetic markers have been based on the following assumptions, summarized by Zhuchenko et al. (1979):

1. The quantitative traits are normally distributed.
2. Both the QTL and the genetic markers have "Mendelian" inheritance. That is, both loci are located on nuclear DNA that undergoes meiosis; except for genetic linkage, allele distributions are random, and the probability of receiving each parental allele is one-half.
3. The distribution of QTL throughout the genome is sparse. Thus, if a significant marker-linked effect is found, it is assumed to be due to a single QTL.
4. Only two QTL alleles are segregating in the population.
5. There are no pleiotropic effects of the genetic markers on the quantitative traits.
6. There are no interactions of QTL alleles with other loci (epistasis), with the environment, or with phenotype.

Several studies have removed or tested several of these assumptions, and these will be considered below.

The genetic marker will detect an effect of a quantitative trait only if the marker is in linkage disequilibrium with a QTL. Thus, all experimental designs are based on within-family analyses, in which linkage disequilibrium can be expected due to a specific linkage relationship in the "founder" individuals. Most early studies were based on crosses between inbred lines, each homozygous for a different allele of the genetic marker and an assumed linked QTL to produce an F_1 population heterozygous for both loci (Sax, 1923; Soller et al., 1976; Zhuchenko et al., 1979). The F_1 was then either crossed on itself to produce an F_2, backcrossed to one of the parental lines (BC), or crossed to a third, "tester," strain (TC). In the case of the F_2, all three genotypes for the genetic marker, denoted MM, Mm, and mm, will be represented, and, except for recombinants, these three groups will have QTL genotypes denoted QQ, Qq, and qq, respectively. In the F_2 generation, all environmental factors and all genetic factors not linked to the marker locus are assumed to be randomly distributed. Thus, a significant difference between the means among the marker genotypes for the quantitative trait can be assumed to be due to a linked QTL for which the parental strains expressed different alleles. The expected contrast between the homozygotes for the marker locus will be (1–2r)2a, where a is half the difference between the QTL homozygotes for quantitative trait, and r is the recombination frequency between the QTL and the genetic marker. The BC and TC differ from the F_2 in that only two marker genotypes will be represented, say Mm and mm. The contrast for the quantitative trait for these two genotypes in the BC will be (1–2r)(a–d) where d is difference between the midpoint of the two homozygotes and the heterozygote.

Although inbred lines are common for many selfing plant species, they are not a viable alternative for most agricultural animal species. Furthermore, it is generally not economically feasible to produce test populations for QTL detection. Rather, it is necessary to analyze existing populations. Two specific population structures have been considered in detail with respect to QTL detection: a sample of a few large half-sib families and a sample of many small full-sib families.

The half-sib or "daughter design" was first considered in detail by Neimann-Sorensen and Robertson (1961) using blood groups as genetic markers in dairy cattle. In this design, the progeny of several sires, each heterozygous for the same genetic marker, are analyzed. The progeny of each sire are divided into two groups, based on which marker allele was received from their common sire. However, even if a segregating QTL is linked to the genetic marker, linkage phase may be different in different families; thus, the marker effect cannot be pooled across families, and a nested ANOVA analysis is required, with the effect of sire allele nested within family (Soller and Genizi, 1978). Alternatively, data can be analyzed by χ^2. The sire allele substitution effect in each family is considered as a single normal deviate, and, under the null hypothesis, their sum of squares has a central χ^2 distribution (Neimann-Sorensen and Robertson, 1961; Weller et al., 1990).

The daughter design is most powerful for very large families, and is, therefore, applicable only to a few agricultural species, dairy cattle in particular. Haseman and Elston (1972) described in detail an alternative design, applicable to a population structure consisting of many small full-sib families. This design was proposed with humans in mind, but it could also be applied to swine and the smaller ruminants. They specifically considered the case of two full-sibs per family. Assume that both parents are heterozygous for the genetic marker, and that at least three different alleles are represented. For example, the sire has alleles "1" and "2", and the dam has alleles "1" and "3". It is then possible to determine allele origin for each offspring. The two offspring can have 0, 1, or 2 marker alleles identical by descent (IBD). If this marker locus is linked to a segregating QTL, then the more alleles the two sibs have IBD, the more similar their expected phenotypes for the quantitative trait. Let σ_g^2 = the variance due to the segregating QTL, and σ_e^2 = the residual variance. Let Y_j = the squared difference between the quantitative trait records, and π_j = the fraction of marker alleles IBD for the jth sib pair, then:

$$E(Y_j|\pi_j) = \alpha + \beta\pi_j \qquad [1]$$

where $\alpha = \sigma_e^2 + 2\sigma_g^2$ and $\beta = -2(1-2r)^2\sigma_g^2$. Thus, $Y_j|\pi_j$ is a linear function of π_j, which can have values of 0, 1/2, or 1. α and β can then be estimated by linear regression over all the families in the sample. In the absence of a segregating linked QTL, $E(\beta) = 0$. With more than two sibs per family, Equation [1] can be applied to all the possible pair-wise combinations. For the same number of genotypes, this method will be less powerful than the methods described previously.

As will be seen below, statistical power to detect QTL can be increased by scoring quantitative trait values in future generations. For crosses between inbred lines, it is possible to selfbreed the BC, TC, or F_2 to produce recombinant inbred lines (RIL), while for large half-sib families, it is possible to genotype progeny of the heterozygous sire, and phenotype the grandprogeny. This design has been denoted the "granddaughter design" (Weller et al., 1990).

The model of Fernando and Grossman (1989) is suitable for any population structure and also can incorporate non-linked polygenic effects and other "nuisance" effects, such as herd or block. Their method is based on Henderson's mixed-model equations for the individual animal model (Henderson, 1984; Wiggans et al., 1988). They assume that the QTL is a "random" effect with a known variance. Thus, each individual with unknown ancestors has two unique alleles for the QTL, which are "sampled" from an infinite population of alleles. For each individual, they propose a gametic model for the QTL, with separate effects for the sire and dam alleles. For animals that are not genotyped, the probability of receiving either allele from either parent will be equal, but if both the parent and progeny are genotyped for a linked genetic marker, then the probability of receiving a specific parental allele for a QTL linked to the genetic marker will be a function of the progeny marker genotype and r. Based on these probabilities, Fernando and Grossman (1989) demonstrated how a variance-covariance matrix can be constructed for the QTL gametic effects. They further describe a simple algorithm to invert this matrix analogous to Henderson's method for inverting the numerator relationship matrix. Thus, an augmented set of mixed-model equations can be constructed with an additional 2n equation where n is the number of animals included in the analysis. This method has an additional advantage that by assuming random QTL effects, as opposed to fixed effects, the effects of the QTL with the largest estimates are not biased upward (Georges et al., 1995). Theoretically, this method can be extended to handle multiple markers and traits (Goddard, 1992), although the complexity becomes mind-boggling.

Despite the attractive properties of this method, it has not as yet been applied to actual data. In addition to the huge computational requirements, this model assumes that both r and the variance due to the QTL are known *a priori*. Studies on simulated data demonstrate that although restricted maximum likelihood methodology can be used to estimate these parameters, they are completely confounded for a single marker locus (van Arendonk et al., 1994). Furthermore, since each individual with unknown parents is assumed to have two unique alleles, the prediction error variances of the effects for any individual will be quite large, and therefore not very informative. Finally, the assumption of a normal distribution of possible QTL allele effects may not be realistic. However, it is also unclear what will be the effect of assuming only two QTL alleles segregating in the population if the actual number is greater.

The relative statistical power and other properties of the different experimental designs are summarized in Table 1. The optimal experimental design for a given situation will depend on the relative cost of genotyping vs. scoring quantitative traits, what individuals are available for analysis, and what type of mating is possible.

Methods for QTL Parameter Estimation

Estimation of QTL parameters is non-trivial for all the models described above, first because of the effect of recombination. Even for the simplest designs, the contrast between the marker genotypes will be both a function of the QTL allele substitution effect and r. These two parameters will be confounded in a linear model analysis of marker genotypes. As noted above, most studies have assumed only two segregating QTL alleles. In this case, the QTL genotypes have been treated as fixed effects, and the estimable parameters are a and d. For models consisting of more than one family, such as the daughter design, it is also necessary to estimate the frequencies of the QTL alleles in the population. For those models that assume a random distribution of QTL effects, the variance due to the QTL must be estimated instead of a and d. For all models it is necessary to estimate the residual variance. A few studies have assumed QTL variance effects (Zhuchenko et al., 1979; Weller, 1986; Weller and Wyler, 1992). If the QTL genotype affects the variance of the quantitative trait, it is necessary to estimate a different residual variance for each genotype.

Five basic methods have been suggested to estimate QTL parameters: method of statistical moments, nonlinear least squares, maximum likelihood, Bayesian estimation, and Gibbs sampling. The moments method, although easy to apply, does not utilize all the information in the sample. Furthermore, parameter estimates may not fall within the parameter space (Zhuchenko et al., 1979). Nonlinear least squares has the advantages that it is available in existing statistical software packages, it can accommodate additional "nuisance" variables, and gives results similar to linear least squares, which are most familiar to most researchers. The main disadvantage is that this method is limited to certain models. Maximum likelihood (ML) is much more flexible but requires rather complex programming, except for models that can be analyzed by available software, such as program LE of BMDP (Elkind et al., 1994). Bayesian estimation has the advantage that it accounts for prior knowledge about

Table 1. Summary of experimental designs for QTL detection

	Experimental designs[a]							
	Inbred line				Segregating populations			
Characteristics	BC	TC	F-2	RIL	FS	HS	GDD	AM
Power per trait record[b]	+++	+++	++	+	-	-	--	---
Power per genotyping	++	++	++	+++	--	-	+	---
High h^2	+	+	+	-	+	+	-	-
Low h^2	+	+	+	++	+	+	++	++
Dominance	No	No	Yes	No	No	No	No	No
Use available records	No	No	No	No	Some	Some	Some	All
Mating system	Either	Either	Selfer	Selfer	Cross	Cross	Cross	Cross

[a]BC, backcross; TC, testcross; RIL, recombinant inbred lines; FS, full sibs; HS, half sibs; GDD, granddaughter design; AM, animal model.
[b]+++, very high; ++, high; +, moderate; -, moderately low; --, low; ---, very low.

the variables; however, in most realistic situations, this knowledge will be very limited, and computation is more complex than ML (Hoeschele and VanRaden, 1993). Furthermore, if the quantity of new data is "large," then the prior knowledge will have virtually no effect on the estimates. Gibbs sampling has been applied recently to other complex problems in parameter estimation, and it also has been applied recently to simple models for QTL parameter estimation (van Arendonk et al., 1994). In the future, this method may be able to solve complex models that are intractable to the other methods. Tests of significance can be performed for least squares, ML, and Bayesian estimation.

Maximum likelihood and nonlinear least squares have been used much more widely than the other methods, and only these two methods have been applied to actual data (Weller, 1986; Haley and Knott, 1994). These methods will, therefore, be considered in detail, using the simple example of a BC and an assumed QTL between two linked markers. In this design, two inbred lines are crossed to produce F_1 progeny that then are backcrossed to one of the parental strains. The parental strains are homozygous for different alleles of two linked genetic markers, M and N, with recombination frequency R; and for a QTL, Q, located between them. Recombination frequencies between M and Q and Q and N are denoted r_1 and r_2, respectively. Parental genotypes are denoted MQN/MQN and mqn/mqn. The F_1 genotype is therefore MQN/mqn. The F_1 is mated to the latter parental strain, so eight BC genotypes are possible: the non-recombinants with genotypes MQN/mqn and mqn/mqn; the single recombinants MQn/mqn, mqN/mqn, mQN/mqn, and Mqn/mqn; and the double recombinants MqN/mqn and mQn/mqn. With respect to the genetic markers, only four genotypes are observed.

There are three steps in ML parameter estimation:

1. Defining the assumptions on which the statistical model is based. In addition to the "usual" assumptions listed above, zero crossing-over interference between the segments MQ and QN is assumed.
2. Constructing the likelihood function, which is the joint density of the observations conditional on the parameters.
3. Maximizing the likelihood function with respect to the parameters.

Following Lander and Botstein (1989) and Elkind et al. (1994), the likelihood function for this model is described as follows:

$$L = \prod^{n_{MN}} f_{MN} \prod^{n_{Mn}} f_{Mn} \prod^{n_{mN}} f_{mN} \prod^{n_{mn}} f_{mn} \qquad [2]$$

where n_{MN}, n_{Mn}, n_{mN}, n_{mn} are the number of individuals with genotypes MN/mn, Mn/mn, mN/mn, and mn/mn, respectively, $\prod$ denotes multiplicative sum, and f_{MN}, f_{Mn}, f_{nM}, and f_{mn} are the density functions for the four possible marker genotypes. Since all individuals received an mn chromosomal segment from the recurrent parent, only the chromosomal segment received from the F_1 is indicated. The density functions for the possible marker genotypes are computed as follows:

$$f_{MN} = (1 - \alpha)f(Q) + \alpha f(q) \qquad [3]$$
$$f_{Mn} = (1 - \beta)f(Q) + \beta f(q) \qquad [4]$$
$$f_{mN} = (1 - \beta)f(q) + \beta f(Q) \qquad [5]$$
$$f_{mn} = (1 - \alpha)f(q) + \alpha f(Q) \qquad [6]$$

where $\alpha = r_1r_2/(1 - R)$, $\beta = r_1(1 - r_2)/R$, and f(Q) and f(q) are the normal density functions for each observation with standard deviations of σ, and means of μ_Q for the Qq genotype and μ_q for the qq genotype. Thus, the likelihood can be computed by calculating the appropriate density function for each individual, depending on its marker genotype, and multiplying. Assuming zero recombination interference, for given values of R and r_1, r_2 can be estimated as follows:

$$r_2 = (R - r_1)/(1 - 2r_1) \qquad [7]$$

Thus, it is necessary to derive ML estimates for only four parameters, r_1, σ, μ_Q, and μ_q.

In theory, this can be done by taking the partial derivatives of the likelihood function or its log with respect to each parameter, setting each derivative equal to zero, and solving this system of equations. Although it is possible to differentiate the log of the likelihood function with respect to these four parameters, the resultant system of equations cannot be solved analytically. Approximate solutions can be derived by iterative methods.

Numerous methods have been applied to maximize likelihood functions (Weller, 1986; Jenson, 1989; Simpson, 1989; Darvasi et al., 1993; Bovenhuis and Weller, 1994). The program LE of the statistical package BMDP uses Newton-Raphson iteration (Jenrich and Robinson, 1969) to derive parameter estimates (Elkind et al., 1994). The expectation-maximization (EM) algorithm is often preferred because it is guaranteed to converge, provided a maximum exists within the parameter space (Dempster et al., 1977). Several studies have used partial EM for QTL detection, applying the EM algorithm to estimate QTL means and variances, but not recombination frequency (Lander and Botstein, 1989; Luo and Woolliams, 1993). Jansen (1992) developed a general technique to estimate all QTL parameters via the EM algorithm. The log likelihood function is decomposed into two terms, so that parameters related to recombination frequency and parameters related to QTL means and variances can be estimated separately. This technique is amenable to analysis with standard statistical packages and can be applied also to estimate fixed effects.

Two methods can be used to test hypotheses with respect to ML parameter estimates. In likelihood ratio tests, the ratio of the likelihood maximized for all parameters and the likelihood maximized with some parameters fixed is computed. Under the null hypothesis that the true parameter values are not different from the fixed values, the log of the likelihood ratio has a $\frac{1}{2}\chi^2$ distribution, with degrees of freedom equal to the number of parameters fixed. In the example given above, the presence of a linked QTL can be tested by fixing $\mu_Q = \mu_q$. This reduces the number of parameters by one, and under the null hypothesis the log of the likelihood ratio will have a $\frac{1}{2}\chi^2$ distribution with one degree of freedom (Simpson, 1989; Bovenhuis and Weller, 1994). Alternatively, it is possible to estimate confidence intervals for the parameter estimates. As sample size increases, the inverse of the matrix of second differentials with respect to the parameters at their ML values, times –1, becomes asymptotically equal to the prediction error variance matrix of the parameter estimates. These values then can be used to make probability statements about the parameter estimates, for example the probability that $\mu_Q - \mu_q \neq 0$ given the prediction error variance of this difference (Jenson, 1989; Darvasi et al., 1993).

These parameters can also be estimated by nonlinear least squares using the following model derived by Martinez and Curnow (1992) and Haley and Knott (1994). For the backcross design with two flanking markers, it is possible to write the following model:

$$Y_{ij} = \mu_Q + (\mu_q - \mu_Q)p_i + e_{ij} \qquad [8]$$

where Y_{ij} is the production record of the jth individual with marker genotype i, p_i is the probability that an individual with marker genotype i has genotype qq, e_{ij} is the random residual, and the other terms are as defined previously. p_i for each of the four marker genotypes are given above in Equations [3] through [6]. If r_1 were known, it would be possible to substitute these values into Equation [8], and then solve as a simple linear regression, with μ_Q as the y-intercept and $\mu_q - \mu_Q$ and the slope. Since r_1 is not known, it is necessary to solve for all three parameters by finding the values that minimize the residual sum of squares. This model readily can be solved by a nonlinear least squares algorithm, such as PROC NLIN of SAS (SAS, 1988).

The main advantages of this method are that it can be performed by more statistical packages, and significance of the QTL effect can be tested by an *F*-test of the model mean squares against the residual mean squares, which is more familiar to most researchers than a likelihood ratio test. The disadvantage of this method is that it is applicable only in certain situations. It cannot be applied to estimate recombination frequency between a QTL and a single marker, used to estimate QTL variance effects, or applied to segregating populations. All of these questions have been addressed by ML (Weller, 1986; Bovenhuis and Weller, 1994).

Statistical Power for QTL Detection

Statistical power to detect segregating QTL will depend on the sample size, the magnitude of the Type I error allowed, the effect of the segregating QTL in comparison to the genetic and environmental variances, the recombination distance between the QTL and the genetic marker, the specific experimental design employed, and the method of statistical analysis. Since the number of possible combinations described is quite large, I will present only a few examples from the literature, and describe in general terms the effect of various parameters on the statistical power of the experiment.

Different studies that estimate the power of QTL detection considered the substitution effect of QTL alleles in terms of either the phenotypic, the residual, or the genetic standard deviation. However, since residual and genetic variances only are known *a posteriori*, QTL effects will be given in units of the phenotypic standard deviation (SDU). Statistical power of .9 to detect a tightly marker-linked QTL that accounts for 1% of the phenotypic variance in the F_2 design, with a type I error of .05, is obtained with sample size of about 1,000 individuals (Soller et al., 1976). The mean difference in trait values between alternative homozygotes for a QTL of this magnitude will be .282 SDU (i.e., a substitution effect of .141 SDU). For a power of .5, and .1 recombination, about 500 F_2 recombinant inbred lines are required to detect a QTL of similar magnitude, if only one individual is scored per inbred line (Simpson, 1989). For the daughter design, power of .7, with a Type I error of .01, is obtained for a QTL with a substitution effect of .2 SDU if 400 daughters of each of 10 sires are analyzed for a trait with heritability of .2 (Weller et al., 1990). This entails genotyping 4,000 individuals.

Statistical methods that utilize more information in the data and make more extensive assumptions will be more powerful than methods that utilize less information and make fewer assumptions. Thus, ML, which utilizes all information in the data, should be more powerful than ANOVA, which utilizes only the mean and variance of the distributions. Simulation results to this effect were obtained (Simpson, 1989), but later retracted (Simpson, 1992).

The effects of the magnitude of the QTL and the proportion of recombination between the marker and the QTL on sample size to achieve a given power will be quadratic. That is, for an effect of half the magnitude, it will be necessary to increase the number of individuals scored fourfold to achieve the same power. In either the F_2 or BC designs, the magnitude of the effect measured will decrease proportional to $1 - 2r$, as compared to complete linkage. Thus, to achieve power equal to the case of complete linkage, it will be necessary to increase the experiment size by a factor of $1/(1 - 2r)^2$ (Soller et al., 1976).

For QTL bracketed by two markers, the effect measured will not be reduced by recombination (except for double cross-overs), but, in a simple linear model analysis, recombinant individuals will be deleted. The proportion of recombinants for the F_2 and BC designs will be $(1 - R)^2$ and $(1 - R)$, respectively, where R is the recombination frequency between the markers. Power with a marker bracket, therefore, will be reduced by this factor relative to complete linkage. Assuming $R = 2r_1$, which is the optimum case for a marker bracket, power with a marker bracket will be increased by $(1 - R)$ for the BC design and will be equal to a single-marker analysis for the F_2 design (Weller, 1992). For a sample of 500 to 1,000 individuals, ML analysis with genetically linked markers increases the accuracy of QTL detection over the range of .5 to .2 Morgans. Further "saturation" of the genetic map has no effect, unless the number of individuals genotyped is increased to several thousand (Darvasi et al., 1993).

Most studies have analyzed several quantitative traits using several markers but performed a separate statistical analysis for each marker-trait combination. This will result in a large number of spurious "significant" effects. For example, if five markers are used to analyze 10 traits, there are 50 marker-trait combinations. If each combination is analyzed separately, two to three should display "significance" at the 5% level purely by chance. Although this problem has been noted by several studies (Weller et al., 1990; Lander and Botstein, 1989; Hoeschele and VanRaden, 1993), there is no complete solution to the problem. Lander and Botstein (1989) suggested increasing the nominal significance level to the level required to obtain a significant result by chance among any of the individual marker-trait comparisons. In this case many true effects will not be detected. VanRaden and Weller (1994) suggested a nonlinear regression method suitable for an analysis of a cross "saturated" with many closely linked genetic markers. Rather than analyzing each marker, the analysis is based on events of recombination, which are assumed known, and only a single test is performed for each chromosome. Hoeschele and VanRaden (1993) suggested a Bayesian analysis that takes account of prior information. In this case, the combined variance of the effects detected must be less than the total genetic variance of the trait.

Korol et al. (1995) proposed a multitrait ML analysis for estimating QTL parameters on several correlated traits. They showed that power of QTL detection can be increased if the effects of the QTL on the correlated traits differs from the residual correlation. However, their method becomes extremely complicated for more than two traits. Another alternative would be to analyze a set of canonical variables derived from the actual traits. The canonical variables will be by definition uncorrelated, and thus overall significance levels readily can be computed. It also will be possible to determine whether an effect found associated with two correlated traits is due to a single QTL or to two different loci.

Optimization of Experimental Designs

The major cost elements of QTL detection are producing the individuals for analysis, scoring the quantitative traits, genotyping for the genetic markers, and data analysis. Optimization of experimental designs to obtain maximum power per unit cost will depend on the relative costs of these factors. Genotyping individuals for the genetic markers is often the most expensive part of the experiment. This is especially true for DNA-level markers such as restriction length polymorphisms (RFLP) or DNA-microsatellites (Beckmann and Soller, 1983). Furthermore, if the analysis is based on existing records, marker genotyping is the only significant expense. If both the numbers of individuals and markers available for genotyping are unlimited, then marker spacing of close to 50 cM will give maximum statistical power per unit cost (Darvasi and Soller, 1994a). Several techniques have been proposed to increase statistical power to detect segregating QTL as a function of the number of genotype assays performed: replicate progeny, selective genotyping, sample pooling, and sequential sampling. All of these techniques require increasing the number of individuals produced and scored for quantitative traits, as compared to designs in which all individuals scored for the quantitative traits are also genotyped.

Power per sample genotyped can be increased fourfold by application of techniques based either on increasing the number of generations or the number of individuals scored for the quantitative traits (Soller, 1990, 1991). For example, instead of genotyping F_2 individuals, it is possible to first selfbreed for several generations to produce recombinant inbred lines. All individuals within each line will then have the same genotype, and it will be necessary to genotype only a single individual of each line, even though the phenotypic performance of all individuals can be used to determine the QTL effect. This reduces the residual variance by a factor of n, where n is the number of individuals measured in each line, and assuming that the other sources of variance are small, increases statistical power by a factor of $\sqrt{n}$. Since only the residual variance is decreased by either of these designs, the advantage of these techniques decreases as a function of heritability and recombination frequency between the QTL and the genetic marker. With minimal recombination, power equal to an F_2 design can be obtained with only a third the number of individuals genotypes for traits with heritability in the range of .2 to .3 (Soller and Beckmann, 1990). However, this advantage decreases to only twofold if either heritability is greater than .5 or recombination frequency above .2.

In outbreeding populations, such as dairy cattle, marker genotypes can be determined on a sample of progeny of a heterozygous parent, and the quantitative traits can be scored on the grandprogeny (i.e., the progeny of the marker-genotyped offsprings of the heterozygous parent). This design is termed a "granddaughter" design, as opposed to the "daughter" design described above (Weller et al., 1990). A segregating marker-linked QTL can be detected by analysis with the following linear model:

$$Y_{ijkl} = GS_i + G_{ij} + SO_{ijk} + e_{ijkl} \qquad [9]$$

Where Y_{ijkl} is the trait record for granddaughter l of son k that inherited marker allele j from grandsire i, GS_i is the effect of the ith grandsire, G_{ij} is the effect of marker allele j inherited from grandsire i, SO_{ijk} is the effect of the kth son that inherited marker allele j from grandsire i, and e_{ijkl} is the random residual. As in the daughter design, significant marker-allele effect will indicate a linked QTL.

Straightforward analysis by Equation [9] is generally not practical because data will be collected over many herds, the granddaughters have multiple records, and individuals are related. Therefore, several studies have suggested analysis of the genetic evaluations of the sons, or their "daughter yield deviations", both of which are readily available (Andersson-Eklund and Rendel, 1993; Hoeschele and VanRaden, 1993). In either case, the model then includes only the effects of grandsire and marker allele nested within grandsire. However, variances and covariances among genetic evaluations or daughter yield deviations are dependent on the sons' reliabilities, which will be different for each individual. Furthermore, genetic evaluations are regressed, and thus estimates of QTL effects derived by analysis of evaluations will be biased downward.

Increasing the number of granddaughters will reduce the residual variance, but not between-son genetic variance. Thus, the advantage of the granddaughter design is greater for low heritability traits. With heritability of .2 and a Type I error of .01, power is .74 to detect a segregating QTL with a substitution effect of .2 SDU if genetic markers are analyzed on 100 sons of each of 10 grandsires, with 50 quantitative trait-recorded granddaughters per son (Weller et al., 1990). Comparing this example to the example above for the daughter design, greater power is obtained to detect an effect of the same magnitude with the granddaughter design even though only one-fourth as many individuals are genotyped (4,000 vs. 1,000). The disadvantages of replicate progeny techniques are that many more individuals must be scored for the quantitative trait, and it may take longer until records become available. The relative power of the different experimental designs that have been considered is summarized in Table 1.

Selective genotyping was first proposed by Lebowitz et al. (1987) and elaborated by Lander and Botstein (1989) and Darvasi and Soller (1992). Most of the information with respect to QTL detection can be derived from the individuals with the extreme phenotype values for the quantitative trait. Thus, if the sample of individuals recorded for the quantitative trait is large, power per individual genotyped can be increased by selectively genotyping those individuals with the highest and lowest phenotypes. With selective genotyping it is possible to obtain the same statistical power by genotyping only one-third as many individuals as compared to a random sample. The disadvantages of this method are, first, that a much larger sample of individuals must be scored for the quantitative trait. Second, if there are several quantitative traits of interest, it will be necessary to genotype a different sample for each trait. Thus, this technique is only useful if the number of traits of interest is low. Third, estimates of the QTL effect will be biased, and finally, QTL variance effects cannot be detected (Weller and Wyler, 1992).

The number of genotypings that must be performed can be reduced further by up to two orders of magnitude by sample pooling (Plotsky et al., 1993; Darvasi and Soller, 1994b). That is, instead of genotyping each individual separately, genetic material from several individuals with similar phenotypes for the quantitative trait can be combined prior to assay. In this case a linked QTL is detected by band intensity when pools of individuals with "high" and "low" phenotypes are compared. For this method to be effective, it must be possible to determine accurately the number of individuals of each genotype in a pool from the band intensity. Sample pooling must be applied together with selective genotyping. Thus, similar to selective genotyping, this method is only useful if the number of quantitative traits of interest is relatively low. Since there will be some degradation of information when genetic material of individuals is pooled, more individuals must be scored for the quantitative trait, as compared to selective genotyping without sample pooling to obtain the same statistical power.

Finally, Motro and Soller (1993) suggested sequential sampling as a further tool to reduce the number of individuals genotyped. Rather than genotyping a sample large enough to obtain the desired statistical power, a smaller sample is genotyped. Further genotyping will not be required for those markers that either clearly show no significant effect or that show a significant effect. Additional individuals will be genotyped only for those markers that display "borderline" significance. By this method it is possible to reduce the total number of genotypings required by nearly half for a single trait. However, if the number of traits is large, nearly every marker will have borderline significance for at least one quantitative trait. Thus, like selective genotyping and sample pooling, this method is useful only if the number of traits under consideration is small.

Implementation and Expected Gains from Marker-Assisted Selection

Marker-assisted selection can be applied both to crossbreeding programs and to selection within a breed. In general, MAS will make the greatest impact in those situations in which trait-based selection is not effective. I will briefly discuss application of MAS to crossbreeding, and then discuss application of MAS within a breed using the example of dairy cattle.

By crossbreeding it is possible to combine desirable genes from two sources, and to utilize heterosis (Soller and Beckmann, 1983). The classic example of the first objective is to introduce genes for resistance from a wild or "primitive" strain into an "advanced" agricultural strain. The breeding program will be based on a cross between the strains, followed by a series of backcrosses to the advanced agricultural strain. Unless the desired genes from the wild strain can be identified readily, it is very difficult to determine which progeny should be selected. By use of linked genetic markers, QTL genotypes can be identified and individuals with the desired genotypes can be selected even without screening phenotypic expression of the traits of interest. Similarly, if the objective of the cross is to realize heterosis, based only on phenotypic records, all that can generally be determined is that a certain cross is "better" or "worse" than an alternative cross. It cannot be determined whether the overall heterosis is due to interactions between alleles (overdominance) or to interactions between loci (epistasis). Once a desirable combination is found, it is, therefore, not clear how selection should proceed. By identifying the individual loci affecting the traits of interest, it is possible that these questions could be answered.

Within a line, MAS can increase genetic progress in three ways (Weller and Fernando, 1991): 1. increase the accuracy of genetic evaluations; 2. increase the selection intensity; and 3. shorten the generation interval.

Most modern dairy cattle breeding programs are based on a progeny test of young sires, sons of elite cows. Between 50 and 200 daughters are produced from each candidate sire. Daughters of candidate sires constitute only a small fraction of the entire dairy population. Once the daughter records become available, the best of these sires are selected for use in the general population. Thus, the rates of genetic gain per generation and the generation intervals are different along the four paths of inheritance, sire to son, sire to daughter, dam to son, and dam to daughter. The annual rate of genetic gain is equal to the sum of the rates of genetic gain per generation along each path of inheritance divided by the sum of the generation intervals (Owen, 1975). The rate of genetic gain per generation is a function of the selection intensity and the accuracy of the evaluations of the candidates for selection.

Information provided by genetic markers should be combined with phenotypic and pedigree information to derive the most accurate genetic evaluations possible. A first step in this direction was the model of Fernando and Grossman (1989) that uses a modified animal model to estimate both marker-linked QTL effects and unlinked polygenic variance. However, as noted previously, for this method to be practical it is necessary to bracket the QTL within a pair of segregating genetic markers, as described by Goddard (1992). As more information becomes available, it should be possible to locate the QTL within narrow chromosomal regions and to identify the linkage relationships for individuals with many progeny (Soller, 1994). In this case, each QTL allele with its linked marker alleles can be considered a "haplotype" that can be followed through a pedigree. It is then possible to estimate the effects of these "haplotypes" by incorporating their effects into an animal model evaluation of the entire population. However, only a small fraction of the population will be genotyped for genetic markers, and these animals will likely be both individuals with and without production records. Once a significant amount of information accumulates on a particular haplotype, the effects of this haplotype then can be considered a fixed effect, which significantly simplifies the computation of solutions.

Most studies have considered the possible gain from MAS along the sire to son and sire to daughter paths. In current breeding programs, with 50 to 200 daughters, accuracy of sire evaluations is quite high. Therefore, there is little scope for improvement via MAS through increasing the accuracy of sire evaluations (Meuwissen and van Arendonk, 1992). In order to increase the selection intensity, it is necessary that more young males be considered as potential sire candidates. Kashi et al. (1990) considered a two-step procedure in which many young bulls are first selected based on genetic markers, and then a selected group of sires is progeny-tested. This method potentially could increase annual genetic gain by 15%, assuming efficient application of multiple ovulation and embryo transfer to produce a large pool of candidate calves without increasing the number of elite cows used as bull dams.

Alternatively, the generation interval could be shortened by elimination of the progeny test. This could potentially shorten the generation interval along the sire to dam path by up to 4 years, and shorten the sum of the generation intervals by 20%. Without MAS, the accuracy of the sire evaluations would be only half, but with MAS it is possible that accuracies could be increased to the level obtained with a progeny test of 50 daughters. However, farmer opposition to breeding with bulls that have not been progeny-tested is a serious obstacle to implementation of this scheme. More radical strategies to further reduce the generation interval are possible, but not with currently available technology (Georges and Massey, 1991).

Implications

Seventy years after Sax's first experiment, QTL mapping is an idea whose time has come. Traditional limitations to practical application of QTL mapping and marker-assisted selection have been removed, and the only limiting factors now are costs. The number of markers and traits has become so large for many species that joint analysis of all data derived from a single experiment is necessary. The challenges presented by QTL analysis for segregating populations have been only partially met at present. Application of marker-assisted selection potentially can increase rates of genetic gain by up to 20% but will require significant changes in breeding programs.

Literature Cited

Andersson-Eklund, L., and J. Rendel. 1993. Linkage between amylase-1 locus and a major gene for milk fat content in cattle. Anim. Genet. 24:101.

Beckmann, J. S., and M. Soller. 1983. Restriction fragment length polymorphisms in genetic improvement: Methodologies, mapping and costs. Theor. Appl. Genet. 67:35.

Bishop, M. D., S. M. Kappes, J. W. Keele, R. T. Stone, S.L.F. Sunden, G. A. Hawkins, S. S. Toldo, R. Fries, M. D. Grosz, J. Yoo, and C. W. Beattie. 1994. A genetic linkage map for cattle. Genetics 136:619.

Bovenhuis, H., and J. I. Weller. 1994. Mapping and analysis of dairy cattle quantitative trait loci by maximum likelihood methodology using milk protein genes as genetic markers. Genetics 136:267.

Darvasi, A., and M. Soller. 1992. Selective genotyping for determination of linkage between a marker locus and a quantitative trait locus. Theor. Appl. Genet. 85:353.

Darvasi, A., and M. Soller. 1994a. Optimum spacing of genetic markers for determining linkage between marker loci and quantitative trait loci. Theor. Appl. Genet. 89:351.

Darvasi, A., and M. Soller. 1994b. Selective DNA pooling for determination of linkage between a marker locus and a quantitative trait loci. Genetics 138:1365.

Darvasi, A., A. Vinreb, V. Minke, J. I. Weller, and M. Soller. 1993. Detecting marker-QTL linkage and estimating QTL gene effect and map location using a saturated genetic map. Genetics 134:943.

Dempster, A. P., N. M. Laird, and D. B. Rubin. 1977. Maximum likelihood from incomplete data via the EM algorithm. J. R. Stat. Soc. (Series B) 39:1.

Elkind, Y., B. Nir, and J. I. Weller. 1994. Maximum likelihood estimation of quantitative trait loci parameters with the aid of genetic markers using a standard statistical package. Comput. Appl. Biosci. 10:513.

Fernando, R., and M. Grossman 1989. Marker assisted selection using best linear unbiased prediction. Genet. Sel. Evol. 21:467.

Georges, M., and J. M. Massey. 1991. Velogenetics, or the synergistic use of marker assisted selection and germ-line manipulation. Theriogenology 35:151.

Georges, M., D. Nielsen, M. Mackinnon, A. Mishra, R. Okimoto, A. T. Pasquino, L. S. Sargent, A. Sorensen, M. R. Steele, X. Zhao, J. E. Womack, and I. Hoeschele. 1995. Mapping quantitative trait loci controlling milk production in dairy cattle by exploiting progeny testing. Genetics 139:907.

Goddard, M. E. 1992. A mixed model for analyses of data on multiple genetic markers. Theor. Appl. Genet. 83:878.

Haley, C. S., and S. A. Knott. 1994. Interval mapping. Proc. 5th World Congr. Genet. Appl. Livest. Prod. Guelph, ON, Canada. 21:25.

Haseman, J. K., and R. C. Elston 1972. The investigation of linkage between a quantitative trait and a marker locus. Behav. Genet. 2:3.

Henderson, C. R. 1984. Applications of Linear Models in Animal Breeding. Univ. of Guelph, Guelph, ON, Canada.

Hoeschele, I., and P. M. VanRaden. 1993. Bayesian analysis of linkage between genetic markers and quantitative trait loci. I. Prior knowledge. Theor. Appl. Genet. 85:953.

Israel, C. 1993. Animal model, a memory efficient computational strategy applied to the genetic evaluation of the Israeli dairy cattle population. M. S. Thesis. The Hebrew Univ. of Jerusalem, Jerusalem, Israel.

Jansen, R. C. 1992. A general mixture model for mapping quantitative trait loci by using molecular markers. Theor. Appl. Genet. 85:252.

Jenrich, R. I., and S. M. Robinson. 1969. A Newton-Raphson algorithm for maximum likelihood factor analysis. Psychometrika 34:111.

Jenson, J. 1989. Estimation of recombination parameters between a quantitative trait locus (QTL) and two marker gene loci. Theor. Appl. Genet. 78:613.

Kashi, Y., E. Hallerman, and M. Soller. 1990. Marker-assisted selection of candidate bulls for progeny testing programmes. Anim. Prod. 51:63.

Korol, A. B., Y. I. Ronin, and V. M. Kirzhner. 1995. Interval mapping of quantitative trait loci employing correlated trait complexes. Genetics 140:1137.

Lander, E. S., and D. Botstein. 1989. Mapping Mendelian factors underlying quantitative traits using RFLP linkage maps. Genetics 121:185.

Lebowitz, R. J., M. Soller, and J. S. Beckmann. 1987. Trait-based analyses for the detection of linkage between marker loci and quantitative trait loci in crosses between inbred lines. Theor. Appl. Genet. 73:556.

Lee, K. L., A. E. Freeman, and L. P. Johnson. 1985. Estimation of genetic change in the registered Holstein cattle population. J. Dairy Sci. 68:2629.

Luo, Z. W., and J. A. Woolliams. 1993. Estimation of genetic parameters using linkage between a marker gene and a locus underlying a quantitative character in F_2 populations. Heredity 70:245.

Martinez, O., and R. N. Curnow. 1992. Estimating the locations and the sizes of the effects of quantitative trait loci using flanking markers. Theor. Appl. Genet. 85:480.

Meuwissen, T.H.E., and J.A.M. van Arendonk. 1992. Potential improvements in rate of genetic gain from marker-assisted selection in dairy cattle. J. Dairy Sci. 75:1651.

Motro, U., and M. Soller. 1993. Sequential sampling in determining linkage between marker loci and quantitative trait loci. Theor. Appl. Genet. 85:658.

Neimann-Sorensen, A., and A. Robertson. 1961. The association between blood groups and several production characters in three Danish cattle breeds. Acta Agric. Scand. 11:163.

Nicholas, F. W., and C. Smith. 1983. Increased rates of genetic change in dairy cattle by embryo transfer and splitting. Anim. Prod. 36:341.

Owen, J. B. 1975. Selection of dairy bulls on half-sister records. Anim. Prod. 20:1.

Plotsky, Y., A. Cahaner, A. Haberfeld, U. Lavi, S. J. Lamont, and J. Hillel. 1993. DNA fingerprint bands applied to linkage analysis with quantitative trait loci in chickens. Anim. Genet. 24:105.

Robinson, J.A.B., J. P. Chesnais, P. A. Abraham, T. D. Osterhout, and D. K. Luu. 1994. Canadian genetic evaluation for dairy cattle production traits. Proc. 5th World Congr. Genet. Appl. Livest. Prod. Guelph, ON, Canada. 17:38.

Rohrer, G. A., L. J. Alexander, J. W. Kelle, T. P. Smith, and C. W. Beattie. 1994. A microsatellite linkage map of the porcine genome. Genetics 136:231.

SAS. 1988. SAS/STAT® User's Guide (Release 6.03). SAS Inst. Inc., Cary, NC.

Sax, K. 1923. The association of size differences with seed-coat pattern and pigmentation in *Phaseolus vulgaris*. Genetics 8:552.

Simpson, S. P. 1989. Detection of linkage between quantitative trait loci and restriction fragment length polymorphisms using inbred lines. Theor. Appl. Genet. 77:815.

Simpson, S. P. 1992. Correction: Detection of linkage between quantitative trait loci and restriction fragment length polymorphisms using inbred lines. Theor. Appl. Genet. 85:110.

Soller, M. 1990. Genetic mapping of the bovine genome using DNA-level markers with particular attention to loci affecting quantitative traits of economic importance. J. Dairy Sci. 73:2628.

Soller, M. 1991. Mapping quantitative trait loci affecting traits of economic importance in animal populations using molecular markers. In: L. B. Schook, H. A. Lewin, and D. G. McLaren (Ed.) Gene Mapping: Strategies, Techniques and Applications. p 21. Marcel Dekker, New York.

Soller, M. 1994. Marker assisted selection—an overview. Anim. Biotech. 5:193.

Soller, M., and J. S. Beckmann. 1983. Genetic polymorphism in varietal identification and genetic improvement. Theor. Appl. Genet. 67:25.
Soller, M., and J. S. Beckmann. 1990. Marker-based mapping of quantitative trait loci using replicated progeny. Theor. Appl. Genet. 80:205.
Soller, M., and A. Genizi. 1978. The efficiency of experimental designs for the detection of linkage between a marker locus and a locus affecting a quantitative trait in segregating populations. Biometrics 34:47.
Soller, M., A. Genizi, and T. Brody. 1976. On the power of experimental designs for the detection of linkage between marker loci and quantitative loci in crosses between inbred lines. Theor. Appl. Genet. 47:35.
van Arendonk, J.A.M., H. Bovenhuis, S. van der Beek, and A. F. Groen. 1994. Detection and exploitation of markers linked to quantitative traits in farm animals. Proc. 5th World Congr. Genet. Appl. Livest. Prod. Guelph, ON, Canada. 21:193.
VanRaden, P. M., and J. I. Weller. 1994. A simple method to locate and estimate effects of individual genes with a saturated genetic marker map. J. Dairy Sci. 77(Suppl. 1):249 (Abstr.).
Van Vleck, L. D. 1986. Evaluation of dairy cattle breeding programs: Specialized milk production. Proc 3rd World Congr. Genet. Appl. Livest. Prod. 9:141.
Weller, J. I. 1986. Maximum likelihood techniques for the mapping and analysis of quantitative trait loci with the aid of genetic markers. Biometrics 42:627.
Weller, J. I. 1992. Statistical methodologies for mapping and analysis of quantitative trait loci. In: J. S. Beckmann and T. C. Osborn (Ed.) Plant Genomes: Methods for Genetic and Physical Mapping. p 181. Kluwer Academic Publishers Group, Dordrecht, The Netherlands.
Weller, J. I. 1994. Economic Aspects on Animal Breeding. Chapman and Hall, London.
Weller, J. I., and R. L. Fernando 1991. Strategies for the improvement of animal production using marker assisted selection. In: L. B. Schook, H. A. Lewin, and D. G. McLaren (Ed.) Gene Mapping: Strategies, Techniques and Applications. p 305. Marcel Dekker, New York.
Weller, J. I., Y. Kashi, and M. Soller. 1990. Power of "daughter" and "granddaughter" designs for genetic mapping of quantitative traits in dairy cattle using genetic markers. J. Dairy Sci. 73:2525.
Weller, J. I., and M. Ron. 1989. Trends in secondary traits in the dairy cow population in Israel. 40th Annu. Mtg. EAAP 2:49.
Weller, J. I., and M. Ron. 1994. Detection and mapping quantitative trait loci in segregating populations, theory and experimental results. Proc. 5th World Congr. Genet. Appl. Livest. Prod. Guelph, ON, Canada. 21:213.
Weller, J. I., and A. Wyler. 1992. Power of different sampling strategies to detect quantitative trait loci variance effects. Theor. Appl. Genet. 83:582.
Wiggans, G. R., I. Misztal and L. D. Van Vleck. 1988. Implementation of an animal model for genetic evaluation of dairy cattle in the United States. J. Dairy Sci. 71(Suppl. 2):54.
Zhuchenko, A. A., A. B. Korol, and V. K. Andryushchenko. 1979. Linkage between loci of quantitative characters and marker loci. Genetika 14:771.

Mapping Genes Influencing Meat Characteristics in Pigs

Leif Andersson

Department of Animal Breeding and Genetics, Swedish University of Agricultural Sciences, Box 597, S-751 24 Uppsala, Sweden

Abstract The development of powerful tools for genome analysis opens up possibilities for genetic dissection of complex phenotypic traits. In this paper the current state of the art regarding the mapping and identification of genes controlling meat traits in the pig is reviewed. Two major genes affecting meat characteristics have been identified, the halothane (HAL) and the RN gene. The HAL mutation causes the porcine stress syndrome and is associated with leanness. The HAL-associated phenotype is due to a point mutation in the Calcium Release Channel gene. The RN mutation occurs in the Hampshire breed and has a large effect on the glycogen content in white muscle. The RN gene has recently been mapped to pig chromosome 15. Crosses between divergent populations have been utilized to identify major genes for fatness traits. A major gene for intramuscular fat has been identified by statistical analysis of phenotypic records in Meishan/Large White intercrosses. A quantitative trait locus (QTL) on pig chromosome 4 with large effect on fatness was recently detected by linkage mapping using a wild pig/Large White intercross. The future development of this area of research is discussed.

Key Words: Pig, Gene Mapping, Meat Quality, Linkage

Introduction

Meat quality is a highly complex and also subjective trait. It can be broken down into several interrelated components such as fat content (including intramuscular fat), pH, tenderness, juiciness, and flavor. It is obvious that each of these subcomponents has a complex genetic background involving many genes. Most gene mapping studies in this field have so far focused on the more easily measured meat characters such as leanness and glycogen content. This review on genes influencing meat characteristics in pigs will be divided into two parts. First, I will discuss two loci, the halothane and the RN locus, with major effects on meat quality that have been found to be segregating within commercial populations. Second, I will focus on crossbreeding experiments between divergent lines as a strategy to identify quantitative trait loci (QTL) with major or minor effects on meat characteristics.

Major Genes Segregating Within Population

For more than 20 years the halothane (HAL) locus has been known to control important phenotypic traits in pigs (for review see Archibald, 1991). Individuals homozygous for the recessive *HAL n* allele are associated with the porcine stress syndrome (PSS), which involves sudden stress deaths; pale, soft, exudative (PSE) meat; and sensitivity to the anesthetic gas halothane. However, the recessive allele is also associated with beneficial effects on lean content of meat and the selection for lean growth maintained the allele at a high frequency in many commercial populations. It is now well established that the PSS syndrome is due to a missense mutation at codon 615 (Arg→Cys) in the Calcium Release Channel (CRC) gene (Fuji et al., 1991). Testing for this point mutation is now the most widely used DNA test applied to pig breeding.

It is not yet clear whether the effects on meat quality traits are due to pleiotropic effects of the HAL mutation or strong linkage disequilibrium to other mutations in the CRC region. A recent study from our group sheds some light on this important question. We have studied the segregation of CRC alleles in relation to meat characteristics in an intercross between the European wild pig and Large White domestic pigs (Lundström et al., 1995). We reasoned that if the CRC locus or linked loci have a major influence on meat characteristics we expect to find significant differences between the effects of CRC alleles derived from wild and domestic pigs. Moreover, we were able to compare the effect of these alleles to the effect of a CRC allele carrying the HAL mutation since one of the founder animals was a HAL carrier. The results showed that the effect of the HAL mutation on meat quality (measured, for example, as reflectance value and protein extractability) was by far larger than the difference between CRC alleles derived from the wild and domestic pigs. The result does not support the view that the CRC chromosome region, besides harboring the HAL mutation, has a major influence on meat characteristics.

The second major locus affecting meat quality in pigs that has been described is the so-called RN locus (for review see Fernandez and Monin, 1994). The history of the detection of this locus is that Monin and Sellier (1985) first pointed out that meat from Hampshire pigs often had an unusually low pH and a marked decrease in the yield of cured-cooked ham. Naveau (1986) postulated that this "Hampshire" effect was due to the presence of a dominant gene, which was denoted RN. The presence of a major gene in the Hampshire breed was subsequently confirmed by segregation analysis (LeRoy et al., 1990). It has been shown that the fully dominant RN^- allele increases the amount of glycogen in white muscle by about 70% (Estrade et al., 1993a,b).

The large phenotypic difference between RN genotypes makes it possible to score the segregation at this locus as a monogenic trait. We (Mariani et al., 1996) and others (Milan et al., 1995) have recently mapped the RN locus to chromosome 15. This finding makes it now possible to use marker-assisted selection to modify RN allele frequencies. It is also the first step toward the molecular characterization of the RN locus. The possibilities to clone the RN gene should be quite good due to the accurate scoring of RN genotypes and the rather well-defined function and tissue expression. Comparative mapping may be used to identify putative candidate genes previously mapped in humans or mice. In this context it should be noted that the RN homologue in humans most likely is located on chromosome 2q (Mariani et al., 1996).

Genes Detected in Cross-Breeding Experiments

Linkage mapping in crosses between divergent lines of domestic animals is a powerful strategy for mapping QTL controlling phenotypic traits (Andersson et al., 1994a).

Quantitative trait loci with large effects fixed within lines may be segregating in such crosses. In the field of pig gene mapping, two types of intercrosses have been made for this purpose. These are crosses between the European wild pigs and European domestic pigs and crosses between European domestic pigs and Chinese domestic pigs.

We have crossed two wild pig males with eight Large White sows and generated a three-generation family comprising altogether 200 F_2 offspring (Andersson et al., 1994b). The large phenotypic differences between the founder populations imply that this pedigree will be a rich source for detecting QTL. Consequently, phenotypic records for a variety of traits were collected on all F_2 animals. Data on meat characteristics include, for instance, various fatness measurements, pH, pigment content, drip loss, tenderness, and protein extractability. A large number of genetic markers have also been analyzed in the complete pedigree. A primary linkage map based on 128 genetic markers has been published (Ellegren et al., 1994) and since then about 100 markers have been added (Marklund et al., 1996). Quantitative trait loci analysis has so far only been completed for the fatness traits but is underway for other traits related to meat quality.

We have found a QTL on pig chromosome 4 with large effect on fatness, measured as abdominal fat and back fat (Andersson et al., 1994b). The QTL explained about 20% of the phenotypic variance for the two fatness traits in the F_2 generation of the wild pig intercross. The effect of the QTL was largely additive and the estimated differences between animals homozygous for the wild pig and Large White alleles were +.76% for abdominal fat percentage and +4.6 mm for average back fat depth. Thus, in both cases the estimated effects were consistent with the differences between the two founder lines, namely that the wild pigs apparently deposit more fat than purebred Large White animals. The latter population has been exposed to intense selection for lean growth. It is currently not known whether this huge QTL effect is due to a single gene or to several linked genes forming a haplotype effect. This will be investigated by following the segregation at this locus in subsequent generations. Moreover, comparative gene mapping of pig chromosome 4 will be of great interest. The comprehensive human and mouse maps may then be used to identify putative candidate genes located in a homologous region that may play a crucial role in the genetic control of fat metabolism in mammals.

No comprehensive QTL mapping experiments involving Chinese and European domestic pigs have yet been reported. Janss et al. (1994) recently reported that they have identified a major gene for intramuscular fat in a Meishan/Large White crossbred population. The gene was detected by statistical analysis using Gibbs sampling and the effect of the QTL was very large, about three phenotypic standard deviations. It should be straightforward to map a gene with such a large effect. Furthermore, an association of polymorphism at the PIT1 locus, located on chromosome 13, with early growth and carcass traits has recently been found utilizing crosses between Chinese and European breeds (Yu et al., 1995). Interestingly, the same region on chromosome 13 showed a significant QTL effect on early growth in our wild pig/Large White intercross (Andersson et al., 1994). It is thus possible that the two studies have revealed the same QTL, which may be identical to the PIT1 locus.

Future Developments

Linkage mapping has already proven to be a powerful strategy for the identification of genes affecting meat quality traits in pigs as illustrated in this review. However, there are a number of important improvements that can be made to make genome analysis an even more efficient strategy for genetic dissection of complex traits.

Improved Phenotypic Records A prerequisite for the successful use of gene mapping strategies is the presence of accurate phenotypic records. As stated above, meat quality is a highly complex trait and there is a strong need for adequate phenotypic measurements. Moreover, to map genes with a moderate or small effect segregating in outbred populations we need to analyze large pedigrees (thousands of animals), and thus methods suitable for large scale screenings must be used.

An Improved Porcine Genetic Map A good, low-resolution pig genetic map with markers every 5 to 10 cM is now available (Ellegren et al., 1994; Rohrer et al., 1994; Archibald et al., 1995). This map has a sufficient resolution for the initial mapping of trait loci. However, a high-resolution genetic map (1 cM spacing between markers) is necessary for more precise mapping. This is well illustrated by the recent mapping of the RN gene to pig chromosome 15, which calls for the development of additional microsatellite markers on this chromosome. The mapping of the RN gene to a 1 cM interval would facilitate the utilization of marker-assisted selection as well as the eventual cloning of the gene.

An Improved Comparative Map Comparative mapping will become more important and more powerful in the future. It is expected that large-scale sequencing of the human genome will be started in 1996 and that the sequence will be completed by the year 2001. There are also several large-scale human cDNA sequencing projects that currently generate a huge amount of new information on expressed sequences. By combining the information from genome and cDNA sequence data it will be possible to construct high-resolution maps of the great majority of human coding sequences. Information on tissue expression patterns as well as other gene characteristics will also be rapidly accumulating. This information will be an invaluable tool also for animal geneticists when high-resolution comparative maps have been developed.

Improved Genotyping Methods Large pedigree materials (thousands of animals) will be needed to map genes with moderate or small phenotypic effects in commercial, outbred populations. There is therefore a strong need to further improve methods for large-scale genotyping analyses to make such studies feasible. This can be done by automatization of several of the steps in microsatellite analysis, including interpretation and collection of genotype information from electrophoretograms. New types of genetic markers that are not based on the electrophoretic detection of polymorphism may also be developed.

Improved Strategies for Positional Cloning The ultimate goal in many QTL mapping experiments is to unravel the molecular nature of the locus. Information on the nature of QTL will advance our knowledge of the genetic background of polygenic traits and may have profound effects on future animal breeding practices. We therefore need improved methods to move from a linked marker to the gene itself (i.e., positional cloning). The development of good comparative maps is one step toward this goal since it will increase our chances to identify candidate genes in a given chromosomal region. Novel methods such as bulked segregation analyses (e.g., Horvat and Medrano, 1994) and Genetically Directed Representational Difference Analysis (GDRDA; Lisitsyn et al., 1994) may also play a role for detecting DNA polymorphism close to a trait locus.

Literature Cited

Andersson, L., I. Edfors-Lilja, H. Ellegren, M. Johansson, and L. Marklund. 1994a. Mapping trait loci by crossbreeding genetically divergent populations of domestic animals. Anim. Biotechnol. 5:225.

Andersson, L., C. S. Haley, H. Ellegren, S. A. Knott, M. Johansson, K. Andersson, L. Andersson-Eklund, I. Edfors-Lilja, M. Fredholm, I. Hansson, J. Håkansson, and K. Lundström. 1994b. Genetic mapping of quantitative trait loci for growth and fatness in pigs. Science (Wash., DC) 263:1771.

Archibald, A. L. 1991. Inherited halothane-induced malignant hyperthermia in pigs. In: J. B. Owen and R.F.E. Axford (Ed.) Breeding for Disease Resistance in Farm Animals. p 449. C.A.B. International, London.

Archibald, A. L., C. S. Haley, J. F. Brown, S. Couperwhite, H. A. McQueen, D. Nicholson, W. Coppieters, A. Van de Weghe, A. K. Winterö, M. Fredholm, N. J. Larsen, V. H. Nielsen, D. Milan, N. Woloszyn, A. Robic, M. Dalens, J. Riquet, J. Gellin, J-C. Caritez, D. Hue, G. Burgaud, L. Ollivier, J-P. Bidanel, M. Vaiman, C. Renard, H. Gelderman, R. Davoli, D. Ruyter, E.J.M. Vestege, M.A.M. Groenen, W. Davies, B. Höyheim, A. Keiserud, L. Andersson, H. Ellegren, M. Johansson, L. Marklund, R. J. Miller, D. V. Anderson-Dear, E. Signer, and A. J. Jeffreys. 1995. The PigMaP Consortium linkage map of the pig *(Sus scrofa)*. Mamm. Genome 6:157.

Ellegren, H., B. P. Chowdhary, M. Johansson, L. Marklund, M. Fredholm, I. Gustavsson, and L. Andersson. 1994. A primary linkage map of the porcine genome reveals a low rate of genetic recombination. Genetics 137:1089.

Estrade, M., X. Vignon, and G. Monin. 1993a. Effect of the RN^- gene on ultrastructure and protein fractions in pig muscle. Meat Sci. 35:313.

Estrade, M., X. Vignon, E. Rock, and G. Monin. 1993b. Glycogen hyperaccumulation in the white muscle fibres of RN^- carrier pigs. A biochemical and ultrastructural study. Comp. Biochem. Physiol. 104B(2):321.

Fernandez, X., and G. Monin. 1994. A major gene affecting pork quality: The RN gene. Meat Focus International. August 1994:332.

Fuji, J., K. Otsu, F. Zorzato, S. de Leon, V. K. Khanna, J. Weiler, P. J. O'Brien, and D. H. MacLennan. 1991. Identification of a mutation in the porcine ryanodine receptor that is associated with malignant hyperthermia. Science (Wash., DC) 253:448.

Horvat, S., and J. F. Medrano. 1994. Targeting a specific genomic interval to identify RAPD markers linked to the high growth (*hg*) locus in mice. Proc. 5th World Congr. Genet. Appl. Livest. Prod. Guelph, ON, Canada 21:71.

Janss, L.L.G., J.A.M. Van Arendonk, and E. W. Brascamp. 1994. Identification of a single gene affecting intramuscular fat in Meishan crossbreds using Gibbs sampling. Proc. 5th World Congr. Genet. Appl. Livest. Prod. Guelph, ON, Canada 18:361.

Le Roy, P., J. Naveau, J. M. Elsen, and P. Sellier. 1990. Evidence for a new major gene influencing meat quality in pigs. Genet. Res. Camb. 55:33.

Lisitsyn, N. A., J. A. Segre, K. Kusumi, N. M. Lisitsyn, J. H. Nadeau, W. N. Frankel, M. H. Wigler, and E. S. Lander. 1994. Direct isolation of polymorphic markers linked to a trait by genetically directed representational difference analysis. Nat. Genet. 6:57.

Lundström, K., A. Karlsson, J. Håkansson, I. Hansson, M. Johansson, L. Andersson, and K. Andersson. 1995. Production, carcass and meat quality traits of F_2-crosses between European Wild pigs and domestic pigs including halothane gene carriers. Anim. Sci. 61:325.

Mariani, P., K. Lundström, U. Gustafsson, A.-C. Enfält, R. K. Juneja, and L. Andersson. 1996. A major locus (RN) affecting muscle glycogen content is located on pig chromosome 15. Mamm. Genome 7:52.

Marklund, L., W. Davies, H. Ellegren, M. Fredholm, B. Höyheim, M. Johansson Moller, K. Juneja, P. Mariani, W. Coppetiers, and L. Andersson. 1996. A comprehensive pig linkage map based on a wild pig - Large White intercross. Anim. Genet. 27 (In press).

Milan, D., P. Le Roy, N. Woloszyn, J. C. Caritez, J. M. Elsen, P. Sellier, and J. Gellin. 1995. The RN locus for meat quality maps to pig chromosome 15. Genet. Sel. Evol. 27:(In press).

Monin, G., and P. Sellier. 1985. Pork of low technological quality with a normal rate of muscle pH fall in the immediate post-mortem period: The case of the Hampshire breed. Meat Sci. 13:49.

Naveau, J. 1986. Contribution à l'étude de déterminisme génétique de la qualité de la viande porcine. Héritabilité du rendement technologique Napole. J. Rech. Porcine Fr. 18:265.

Rohrer, G. A., J. A. Leeson, J. W. Keele, T. P. Smith, and C. W. Beattie. 1994. A microsatellite linkage map of the porcine genome. Genetics 136:231.

Yu, T.-P., C. K. Tuggle, C. B. Schmitz, and M. F. Rothschild. 1995. Association of PIT1 polymorphisms with growth and carcass traits in pigs. J. Anim. Sci. 73:1282.

Improving Dairy Cattle Using Marker Data

Margaret R. Dentine

Department of Dairy Science, University of Wisconsin, Madison, WI 53706, U.S.A.

Abstract Use of molecular data will require dense genomic maps of the bovine with sufficiently polymorphic markers to track quantitative trait loci (QTL). Use of these maps for dairy cattle improvement is facilitated by pedigree structures over several generations and organized performance testing. Once QTL-marker associations are accurately estimated, young animals with favorable allelic combinations can be selected. Continued performance recording and evaluation of marker-QTL associations will be required for several reasons. Initial associations of markers and QTL will be family-specific and will need estimation in each generation to detect recombination. Even very close markers that do not recombine with QTL will acquire other positive or negative linkages due to recombinations away from initial marker-QTL position. Second, as some QTL alleles reach high frequencies, marker-assisted selection will shift to other loci not yet utilized. Third, traits under selection are likely to change so that the new selection goals or traits will require estimates of desirable marker-QTL combinations with different economic weighting. In each case, marker data will be combined with performance data on the individual and on relatives to make optimal progress. Marker-assisted selection will provide additional progress provided marker information can be accurately and efficiently included in ongoing selection programs.

Key Words: QTL, Marker-Assisted Selection

Introduction

Recent genetic mapping efforts in cattle have developed polymorphic molecular markers well distributed across the bovine genome (Barendse et al., 1994; Bishop et al., 1994; Georges et al., 1995). Individual genes with effects on performance may not yet be mapped, but linkage relationships allow tracking of segregation of large segments of chromosomes that do have significant effects on production and health traits (Cowan et al., 1990; Hoeschele and Meinert, 1990; Van Eijk et al., 1992; Andersson-Eklund and Rendel, 1993; Dentine, 1994; Ron et al., 1994; Georges et al., 1995). Maps with polymorphic markers at even 20 cM spacing have been suggested in marker-assisted selection in plants (Hospital et al., 1992; Edwards and Page, 1994). Use of flanking markers can characterize the interval between markers and can facilitate the passage of whole segments to progeny by rejecting those progeny that have recombination events in the interval (Lander and Botstein, 1989; Lande and Thompson, 1990). If finer mapping is desired or if polymorphic markers are not located close to QTL, methods to develop markers in target areas are available (Parrish and Nelson, 1993). Previous studies of marker utility have assumed that parents may not be polymorphic at close markers (Smith and Simpson, 1986). New marker systems ensure that every parent will have some linked heterozygous marker that can be followed to progeny. Marker-assisted selection of dairy cattle has become a reality and is already being used for some selections of young bulls.

Advantages to Marker-Assisted Selection

Dairy cattle genetic improvement has some particular advantages in considering use of molecular markers. Historical pedigree structures are well-documented and include performance records in a consistent national format. Accurate evaluations of genetic merit are available from use of phenotypic records in an animal model. Substantial power to detect quantitative trait loci (QTL) can be gained by use of three generation pedigrees (Weller et al., 1990; Moreno-Gonzalez, 1993). Segregations of markers from grandsire to sons can be tracked and performance records of granddaughters in progeny test can be used to estimate allele substitution effects (Fernando and Grossman, 1989; Hoeschele and VanRaden, 1993). Additional generations or collateral relationships provide independent tests of estimates for confirmation. These estimates from studies over several generations require less genotyping and concentrate investigations on young bulls in artificial insemination where DNA from semen is available. These bulls have been selected on pedigree but would not be selected for marker segregation from the sire, thus providing an estimate with less bias than that in the population of bulls already selected on progeny test results. (Mackinnon and Georges, 1992).

Current selection schemes utilize very few sires of sons and selection of these sons is one of the most influential decisions for genetic progress (Van Vleck, 1977). Thus, knowledge of marker-QTL relationships in a few key individuals may provide additional genetic progress even if the rest of the population is not genotyped (Beckmann and Soller, 1983; Kashi et al., 1990; Brascamp et al., 1993). High selection intensity in sires of sons also increases the allele frequency of any genetic defects carried by these few individuals. Marker data can be used to eliminate these defects as the causative loci are detected and mapped.

Use of pedigrees over several generations implies that the original marker-QTL estimates may be historical and that information will be utilized in sons or daughters of the original bulls where the estimates are made. Critics of this detection scheme point out that information will be known too late for use in selecting progeny of the original bull. Use of these same key individuals as sires of bull-dams or grandsires of young bulls will allow use of these marker-QTL estimates in subsequent generations.

Selection of young bulls for progeny testing can take advantage of selection among multiple offspring of the same mating through embryo transfer. These individuals can be ranked easily on genetic merit by using simple adjustments due to marker information because pedigrees within full-sib families would be identical. Use of marker information has the highest utility at this stage of selection (Dentine, 1992; Meuwissen and van Arendonk, 1992), and segregation information is independent of other performance data on relatives. With few sires of sons being used, genetic defects from these sires can also be eliminated at this stage to remove carriers from the population while retaining other superior genes.

Complications to Use of Molecular Data

Differentiating between young bulls of equal pedigree merit also illustrates one of the limitations of marker data. If molecular data are used to discriminate between full-sibs, the number of available genotypes will be limited by family size. Even if only the few largest marker-QTL associations are utilized due to costs of genotyping, no individual of the most favorable genotype may be available (Zehr et al., 1992). Compromises to select better individuals from the available array of genotypes will involve use of both marker data and phenotypic records on relatives to estimate predicted transmitting ability (Lande and Thompson, 1990; van Arendonk et al., 1994). Schemes for preselecting embryos ("Velogenetics", Georges and Massey, 1991) have the advantage of being able to search more offspring for favorable genotypes but will increase costs. Use of individuals with the most favorable genotypes for cloning has been suggested as a method to recover additional costs (Bishop et al., 1995). Although the small number of candidates may limit marker-assisted selection within family, use of these data represents an additional round of selection otherwise unavailable at an early age. Even low selection intensities can be utilized favorably (Meuwissen and van Arendonk, 1992).

Initially, major loci may not be characterized directly as anonymous microsatellite loci form the largest group of mapped molecular markers (Barendse et al., 1994; Bishop et al., 1994; Georges et al., 1995). These polymorphisms usually occur outside of coding regions of genes and may not directly influence performance. Use of these loci in interval mapping to track segregation will result in good genome coverage and likely detect the chromosome segments containing QTL. However, high linkage disequilibrium is expected due to selection and drift in purebred populations (Dentine, 1992). This disequilibrium will make fine mapping of individual loci less easy and requires larger sample sizes for refining the location of single QTL (Robinson et al., 1991). Additionally, detection of differential performance based on allelic differences is a more challenging task than finding a defective gene (Parish and Nelson, 1993). Larger sample sizes, higher costs, and more candidate genes will make this a difficult task.

Necessary knowledge of locus position and function are quite different when livestock improvement is desired than those details required for positional cloning and human gene therapy. Selection will likely proceed without exact knowledge of genetic differences in most cases. Allele frequencies of large QTL may increase through use of linked markers without knowledge of which genes are involved. Use of linked loci will require separate estimates within individual sires due to different linkage relationships (Geldermann et al., 1985). Marker-QTL associations will be the initial material for selection and in some cases the only criteria needed.

Marker-assisted selection has been suggested as an aid in changing dairy cattle populations rapidly in light of changing economic structure or selection goals. Although markers could provide quicker shifts in directions of selection, one problem exists in changing to goals that include new traits. No marker-QTL associations can be detected without phenotypic records. In the United States, few performance records would be available on traits such as disease resistance, reproduction, or milk quality. Thus, marker-QTL associations would not be known and genetic improvement would not be available until data were collected. This dependence on phenotypic records for marker-assisted selection is also a concern with changing attitudes toward the role of various organizations toward maintaining or sharing records on yields, conformation, and calving difficulty. Changes in the current national data collection schemes could complicate the discovery and utilization of QTL.

New Opportunities

A few genetic situations provide specific advantages to marker-assisted selection. Additional opportunities include genetic improvement using loci that have nonadditive variation. Loci with dominance effects will need continual estimation of allele substitution effects because these will be dependent on population allele frequencies. Epistatic combinations could be selected more easily with molecular markers than with phenotypic records. Markers can be used to select those individuals that break unfavorable linkage relationships such as that between weaver in Brown Swiss cattle and linked QTL for milk yield (Georges et al., 1993). This concept of teasing apart sections of the genome can also be used to circumvent genetic correlations. Markers could identify alleles that do not have pleiotropic effects that follow the average genetic correlation (Cowan et al., 1992; de Koning and Weller, 1994).

Even if marker-assisted selection becomes widespread, phenotypic records will still be important to dairy cattle improvement. Initially, only a small portion of the genome will be tracked with molecular markers. Use of marker data alone will not be superior to current selection systems (Zhang and Smith, 1992). The unmarked portion of genetic variation will be important and will be evaluated with statistical models. Costs of detection for marker-QTL substitution effects will favor screening the genome with moderately spaced markers; initial estimates will not be precise, and further refinement with phenotypic records will be needed. Clusters of genes with smaller effects will be indistinguishable from major loci with the estimation procedures available (Stam, 1986; Dekkers and Dentine, 1991). These loci are valuable in selection programs but will disassociate over time and require reestimation of marker-QTL associations as clusters break up (Gimelfarb and Lande, 1994).

Finally, even individual loci with large effects directly detected with molecular data may acquire new associations (Figure 1). The original markers M and N detect the favorable Q (in contrast to less favorable q alleles detected with m and n). If only

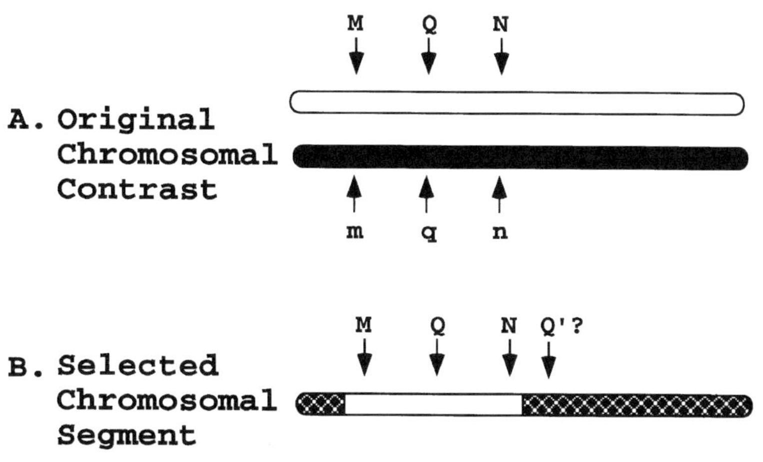

Figure 1. Chromosome configurations in marker-assisted selection. A: Alternate chromosomes showing flanking markers with alleles M and N associated with the more favorable allele Q. B: After selection, M-Q-N relationship maintained, but recombination changes flanking sections, possibly introducing linkage to additional QTL at locus Q′.

individuals with transmission of both M and N alleles are selected, the intact segment containing Q increases in frequency. Outside of the M-Q-N segment, recombinations are occurring that may introduce new associations. After time, some individuals may have new loci, Q′ with positive or negative effects, associated with M and N. This "hitchhiking" effect can occur even if the first QTL is known without errors by direct molecular data (Kaplan et al., 1989; Dekkers and Dentine, 1991).

As some QTL alleles reach high frequencies, marker-assisted selection will shift to other loci not yet utilized. Sources of additional genetic variation may include new clusters of favorable alleles, resorting of favorable alleles originally in unfavorable linkages, favorable alleles previously undetected due to low frequency, mutations, and alleles introgressed from other populations. Genetic variance in dairy cattle has not decreased under intense selection. Marker-assisted selection is likely to be a continuing process of utilizing both marker and phenotypic data in exploiting this variation.

As economics of dairy production and prices for inputs and outputs change, goals of selection will probably shift. Molecular data could change direction of genetic improvement more quickly than phenotypic data alone if records are available on all traits. Desirable marker-QTL associations may vary under different economic weighting of traits but would provide customized selection for those desiring other goals. In each case, marker data will be combined with performance data on the individual and on relatives to make optimal progress. The most likely future will include an interplay of molecular and phenotypic information used for dairy cattle improvement with continual evaluation of relationships between the two types of data. Most efficient genetic progress will include evaluations of genetic merit using all available sources of information.

Literature Cited

Andersson-Eklund, L., and J. Rendel. 1993. Linkage between amylase-1 locus and a major gene for milk fat content in cattle. Anim. Genet. 24:101.

Barendse, W., S. M. Armitage, L. M. Kossarek, A. Shalom, B. W. Kirkpatrick, A. M. Ryan, D. Clayton, L. Li, H. L. Neibergs, N. Zhang, W. M. Grosse, J. Weiss, P. Creighton, F. McCarthy, M. Ron, A. J. Teale, R. Fries, R. Q. McGraw, S. S. Moore, M. Georges, M. Soller, J. E. Womack, and D.J.S. Hetzel. 1994. A genetic linkage map of the bovine genome. Nat. Genet. 6:227.

Beckmann, J. S., and M. Soller. 1983. Restriction fragment length polymorphisms in genetic improvement: Methodologies, mapping and costs. Theor. Appl. Genet. 67:35.

Bishop, M. D., G. A. Hawkins, and C. L. Keefer. 1995. Use of DNA markers in animal selection. Theriogenology 43:61.

Bishop, M. D., S. M. Kappes, J. W. Keele, R. T. Stone, S.L.F. Sunden, G. A. Hawkins, S. S. Toldo, R. Fries, M. D. Grosz, J. Yoo, and C. W. Beattie. 1994. A genetic linkage map for cattle. Genetics 136:619.

Brascamp, E. W., J.A.M. van Arendonk, and A. F. Groen. 1993. Economic appraisal of the utilization of genetic markers in dairy cattle breeding. J. Dairy Sci. 76:1204.

Cowan, C. M., M. R. Dentine, R. L. Ax, and L. A. Schuler. 1990. Structural variation around prolactin gene linked to quantitative traits in an elite Holstein sire family. Theor. Appl. Genet. 79:577.

Cowan, C. M., M. R. Dentine, and T. Coyle. 1992. Chromosome substitution effects associated with κ-casein and β-lactoglobulin in Holstein cattle. J. Dairy Sci. 75:1097.

Dekkers, J.C.M., and M. R. Dentine. 1991. Quantitative genetic variance associated with chromosomal markers in segregating populations. Theor. Appl. Genet. 81:212.

de Koning, G. J., and J. I. Weller. 1994. Efficiency of direct selection on quantitative trait loci for a two-trait breeding objective. Theor. Appl. Genet. 88:669.

Dentine, M. R. 1992. Marker-assisted selection in cattle. Anim. Biotech. 3(1):81.

Dentine, M. R. 1994. Effects of individual loci on performance during lactation. Proc. 5th World Congr. Genet. Appl. Livest. Prod., Guelph, ON, Canada 19:291.

Edwards, M. D., and N. J. Page. 1994. Evaluation of marker-assisted selection through computer simulation. Theor. Appl. Genet. 88:376.

Fernando, R. L., and M. Grossman. 1989. Marker assisted selection using best linear unbiased prediction. Genet. Sel. Evol. 21:467.

Geldermann, H., U. Pieper, and B. Roth. 1985. Effects of marked chromosome sections on milk performance in cattle. Theor. Appl. Genet. 70:138.

Georges, M., A. B. Dietz, A. Mishra, D. Nielsen, L. S. Sargeant, A. Sorensen, M. R. Steele, X. Zhao, H. Leipold, J. E. Womack, and M. Lathrop. 1993. Microsatellite mapping of the gene causing weaver disease in cattle will allow study of an associated quantitative trait locus. Proc. Natl. Acad. Sci. USA 90:1058.

Georges, M., D. Nielsen, M. Mackinnon, A. Mishra, R. Okimoto, A. T. Pasquino, L. S. Sargeant, A. Sorensen, M. R. Steele, X. Zhao, J. E. Womack, and I. Hoeschele. 1995. Mapping quantitative trait loci controlling milk production in dairy cattle by exploiting progeny testing. Genetics 139:907.

Georges, M., and J. M. Massey. 1991. Velogenetics, or the synergistic use of marker-assisted selection and germ-line manipulation. Theriogenology 35:151.

Gimelfarb, A., and R. Lande. 1994. Simulation of marker assisted selection in hybrid populations. Genet. Res. Camb. 63:39.

Hoeschele, I., and T. R. Meinert. 1990. Association of genetic defects with yield and type traits: The weaver locus effect on yield. J. Dairy Sci. 73:2503.

Hoeschele, I., and P. M. VanRaden. 1993. Bayesian analysis of linkage between genetic markers and quantitative trait loci. II. Combining prior knowledge with experimental evidence. Theor. Appl. Genet. 85:946.

Hospital, F., C. Chevalet, and P. Mulsant. 1992. Using markers in gene introgression breeding programs. Genetics 132:1199.

Kaplan, N. L., R. R. Hudson, and C. H. Langley. 1989. The "hitchhiking effect" revisited. Genetics 123:887.

Kashi, Y., E. Hallerman, and M. Soller. 1990. Marker-assisted selection of candidate bulls for progeny testing programmes. Anim. Prod. 51:63.

Lande, R., and R. Thompson. 1990. Efficiency of marker-assisted selection in the improvement of quantitative traits. Genetics 124:743.

Lander, E. S., and D. Botstein. 1989. Mapping Mendelian factors underlying quantitative traits using RFLP linkage maps. Genetics 121:185.

Mackinnon, M. J., and M. Georges. 1992. The effects of selection on linkage analysis for quantitative traits. Genetics 132:1177.

Meuwissen, T.H.E., and J.A.M. van Arendonk. 1992. Potential improvements in rate of genetic gain from marker-assisted selection in dairy cattle breeding schemes. J. Dairy Sci. 75:1651.

Moreno-Gonzalez, J. 1993. Efficiency of generations for estimating marker-associated QTL effects by multiple regression. Genetics 135:223.

Parrish, J. E., and D. L. Nelson. 1993. Methods for finding genes: A major rate-limiting step in positional cloning. Genet. Anal. Tech. Appl. 10(2):29.

Robinson, W. P., A. Cambon-Thomsen, N. Borot, W. Klitz, and G. Thomson. 1991. Selection, hitchhiking and disequilibrium analysis at three linked loci with application to HLA data. Genetics 129:931.

Ron, M., M. Band, A. Yanai, and J. I. Weller. 1994. Mapping quantitative trait loci with DNA microsatellites in a commercial dairy cattle population. Anim. Genet. 25:259.

Smith, C., and S. P. Simpson. 1986. The use of genetic polymorphisms in livestock improvement. J. Anim. Breed. Genet. 103:205.

Soller, M., and J. S. Beckmann. 1990. Marker-based mapping of quantitative trait loci using replicated progenies. Theor. Appl. Genet. 80:205.

Stam, P. 1986. The use of marker loci in selection for quantitative characters. In: C. Smith, J.W.B. King, and T. C. McKay (Ed.) Exploiting New Technologies in Animal Breeding: Genetic Developments. p 170. Oxford University Press, Oxford, U.K.

van Arendonk, J.A.M., B. Tier, and B. P. Kinghorn. 1994. Use of multiple genetic markers in prediction of breeding values. Genetics 137:319.

Van Eijk, M.J.R., J. A. Stewart-Haynes, J. E. Beever, R. L. Fernando, and H. A. Lewin. 1992. Development of persistent lymphocytosis in cattle is closely associated with *DRB2*. Immunogenetics 37:64.

Van Vleck, L. D. 1977. Theoretical and actual genetic progress in dairy cattle. In: E. Pollak, O. Kempthorne, and T. B. Bailey, Jr. (Ed.) Proc. Int. Conf. Quant. Genet. p 543. Iowa State University Press, Ames.

Weller, J. I., Y. Kashi, and M. Soller. 1990. Power of daughter and granddaughter designs for determining linkage between marker loci and quantitative trait loci in dairy cattle. J. Dairy Sci. 73:2525.

Zehr, B. E., J. W. Dudley, and J. Chojecki. 1992. Some practical considerations for using RFLP markers to aid in selection during inbreeding of maize. Theor. Appl. Genet. 83:704.

Zhang, W., and C. Smith. 1992. Computer simulation of marker-assisted selection utilizing linkage disequilibrium. Theor. Appl. Genet. 83:813.

The Use of Markers in Poultry Improvement

R.W. Fairfull[1]

Centre for Food and Animal Research, Agriculture and Agri-Food Canada, Ottawa, Ontario, Canada

Abstract Markers have several uses in poultry improvement: 1) providing a tool for understanding the genome and how it works; 2) establishing genetic relationships; 3) prediction of heterosis; 4) genomic selection; and 5) marker-assisted selection. Expediting learning about the genome and how it works is the most important facet of marker technology because knowledge is the basis of authentic progress. In poultry, useful genes, such as genes to allow "autosexing" at hatch or genes to reduce heat stress, have been introduced into commercial and other important stocks. This required several generations of backcrossing to allow recombination and permit the elimination of unwanted genes with no assurance that all undesirable genes would be eliminated. Genomic selection using markers would greatly accelerate the introgression of useful genes. Flanking markers would ensure that only the desired gene was introduced. Markers may provide a useful basis for predicting heterosis and other genetic effects associated with crossing. Commercial performance is enhanced by taking advantage of heterosis and other crossing effects in many stocks of poultry. Markers provide the possibility of determining pedigrees or assessing inbreeding levels in stocks where that information may not be possible otherwise. Finally, marker-assisted selection may allow opportunities for enhanced improvement of economic traits, especially in cases in which direct selection would not otherwise be possible.

Key Words: Poultry, Markers, Genome, Introgression, Selection, Heterosis, Pedigrees

Introduction

Markers are not new in poultry research and improvement. Markers of various types, from phenotypic markers to proteins, have been used in poultry for some time. Many uses of markers are not novel either; markers have been used for mapping and in selection, although DNA-based techniques have increased the power of markers. Thus, molecular markers are part of a continued refinement of animal genetic technology that will go on to generate new techniques and insights. At some time in the future, markers will be genes controlling characters of interest, not inert road signs. However, what is anticipated now is not necessarily the future. The unexpected, the serendipitous, will contribute new techniques and insights for genetic development. Markers will be most valuable in developing information on what we cannot do, or cannot do well, at present. Markers will be of less benefit in areas that can be addressed effectively by classical breeding techniques.

Poultry have a role as a pilot organism in the development, authentication, and application of molecular genetic knowledge in livestock. To some degree, poultry perform the role of genetic model in developing knowledge and systems for other

[1]Present address: Argo Breeding Company International Inc., P.O. Box 631940, Nacogdoches, TX 75963-1940, U.S.A.

economic species now. Poultry are suited to the role of model for other livestock species for several reasons. They are genetically and physiologically complex and they have intrinsic economic value. Also, poultry have a meat model (the broiler) and a reproduction model (the layer). In addition, poultry have a comparatively short generation interval. Each bird is relatively cheap and the cost of rearing and maintaining large populations is comparatively low. Also, economical empirical testing of new theories and applications quickly should be possible with poultry.

Understanding the Genome and How It Works

Most recent quantitative genetic theory is based on the infinitesimal model. This model is unrealistic in its basic assumptions of the organization of the genome, the size and distribution of genetic effects, and linearity of genetic effects. It is, however, a very convenient model because of its simplicity, allowing one to ignore many otherwise irritating problems. Many "proofs" in quantitative genetics are based on theory using the infinitesimal model.

Science grows degenerate when it banks too much on "thought experiments" without empirical verification. Occasionally, working scientists have trouble discriminating between theory (or theory-based simulation) and reality. New "experiments" are based on results of previous simulations that were in turn based on earlier simulations. Theory is not an end point and theoretical proof is not verification. If it were, we could simply show Nature a robust, mature, rational theory to revise the universe. Of course, then, the universe might not work—except in theory.

Many techniques, including all statistical estimation and simulation, are model-dependent. Models are the result of a series of assumptions that have implications, whether the user understands them or not. Often, conclusions are sensitive to the assumptions. Hence, proofs resting on quantitative genetic theory must be viewed with skepticism about the validity of the underlying model. We do not understand the genome very well at present, but sound conclusions and applications demand knowledge.

Genome Mapping

Gene mapping will give us considerable knowledge quickly and the keys to extending it. Molecular markers are essential tools in genome mapping. Mapping of potential quantitative trait loci (QTL) is proceeding at an astonishing rate with a worldwide cooperative effort (Archibald, 1994). Several strategies have been proposed for mapping QTL. The best of them have advantages over other methods, but they all have weaknesses (Bulfield, 1992; Muir, 1994). As Bulfield (1992) states, "We have, however, to realize that these approaches are imperfect and we have a long way to go before we have a straight-forward method for identifying genes for commercially important traits."

Gene mapping as it now stands has several difficulties. The models used are additive, but it is known that both dominance and epistasis are important for commercial traits in chickens (Fairfull et al., 1987). The size and placement of QTL are uncertain. Genotype-environment interactions (G×E) affect the genome as a whole. How

G×E are distributed across the genome and how G×E affect QTL estimates is not known. Nevertheless, they will be affected. Genetic background may affect the value of a QTL and the estimate of the relationship between marker and QTL (Dunnington et al., 1993). Also, in initial mapping efforts, only the linked segments with the largest effects are going to be detected as QTL. For example, in a recent cooperative mapping effort in pigs, the smallest QTL detectable represented about 7% of the total variation of the trait mapped (Haley et al., 1995). Thus, the initial mapping efforts of poultry biotechnologists are likely to yield little of direct, immediate application to practical poultry improvement because these findings must be applied in high-performance, highly selected lines where QTL of large effects likely have little remaining usable genetic variation. However, if a QTL with large effect is found in a highly selected line, extreme caution may be advisable because the only explanation of such a result is a negative correlation of the QTL with fitness.

All these problems will ultimately be of little consequence to the mapping effort. The detection and specification of QTL with gross effects will allow the mapping of QTL with smaller effects. Mapping errors will result only in a necessary inefficiency in the mapping process. However, the mapping effort will be enhanced by contributions from other molecular techniques and information gained in the mapping of other species. The genes affecting quantitative traits in poultry will be found and cloned (Bulfield, 1992), variants will be classified, and QTL themselves will become the markers, detectable with no greater effort or cost than linked markers. Nonetheless, the problems of genetic background, epistasis, and so on, remain.

Obstacles must be kept in mind concerning mapping. First, the genes for a trait that was not measured in a mapping population cannot be mapped. Support must be found for centers willing to put effort into complex traits that are more costly to assess. Second, any single population used for mapping may not permit the mapping of all QTL associated with any particular trait. Some consideration should be given to the characterization and selection of populations to be used for mapping the more complex and difficult traits.

Finding QTL is not an ending. It is a departure point for another leg of the journey that will allow us finally to begin to understand the genetics and physiology of species. The information generated will form the basis of an authentic revolution in genetic progress and techniques to achieve this progress. Nevertheless, as the amount of information available grows, the more complex a subject and its applications become. Thus, make no mistake: all problems will never be solved. Also, the complexity of estimating the value of the thousands of genes involved in economic and fitness traits and their interactions, and their application in breeding value estimation, will be much greater.

Establish Genetic Relationships

Markers have broad application in various roles associated with identification and pedigree determination. It is now established that markers are highly reliable in individual identification (Hillel et al., 1993a) to such a degree that it is accepted as legal proof of identity in most jurisdictions. Pedigree determination is a simple extension of individual identification, and the use of markers for this purpose has revolutionized the study of feral avian populations in field studies of behavior, genetics, ecology, and related areas. In poultry, this technology can be beneficial in many cases: preventing or correcting pedigree errors in important lines, especially small, special-purpose lines in which pedigree errors could have large negative consequences; recovering pedigrees following an accident; or retrospective analysis of genetic events (where appropriate tissue samples exist). This type of marker technology is also valuable at the population level, where it can be used to determine inbreeding levels (Kuhnlein et al., 1990) and study population phenomena such as genetic drift, mutation rate, and so on.

Prediction of Heterosis

The use of crosses of breeds or strains is standard practice in commercial broilers and layers. Heterosis can be advantageous in parental and commercial stocks, where crosses are the final product. In sire lines, useful heterosis can be present for viability and male reproductive characters. Dam performance can be improved by heterosis for viability, fertility, hatchability, egg production, egg size, feed conversion, and sexual maturity. Commercial layer crosses usually possess worthwhile heterosis for viability, hatchability, egg production, egg size, feed conversion, and sexual maturity, and commercial broilers typically manifest beneficial heterosis for viability, hatchability, growth, and feed conversion.

Heterosis for many traits is substantial in chickens: commonly 10% or more for egg production traits; 3 to 7% for sexual maturity; 2 to 5% for body weight and low, but consistent (1 to 3%), for egg weight (Fairfull and Gowe, 1986). Heterosis for viability is complex and variable, but often high. Both dominance and epistasis are important for commercial traits in chickens (Fairfull et al., 1987).

Heterosis and Genetic Distance

Gavora et al. (1993, 1996) used DNA fingerprints (DFP) to assess the predictability of heterosis for a broad array of traits in chickens. The correlation coefficients between genetic distance (band sharing) of parental populations and heterosis were significant for age at first egg, egg production traits, egg weight and mature body weight in

Table 1. Correlations (%) of heterosis with genetic distance (band sharing) for Leghorn crosses (best probe combination)

	Egg production cycle	
Trait	133 to 496 days	546 to 797 days
Age at first egg, d	87**	—
Hen-housed egg production, no.	–78**	–41
Egg production rate, %	–69**	–28
Egg weight, g	40*	–42*
Body weight, kg	–68**	–71**

*$P < .05$, **$P < .01$.

the first production cycle (133 to 497 days) and for egg weight and body weight in the second cycle (546 to 797 days) (Table 1). Regressions on band sharing explained significant proportions of the variation in heterosis for age at first egg (78.4%), egg production (60.2%) and mature body weight (46.4%).

As might be expected, correlations between genetic distance and heterosis were generally highest in traits with large heterosis, but this was not a linear result. What was surprising was that the highest correlations were obtained with traits in which heterosis involved epistasis. Of course, these were the traits with the greatest heterosis.

Synopsis

The use of DFP probes to predict heterosis in poultry shows promise. The predictability of heterosis varies by trait, by probe, and with genetic background. Predictions can probably be improved by the addition of suitable markers as predictability seemingly increased with number of bands assessed. Nevertheless, these results should be viewed with caution; Bernardo (1994) found similar associations in maize only when results from related hybrids were added to the model.

Genomic Selection

In the past, the need to introduce useful genes into improved populations has developed, and this need will arise at times in future. This must be done quickly, safely, and cheaply with little or no loss of commercial value in the line or lines involved. Inevitably, the source of many of these genes will be donor populations with undesirable characteristics. The difficulty is to introduce the desirable gene into the host genome (introgression) with as little of the undesirable genome as possible while retaining beneficial genetic variation, controlling the rate of inbreeding, and maintaining commercial performance. The introgression problem with transgenic animals is similar (see Hillel et al., 1993b).

The introduction of "autosexing" (slow feathering) genes into commercial chicken stocks illustrates conventional introgression and possible attendant problems. Commercial companies developed "autosexing" lines by introducing the slow feathering gene to some of their lines initially by crossing. This was followed by many generations of backcrossing coupled with phenotypic selection to reduce the undesirable genetic contributions of the donor genotype. When these lines were released, their commercial performance was excellent. After some years on the market some of these lines encountered performance problems associated with avian leukosis virus (ALV) infection. Sales of these lines were restricted with a consequent loss of business. When these slow feathering lines were studied, it was found that a deleterious endogenous viral gene, *ev21*, which was closely linked (or part of the same major gene) to slow feathering (Bacon et al., 1988), had been incorporated into the host genome along with the slow feathering gene. The use of the slow feathering gene to permit feather sexing necessitated eliminating ALV from the parent populations.

Markers and Introgression

The ideal situation for introgression would be to have equally spaced markers throughout the host genome distanced such as to give a high probability of detecting recombinants. Tightly linked flanking markers (TLFM) bracketing the gene of interest in the donor genome would allow recovery of the TLFM system with maximal recombination with the host genome. In contemplating the introgression problem, two types of independently segregating genomic segments (chromosomes or part of chromosomes) need to be considered: 1) those without the gene to be introgressed and 2) that which has the gene to be introgressed. In the backcross generations, host genomic segments of the first type not containing recombinants with the donor would be selected; in the segment containing the TLFM system, the recombinants with the highest proportion of the host genome carrying the TLFM system would be chosen.

The efficiency of introgression is related to the length of time required to obtain a new line with the donor gene satisfactorily incorporated into the host genome. Simulation studies have shown that the use of markers allows more efficient and complete introgression than conventional methods (Groen and Timmermans, 1992; Hospital et al., 1992; Hillel et al., 1993b). The use of genomic selection based on markers speeds the introgression process by at least two generations and can reduce the interval for each generation by 100 to 200 or more days depending on assumptions. Also, markers increase the efficiency and completeness of elimination of the other parts of the donor genome. Backcrossing is the most rapid way to recover the average genetic value of the host genome. This, of course, assumes no phenotypic selection during the early introgression process because that would lengthen the time interval of the introgression process and its benefit would not compensate for the loss of time.

Marker-Assisted Selection (MAS)

In practical selection programs, the problem is to find the optimal selection program, not the maximal one. Selection programs must be cost-effective because business is built on profits (Emsley, 1993), not simply genetic gain (for a more comprehensive discussion see Muir, 1994). In the short term, advanced technologies may have little impact on poultry breeding practices if they cannot be shown to be both effective and profitable.

The optimal selection program will depend on many factors, including competitive needs for improving traits, genetic parameters, measurement costs, performance levels of traits in the stock and others used in the final product, and so on. Rate of progress is not the only consideration. Based on present technology, the rate of progress could be increased in most selection programs. However, a stock's value could be diminished by over-selection with present technologies. Extreme selection to solve short-term problems has occurred in poultry breeding and resulted in the loss of commercial value (see Gowe and Fairfull, 1982). On the other hand, progress seemingly has no foreseeable end with balanced selection (Gowe and Fairfull, 1985; Fairfull and Gowe, 1990). The same concerns are mirrored in MAS. In a simulation study, Gibson (1994) found that conventional selection achieved greater long-term responses than MAS, although MAS had better short-term responses. Since an additive infinitesimal model was used, this might be viewed as a best case scenario for MAS.

In a practical breeding situation, using MAS simply to replace or augment conventional selection without a change in age at selection or selected characters, would be premature unless some clear advantage were apparent, especially because empirical results with MAS are not yet available. If MAS is applied commercially, it will be applied in improved populations that have been under selection for decades but retain a high degree of genetic variation (Gowe and Fairfull, 1985). There is likely to be little scope to exploit QTL with gross effects in such populations. These genes with large effects will be the ones detected and characterized first; however, likely little usable allelic variation remains at these loci in highly selected populations (Rocha et al., 1995). Nonetheless, circumstances exist under which MAS could be of immediate advantage: 1) to augment selection for sex-limited traits, 2) to select for nonconventional traits that cannot be effectively or economically selected in any other way, 3) to effectively increase population size, or 4) select at very early ages when the trait is not expressed.

Weaknesses of Marker-Assisted Selection

In theory, there are no problems either technically or mathematically with the inclusion of markers in selection (e.g., Arendonk et al., 1994). The value of markers is usually connected to the fact that in some cases the phenotype may be a poor estimator of breeding value. However, how good an estimator of breeding value (QTL) is a marker? The value of a marker is a statistical estimate (of the effect of linked genes) and initially may not be as good as the direct breeding value estimates of the trait in many populations. Many more parameters must be predicted when markers are added to a model. Also, there is an uncertainty associated with the location of the QTL. Further, the estimate of an effect in a static population may not translate to the "effect" realized under selection. In addition, estimates of the value of a marker may differ between environments and between populations. Over time these problems may disappear in many populations, but they should always be considered.

Linkage Marker-assisted selection is dependent on linkage disequilibrium (Muir, 1994). Over time, recombination will disrupt linkage disequilibriums. The problem is how to establish and maintain, or estimate, the required linkage disequilibrium in a practical poultry improvement program. Various crosses of inbred lines can result in the desired disequilibrium (Darvasi et al., 1993; Moreno-Gonzalez, 1993). While this is useful for mapping, envisioning this as a basis for any poultry improvement is difficult. Any use of crosses to generate a base population for within-line selection would be questionable in the modern poultry breeding situation. Knott and Haley (1992) proposed within-progeny analyses to determine marker-QTL linkages, and this may be feasible with dairy cattle, but family sizes are unlikely to be large enough in a poultry program to use this method. Mixed-model approaches suffer from similar problems (Arendonk et al., 1994). Efficiency and power would be increased if flanking markers were available for QTL (Arendonk et al., 1994). It is likely that many QTL detectable in reasonably sized poultry populations would not be useful in MAS because QTL with gross effect would likely have little useful genetic variation in highly selected poultry populations. In evaluating MAS programs, the benchmark for comparison must be a thoughtful conventional selection program that may make better use of finite resources.

Gene Frequency Selection results in changes in gene frequencies. With changes in gene frequencies, there are changes in genetic correlations (McMillan et al., 1995) and associated changes in the relative values of alleles. The results of Dunnington et al. (1993) suggest that relationships between markers and quantitative traits vary with genetic background. Thus, the value of QTL-markers will probably change with selection because of changing genetic background and changing genetic relationships. Hence, the value of QTL-markers in selection must constantly be updated with "breeding value" estimates from phenotypes.

Effective Marker-Assisted Selection

Sex-Limited Traits Many traits of interest are expressed in only one sex in poultry, and in the opposite sex direct selection on the phenotype is not possible for that trait or traits. The prime instances are male and female reproductive traits. Two important examples of this are egg production in egg-type chickens and male reproductive performance in male parent lines in meat-type chickens. Besides being sex-limited, egg production is partially sex-linked, in that the sex chromosome carries many genes affecting egg production. Marker-assisted selection would allow direct selection on the side of the pedigree not expressing the trait, effectively increasing selection pressure on these traits (Lande and Thompson, 1990; Hillel et al., 1993a) by augmenting indirect selection, usually selection on sibling records in poultry.

The benefit of adding direct selection would be proportional to the difference between indirect and direct selection in the limited sex. For traits of low heritability, such as egg production, selection based on pedigree records (e.g., siblings, parents, etc.) can be very effective. Lande and Thompson (1990) overestimate the benefit of MAS in the sex-limited case by assuming no phenotypic selection in the limited sex; however, indirect selection of males based on female sibling records is very important and effective for egg production. Because males have a greater influence on overall selection pressure in most cases with poultry, augmenting indirect (or sibling selection) with direct selection for a gene (or genes) in males would have a greater benefit than a similar change in females. Cost effectiveness should also be considered and all problems of MAS apply.

Non-Conventional Traits Disease resistance is an ideal candidate for the application of marker selection. First, conventional selection for disease resistance (CSDR) is costly. Second, CSDR is usually indirect, using information from disease challenge populations. These studies are becoming more difficult to conduct and of more limited value as restrictions due to animal welfare concerns become more widespread and more inhibiting. Third, indirect selection using markers does not require exposure to pathogens with the resultant losses of stock, extra costs, and welfare problems. Fourth, most markers available for selection are not simply linked markers; they are directly or indirectly involved in disease resistance. In such a case, problems with artificially generating linkage disequilibrium or estimating linked effects by group every generation do not apply.

There are several candidate systems for disease resistance selection: genes of the major histocompatibility complex (Bacon, 1987; Dietert et al., 1991); viral elements (Crittenden, 1991); and molecular markers or genes (Kuhnlein et al., 1991, and unpublished data). For the most part, these are loci with direct or indirect effects on viability as well as effects on production traits (Gavora, 1990), not inert markers. It should also be noted that although we have some knowledge of the genetics of disease resistance, we need a great deal more knowledge of disease resistance genetics before a program of "marker" selection on disease resistance could be any more than a limited success.

Early Selection and Increased Population Size In multiple-stage selection, selection takes place in steps at various ages accompanied by reductions in population size at many, especially earlier, steps. For example, in broilers, selection for broiler body weight takes place during rearing and only "selected" hens are housed for assessment of egg production and other adult traits. Multiple-stage selection reduces costs for feed, for housing, and for those associated with measurement of traits. Another consequence of multiple-stage selection is to effectively increase population size with regard to the selection of some traits. Multiple-stage selection can increase selection efficiency up to 300% and reduce the cost of a breeding program (Muir, 1994). Like benefits should apply to similar MAS programs (Muir, 1994). The earlier in the life cycle that selection can be applied, the greater the benefits.

Marker-assisted selection, by severing the relationship between phenotypic performance and age of selection, would make it potentially possible to select at very early stages: at embryonic stages, at transfer from setting machines to hatching machines, or at hatch. Maximum benefit from the potential of MAS in poultry would apply to selection at or before hatch. This would minimize investment in each animal and would make feasible economic selection based on much larger population sizes than are presently possible.

Conclusion

Markers have important, immediate applications in poultry breeding and genetics: 1) providing a tool for understanding the genome and how it works, 2) establishing genetic relationships, and 3) genomic selection. Marker-assisted selection needs more development, especially empirically, before it can be considered for practical poultry improvement programs. Similarly, the use of markers to predict heterosis is an immature technology that awaits further verification and refinement before any practical implementation. In all considerations of marker technologies, real costs and relative benefit must be considered, not merely theoretical ones.

Literature Cited

Archibald, A., D. W. Burt, and J. L. Williams. 1994. Gene mapping in farm animals and birds: An overview. Proc. 5th World Congr. on Genet. Appl. Livest. Prod. 21:5.

Arendonk, J.A.M. van, H. Bovenhuis, S. van der Beek, and A. F. Groen. 1994. Detection and exploitation of markers linked to quantitative traits in farm animals. Proc. 5th World Congr. Genet. Appl. Livest. Prod. 21:193.

Bacon, L. D. 1987. Influence of the major histocompatibility complex on disease resistance and productivity. Poult. Sci. 66:802.

Bacon, L. D., E. Smith, and L. B. Crittenden. 1988. Association of the slow feathering (*K*) and an endogenous viral (*ev*21) gene on the Z chromosome of chickens. Poult. Sci. 67:191.

Bernardo, R. 1994. Prediction of maize single-cross performance using RFLPs and information from related hybrids. Crop Sci. 34:20.

Bulfield, G. 1992. Molecular approaches to the improvement of production traits. Proc. 19th World Poult. Congr. 1:479.

Crittenden, L. B. 1991. Retroviral elements in the genome of the chicken: Implications for poultry genetics and breeding. Crit. Rev. Poult. Biol. 3:73.

Darvasi, A., A. Weinreb, V. Minke, J. I. Weller, and M. Soller. 1993. Detecting marker-QTL linkage and estimating QTL gene effect and map location using a saturated genetic map. Genetics 134:943.

Dietert, R. R., R. L. Taylor, and M. F. Dietert. 1991. Biological function of the chicken major histocompatibility complex. Crit. Rev. Poult. Biol. 3:111.

Dunnington, E. A., L. C. Stallard, and P. C. Siegel. 1993. Evaluating linkage between DNA fingerprint bands and quantitative traits in chickens: Interactions. Poult. Sci. 72:1000.

Emsley, A. 1993. Aspirations of breeders in the poultry industry as manipulation of the avian genome continues. In: R. J. Etches and A.M.V. Gibbons (Ed.) Manipulation of the Avian Genome. CRC Press, Boca Raton, FL.

Fairfull, R. W., and R. S. Gowe. 1986. Use of breed resources for poultry egg and meat production. Proc. 3rd World Congr. Genet. Appl. Livest. Prod. 10:242.

Fairfull, R. W., and R. S. Gowe. 1990. Genetics of egg production in chickens. In: R. D. Crawford (Ed.) Poultry Breeding and Genetics. pp 705–759. Elsevier Science Publishers, Amsterdam, The Netherlands.

Fairfull, R. W., R. S. Gowe, and J. Nagai. 1987. Dominance and epistasis in heterosis of White Leghorn strain crosses. Can. J. Anim. Sci. 67:663.

Gavora, J. S. 1990. Disease genetics. In: R. D. Crawford (Ed.) Poultry Breeding and Genetics. pp 805–846. Elsevier Science Publishers, Amsterdam, The Netherlands.

Gavora, J. S., R. W. Fairfull, B. F. Benkel, W. J. Cantwell, and J. R. Chambers. 1993. DNA fingerprints as predictors of heterosis. Proc. 42nd Annual National Breeders Roundtable, St. Louis, MO.

Gavora, J. S., R. W. Fairfull, B. F. Benkel, W. J. Cantwell, and J. R. Chambers. 1996. Prediction of heterosis from DNA fingerprints. Genetics (In press).

Gibson, J. P. 1994. Short-term gain at the expense of long-term response with selection of identified loci. Proc. 5th World Congr. Genet. Appl. Livest. Prod. 21:201.

Gowe, R. S., and R. W. Fairfull. 1982. Some lessons from selection studies in poultry. Proc. World Congr. Sheep and Beef Cattle Breed. 1:261.

Gowe, R. S., and R. W. Fairfull. 1985. The direct response to long-term selection for multiple traits in egg stocks and changes in genetic parameters with selection. In: W. G. Hill, J. H. Manson, and D. Hewitt (Ed.) Poultry Genetics and Breeding. pp 125–146. Longman, New York.

Groen, A. F., and M.M.J. Timmermans. 1992. The use of genetic markers to increase the efficiency of introgression—a simulation study. Proc. 19th World Poult. Congr. 2:523.

Haley, C. S., S. A. Knott, and L. Andersson. 1995. Genetic mapping of quantitative trait loci for growth and fatness in pigs. In: Proc. 44th Annu. Natl. Breeders Roundtable. pp 1–11. St. Louis, MO.

Hillel, J., E. A. Dunnington, A. Haberfeld, U. Lavi, A. Cahaner, O. Gal, Y. Plotsky, H. L. Marks, and P. B. Siegel. 1993a. Multilocus DNA markers: Applications in poultry breeding and genetic analyses. In: R. J. Etches, and A.M.V. Gibbons (Ed.) Manipulation of the Avian Genome. CRC Press, Boca Raton, FL.

Hillel, J., A.M.V. Gibbins, R. J. Etches, and D. M. Shaver. 1993b. Strategies for the rapid introgression of a specific gene modification into a commercial poultry flock from a single carrier. Poult. Sci. 72:1197.

Hospital, F., C. Chevalet, and M. Mulsant. 1992. Using markers in gene introgression breeding programs. Genetics 132:1199.

Knott, S. A., and C. S. Haley. 1992. Maximum likelihood mapping of quantitative trait loci using full-sib families. Genetics 132:1211.

Kuhnlein, U., D. Zadworny, Y. Dawe, R. W. Fairfull, and J. S. Gavora. 1990. Assessment of inbreeding by DNA fingerprinting: Development of a calibration curve using defined strains of chickens. Genetics 125:161.

Kuhnlein, U., D. Zadworny, J. S. Gavora, and R. W. Fairfull. 1991. Identification of markers associated with quantitative trait loci in chickens by DNA fingerprinting. In: Burke, Dolf, Lynch and Wolf (Ed.) DNA Fingerprinting. pp 274-282. Birkhauser, Basel, Switzerland.

Lande, R., and R. Thompson. 1990. Efficiency of marker-assisted selection in the improvement of quantitative traits. Genetics 124:743.

McMillan, I., R. W. Fairfull, G. W. Friars, and M. Quinton. 1995. The effect of simultaneous selection on the genetic correlation. Theor. Appl. Genet. 91:776.

Moreno-Gonzalez, J. 1993. Efficiency of generations for estimating marker-associated QTL effects by multiple regression. Genetics 135:223.

Muir, W. M. 1994. Poultry improvement: Integration of present and new genetic approaches for layers. Proc. 5th World Congr. Genet. Appl. Livest. Prod. 20:5.

Rocha, J. L., J. F. Taylor, J. O. Sanders, S. J. Openshaw, and R. Fincher. 1995. Genetic markers to manipulate QTLs: The additive illusion. In: Proc. 44th Annu. Natl. Breeders Roundtable. pp 12–38. St. Louis, MO.

Databases and Information Systems Needed for Maps and Marker-Assisted Selection

J. W. Keele[1] and J. E. Wray

Roman L. Hruska U.S. Meat Animal Research Center, USDA-ARS, Clay Center, NE 68933-0166, U.S.A.

Abstract Maintaining a current database and information system containing marker sequences and genotypic data is critical to the development and utilization of physically anchored genetic linkage maps. Integration of data contributed by scientists with a variety of specialties will require a highly flexible database and information system. The effort required for positional cloning is dependent on the resolution of the genetic linkage map. A target resolution of 2 cM is desired before initiating positional cloning. A large number of markers and animals is required to achieve a genetic linkage map with 2 cM resolution. The integrity of genetic maps is highly sensitive to genotyping errors. Increasing marker density and animals genotyped increases the probability of finding and correcting genotyping errors. Standardizing marker names and allele sizes is required for data integration. Marker names are usually based on function or chromosomal location. Names change as knowledge changes. Use of arbitrary marker names helps but doesn't solve the problem because the same marker can be developed independently by different investigators. Sequence analysis can be effectively used to identify duplicate markers and develop standardized nomenclature. Critical success factors for the effective implementation of genetic linkage maps are maximizing flexibility through effective conceptual database modeling and normalization, management of genotypic and phenotypic data, achieving a genetic linkage map with 2 cM resolution and maintenance of DNA sequence data to aid in standardizing nomenclature.

Key Words: Genome Mapping, Informatics, Database, Information Systems

Introduction

Development and use of genetic linkage maps for marker-assisted selection (MAS) is facilitated by a database and information system maintaining marker genotypes and trait phenotypes on livestock families as well as DNA sequences of markers and genotyping protocols. Successful construction and utilization of relevant genome maps is critically dependent on integration of data contributed by scientists with a variety of specialties, including molecular biology, quantitative genetics, linkage and complex segregation analysis, cytogenetics, meat and muscle biology, reproductive physiology, nutrition, genetic epidemiology, disease resistance, and informatics. Scientists possessing expertise in more than two of these specialties are rare; hence, an important attribute of a genome database is the capability to integrate data from contributors with diverse expertise. Genome mapping is a dynamic and rapidly expanding field with the continual emergence of new methodology resulting in new kinds of data. Integration of new data with data collected previously opens exciting

[1]Author to whom correspondence should be addressed: P.O. Box 166.

possibilities for addressing relevant questions. Obviously, it is impossible to anticipate these questions beforehand. A database capable of integrating data in this way will require great flexibility, consistent standardized nomenclature, and an ability to evolve with the science. We identify critical success factors associated with a database and information system needed for genome mapping and marker-assisted selection.

Database Flexibility

Database flexibility is maximized by a procedure called normalization. Normalization also minimizes data redundancy, which minimizes the number of changes that need to be made when errors are discovered. The essence of normalization is to provide maximum access to the data for cross-referencing through joining tables (Ullman, 1988). For example, microsatellite genotypic data are presented in normal form in Tables 1 and 3 and in non-normal form in Table 2. With Tables 1 and 3, it is possible to cross-reference markers with other tables using a join, while this is not possible with Table 2. With Tables 1 and 3 markers can be added by adding records to the database, whereas to add markers to Table 2 requires the addition of columns, which is a change in the database design. At first glance, it would seem that Table 3 is more normalized than Table 1 because fewer fragment lengths are stored. However, with Table 3 it is not possible to distinguish between homozygotes and genotypes with missing fragments. Table 1 preserves this capability. Using structured query language (SQL), it is possible to produce a view of Table 1 that looks and functions like Table 3; hence, there is no advantage of Table 3 over Table 1. We consider normalization to be a critical success factor in producing a flexible database capable of evolving with the science because normalization will provide maximum cross-referencing access of current data with future tables containing data generated by future genome mapping technology. Conversely, if we do not normalize our databases, it is unlikely that we will be able to integrate existing data with new data generated from future technologies.

Table 1. Microsatellite genotypic data (normalized)

		Genotype	
Animal	Marker	Fragment 1	Fragment 2
926937	11	266	262
926938	11	266	262
926939	11	266	262
926937	12	118	116
926938	12	118	118
926939	12	118	118
926937	14	192	184
926938	14	184	184
926939	14	184	184

Development of genetic linkage maps is an iterative process involving linkage analysis of genotypic data, detection of genotyping errors using double crossovers, rerunning markers and correcting errors and rerunning linkage analysis to produce an improved map (NIH/CEPH Collaborative Mapping Group, 1992). The process is repeated until no additional genotyping errors are detected. Genotyping errors can have very dramatic effects on the accuracy of high-resolution genetic linkage maps (Buetow, 1991), underscoring the importance of error detection and correction. In concert, the efficiency of error detection using double crossovers is very dependent on marker density (Lincoln and Lander, 1992). Hence, the process of map construction and error detection and correction is likely to span several years. For example, suppose an erroneous genotype was mistakenly entered in 1990. Close markers are subsequently genotyped and entered in 2000 and a double-crossover is discovered and the erroneous genotype is corrected. It is important that the evolving genetic maps correctly reflect corrections in the genotypic data that underlie them. Maintenance of genotypic data is a critical success factor for the continual development of livestock genome maps that improve in resolution and integrity over time.

Need for High-Resolution Genetic Linkage Maps

One important application of genetic linkage maps is coarse mapping of economically important quantitative trait loci (QTL) in preparation for positional cloning. An accurate, high-resolution genetic linkage map is required to achieve this. The resolution within which a QTL has been genetically mapped has a large impact on the effort required for positional cloning. Generally, a target resolution of 2 cM is desired before initiating positional cloning. This indicates that a map useful for locating QTLs for positional cloning needs to contain correctly ordered markers with an interval between adjacent markers of 2 cM or less. At least 220 offspring are required to correctly order markers spaced 2 cM apart if each progeny has one parent that is heterozygous for all markers. If each offspring has two parents heterozygous for all markers then the number of progeny required is cut in half. We regard these requirements as minimal because our capability to correctly order markers is com-

Table 2. Microsatellite genotypic data (not normalized)

	Markers					
	11		12		14	
Animal	Fragment 1	Fragment 2	Fragment 1	Fragment 2	Fragment 1	Fragment 2
926937	266	262	118	116	192	184
926938	266	262	118	118	184	184
926939	266	262	118	118	184	184

promised by marker heterozygosities less than 100%. Integrating and validating genotypic data from sufficient markers and animals to construct a genetic linkage map with 2 cM resolution is a critical success factor for positional cloning of QTLs.

Standardized Nomenclature

Traditionally, names of markers have been based on function or chromosomal location. Examples include FSH for follicle-stimulating hormone or D12S25 for anonymous marker 25 on Chromosome 12. The problem with these conventions is that the name changes as more is known about the function of the marker or its location, disrupting the consistent use of name between and within genome databases. The use of arbitrary unique identifiers has been recommended as a way to circumvent this problem (DOE Informatics Summit, 1993). Presumably, there is no temptation to change an arbitrary name. While this helps it doesn't solve the problem because scientists have independently cloned the same sequences, resulting in duplicate markers. We have successfully used sequence analysis to identify duplicate markers developed at MARC as well as equivalences between markers developed at MARC and elsewhere. Analysis of DNA sequence provides a powerful tool for standardizing marker nomenclature. Increasing the accessibility of primer sequences will increase the use of markers. Maintaining DNA sequences of primers and clones is a critical success factor for standardizing nomenclature and increasing the use of markers.

Additional Needs for Mapping QTLs and Marker-Assisted Selection

Genetic linkage mapping of QTLs requires maintenance of trait phenotypes in addition to the data management requirements for genetic linkage mapping of markers

Table 3. Microsatellite genotypic data (normalized)

Animal	Marker	Fragment
926937	11	262
926937	11	266
926937	12	116
926937	12	118
926937	14	184
926937	14	192
926938	11	262
926938	11	266
926938	12	118
926938	14	184
926939	11	262
926939	11	266
926939	12	118
926939	14	184

(Figure 1). The data management requirements for marker-assisted selection include the requirements for mapping QTLs and markers (Figure 1). In addition, MAS places scheduling requirements on the data management and information system (Figure 2). For example, it is important to collect tissue (e.g., white blood cells), extract DNA, collect genotypes, and calculate breeding values between birth and castration. While the timing of castration is somewhat flexible, delaying castration excessively would be associated with increased cost or reduced returns. A data collection and management system that is able to provide decision support prior to the point that decisions need to be made is a critical success factor for marker-assisted selection.

Associations between phenotypes and markers caused by linkage disequilibrium between QTLs and linked markers are not expected to be stable across different families of livestock unless the markers and QTLs are equivalent. Marker-assisted selection is likely to be useful well before the precise DNA sequence polymorphism responsible for the QTL is known. A quantitative trait locus that is strongly associated

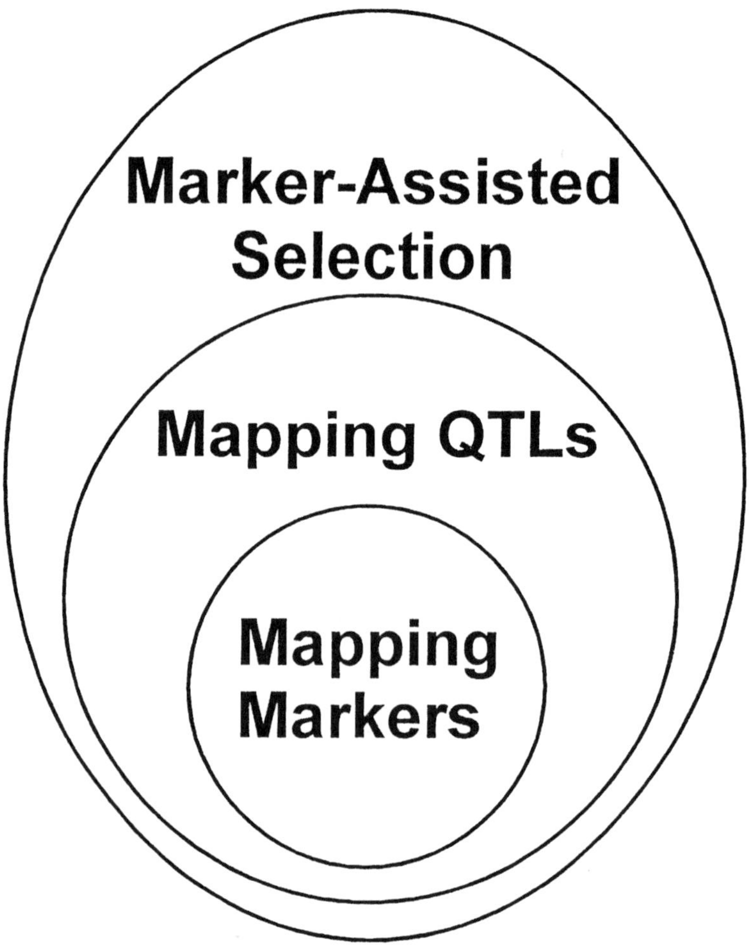

Figure 1. Venn diagram showing relationship between database requirements for mapping markers and quantitative trait loci and marker-assisted selection.

with a marker in one family may be fixed at a specific allele in another family, contributing zero to the variance within that family. For this reason, the usefulness of given markers is family-specific. A critical success factor for marker-assisted selection is whether or not marker genotypes, trait phenotypes, and pedigree databases are complete enough to track chromosomal fragments between generations.

Implications

Development and use of genetic linkage maps for marker-assisted selection is aided by a database and information system maintaining marker genotypes and trait phenotypes on livestock families as well as DNA sequences of markers and genotyping protocols. Critical success factors for the effective implementation of genetic linkage maps are 1) maximizing flexibility through effective conceptual database modeling and normalization, 2) maintenance of sufficient genotypic and phenotypic data within families to be able to determine transmission of chromosomal fragments between generations, 3) achieving a genetic linkage map with 2 cM resolution, and 4) maintenance of DNA sequence data to aid in standardizing nomenclature.

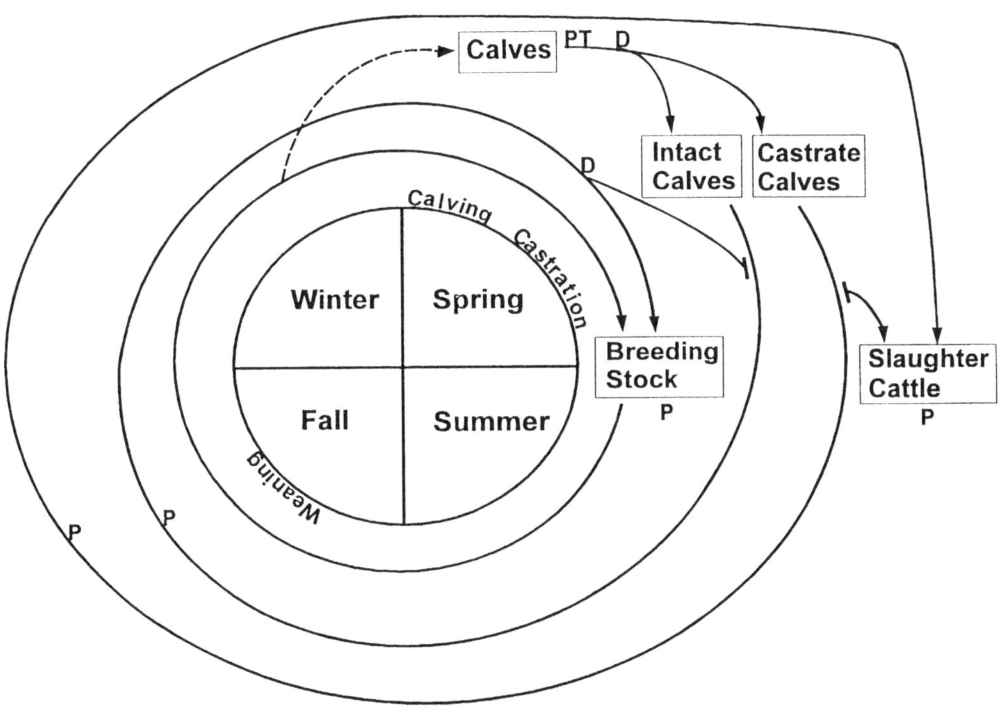

Figure 2. Beef production cycle depicting trait phenotype collection points (P), tissue collection points (T), and decision points (D).

Literature Cited

Buetow, K. H. 1991. Influence of aberrant observations on high-resolution linkage analysis outcomes. Am. J. Hum. Genet. 49:985.

DOE Informatics Summit. 1993. Meeting Report. April 26-27, Baltimore, MD. Available electronically from gopher.gdb.org in the document Math-In-Biology/info_rep.data.

Lincoln, S. E., and E. S. Lander. 1992. Systematic detection of errors in genetic linkage data. Genomics 14:604.

NIH/CEPH Collaborative Mapping Group. 1992. A comprehensive genetic linkage map of the human genome. Science (Wash., DC) 258:67.

Ullman, J. D. 1988. Database and Knowledge-Base Systems. Computer Science Press, Rockville, MD.

Integrating Use of New Markers into the Real World

P. M. VanRaden

Animal Improvement Programs Laboratory, Agricultural Research Service, USDA, Beltsville, MD 20705-2350, U.S.A.

Abstract In theory, all animal breeders could apply genetic marker technology and marker-assisted selection to speed improvement of their populations. In reality, the income from each new biotechnology must exceed its cost before it becomes useful. Fairly small advantages from genotyping will be profitable if reproductive rates are high. Genotyping of most females and of embryos will not be profitable given current reproductive rates. Success of marker-assisted selection depends greatly on the presence of a few genes with moderate or large effect. To estimate allele differences, breeders will genotype siblings in each elite family that have phenotype or progeny-test data. If allele differences are large at a particular locus, breeders will also genotype younger family members without data to speed the selection process. Progeny designs offer faster selection than grandprogeny designs because only one generation separates the tested ancestor from the selected descendants. If families are small, quantitative gene parameters are estimated poorly regardless of how many markers are tested. Estimated effects should be regressed for expected error content before use in breeding values. Industry cooperation will be needed to collect, interpret, and report genotypic data, especially if family members are widely dispersed and if different laboratories use different markers. Genetic markers are a sensible investment, especially when noting the large improvements in animal performance that result from current genetic selection.

Key Words: Biotechnology, Breeding Values, Genetic Markers, Marker-Assisted Selection

Introduction

Biotechnology gives animal breeders the tools to trace genes across generations. According to Mendel's laws, the two parents contribute factors or genes that do not blend but instead remain distinct and traceable from one generation to the next. Until now, most of the important genes affecting profitability could not be traced. Instead of tracing individual genes, animal breeders estimated only the total value of all genes. Statistical methods such as Best Linear Unbiased Prediction and the animal model provided fairly accurate rankings from only performance and pedigree information. Now, a third source of information is available.

Animal breeders have traced individual genes in their populations for many years. Some breeds have been kept separate from each other only on the basis of one or two genes that determine hair color. Hundreds of thousands of cows and bulls have been bloodtyped to verify pedigree information within breeds (Rumler, 1981). Undesirable recessive genes have been discovered and their frequency greatly reduced by a variety of old and new tools. A few quantitative genes with noticeable effects on important traits have been isolated and exploited.

New experiments may discover many more individual genes that merit selection. Then, practical breeders will want answers to basic questions such as these: 1) How many genes with large effect have been found?, 2) How accurate are the estimates of these effects?, 3) Can I genotype my animals at a profit?, and 4) Which animals should be genotyped and when? These questions may not yet have precise answers but must be addressed as the tools of biotechnology become available to the real world.

Estimation

Individual gene models have many parameters, whereas quantitative models may have just a heritability. The number of major loci, their location on the map, the number of alleles at each locus, and each allele's effect on the trait are parameters to be estimated, plus the genetic variance contributed by the minor loci. According to Bayes' theorem, such values can be estimated by blending prior assumptions and current evidence (Hoeschele and VanRaden, 1993a,b). The less evidence the data supply about any parameter, the more important are your assumptions.

Marker-assisted selection (MAS) can increase genetic progress only if major loci, or at least some moderate loci, actually exist. Current genetic evaluations are derived from quantitative theory, in which all loci are assumed to be minor. This prior assumption has been very useful but results in "sophisticated statistical models applied to simplistic genetic models" (Kennedy et al., 1990). Early crossbreeding studies did not detect major loci. Robertson (1949) concluded "There seems no doubt of the answer to the question. . . . The results show that all production characters are controlled by many gene pairs." Since then, however, genes with large effects on important traits have been found in several species (Andersson et al., 1994; Georges et al., 1994; Ron et al., 1994; Rothschild et al., 1994).

Genetic variation must reside somewhere, and heritability comes only from real genes with real effects. These genetic effects may be dispersed among minor loci on each of the chromosomes (Dekkers and Dentine, 1991) or concentrated in a few major loci on particular chromosomes. A particular trait may be associated closely with genotype differences for a particular marker in a particular family. The researcher may then investigate whether one of the quantitative trait locus (QTL) alleles is rare and unfavorable, or rare and favorable, or both are common in the population, or the association is just chance. The researcher may also estimate how far the hypothetical QTL is from the marker, and in which direction.

Exact location of QTL will be difficult to pinpoint. For a given set of families, more and more markers could be tested within a narrower and narrower interval of the map until parameter estimates stopped changing. The estimated location of the QTL, however, does not equal the true location. If more families or family members are added to the data set, the best fit will shift to some other site, and the initial "fine-tuning" will seem to be a wasted effort. The use of marker pairs to flank or bracket a QTL also can be misleading. Obviously, the QTL may be located outside of the bracket or may not exist at all. When searching for QTL, researchers must find a balance between more markers and more animals (Darvasi et al., 1993).

Estimates of parameters such as an individual gene effect usually include the true effect plus a random error. When many markers are tested, the largest estimated effect may be caused by a large true effect or a large error. By considering the probability of each cause in the analysis, a better estimate of the true genetic effect can be obtained. Goddard (1992) and Hoeschele and VanRaden (1993b) presented methods to regress estimates in proportion to their expected error content. After all, breeding values of bulls have been estimated for many years by similar procedures that regress progeny averages by their error content. To use these methods, the variance or distribution of true genetic effects must be roughly known.

Breeding values can be determined more accurately in families that are genotyped. So far, genetic evaluations have used primarily two sources of data: phenotypic and relationship. Now a third data source will be available that identifies the exact genes in common instead of defining probabilities that genes may be in common. Breeding values can be estimated fairly simply for progeny with genotypic but no phenotypic data. If inheritance of markers is known, each progeny's parent average can be updated according to the estimated allele differences within sire and dam and the recombination rate between marker and QTL. Fernando and Grossman (1989) generalized this procedure for animals with both genotypic and phenotypic data and for large populations.

Practical breeders must decide how many of the tested markers will actually be used in MAS. A particular level of significance should not be required (Meuwissen and van Arendonk, 1992). If estimates have been regressed for expected error content, each additional marker will contribute some additional accuracy, even if small. Breeders will start with the markers linked to large effects and genotype more markers as long as the value of information obtained exceeds the cost of genotyping.

Routine genetic evaluations could incorporate marker data when genotyping becomes more routine. Computations will probably be feasible, although the numbers of equations can increase by several times for models that predict a breeding value plus the values of two alleles for several QTL for each individual (Fernando and Grossman, 1989; Hoeschele, 1993). Approximate procedures to adjust breeding values are needed as an aid in both computation and education. Selection index procedures can demonstrate the gains possible when different relatives are genotyped (Hoeschele and Romano, 1993). Genetic principles must be at least partly understood on the farm for breeders to have confidence in the evaluations and in MAS.

Application

Breeders will apply marker-assisted selection as soon as they are convinced it provides better returns than other available investments. The first reports on a new technology may be written by the most optimistic researchers. More skeptical researchers write about some other topic while waiting for the exciting new technology to either mature or fade. Practical breeders must decide whether to continue with the tried and true technologies or to invest in newer, more risky technologies such as nucleus breeding schemes, genetic markers, sexed semen, or transgenics that have not yet become standard. Early estimates of the advantages of nucleus breeding programs for dairy cattle were later revised downward to show little advantage. Will genetic markers gradually mature into a technology as accepted as AI, or will they just fade away?

The benefits of MAS do fade away as the number of generations increase between a tested ancestor and a genotyped descendant (Meuwissen and van Arendonk, 1992; Hoeschele and Romano, 1993). The granddaughter design measures the difference between the two grandsire alleles fairly accurately but provides less information about allele differences for the progeny or grandprogeny that are now being selected. Genotypes of hundreds of offspring may be needed to accurately measure allele differences in parents or grandparents (Weller et al., 1990). Many of these offspring may be poor performers or no longer of interest as candidates for selection. Once allele differences have been measured, candidates for selection will be genotyped to check which allele they inherited. In this second phase of genotyping, offspring might not yet have performance data and might not supply any new information regarding the alleles of their parents or grandparents.

Candidates for selection may be progeny, grandprogeny, great grandprogeny, and so on, of marker-tested grandsires, or may be members of untested families. Relationships between marker-tested grandsires and candidates for selection were determined for the U.S. Holstein population. Grandsires were considered marker-tested if 30 or more sons had evaluations with at least 60% reliability (reliability is the squared correlation of predicted with true transmitting ability) for milk yield. Among the elite (as defined by USDA) Holstein cows, 34% were daughters of marker-tested grandsires. Among yearling bulls being sampled by AI companies, only 3% were sons of marker-tested grandsires. For these 577 yearling bulls (born from July to December 1993), nearly all of their paternal grandsires were marker-tested grandsires, but only 63% of their maternal grandsires had 30 or more evaluated sons. For effective application of MAS, progeny rather than grandprogeny designs will provide more useful information for selection of young animals.

Progeny designs may require twice the numbers of animals genotyped as grandprogeny designs (Weller et al., 1990) to reach the same accuracy. However, if allele differences are detected in sires rather than grandsires, current animals will inherit a tested allele all rather than half of the time and the marker will be a better indicator of the QTL because the genes will have one rather than two chances to recombine. The grandprogeny design is efficient for initial research, but progeny designs can provide information sooner. In routine MAS, all progeny of the best new progeny-tested males could be genotyped for the most promising markers. Allele differences could be estimated less accurately from progeny (Kashi et al., 1990) immediately and then more accurately from grandprogeny (Weller et al., 1990) eventually.

Estimates of allele differences of ancestors and marker genotypes of ancestor and progeny must all be available at times when selection decisions normally occur. Selection usually occurs at a particular point in the reproductive cycle or after the most valuable trait is observed. Traits that arrive later, such as longevity, may increase generation interval by more than accuracy and may not be worth the wait. Selection or MAS will proceed with the available data and future stages of selection will consider the future data.

In some cases, genotype data could come too soon. With embryo transfer, the main goal is to give superior females more progeny. If all embryos are routinely genotyped and many of the embryos discarded, fewer progeny will result from the elite females and the primary goal would be defeated. Genotyping of embryos might make sense only for less expensive embryos or when allele effects are very large. Many other animals will produce too few progeny to repay the initial cost of genotyping. Males that are not likely to be used in AI and females that are not likely to be embryo donors probably also will not be genotyped. In some cases, buyers may request genotyping of embryos or breeding stock before purchase or export just to prevent being sent the poorest of the litter.

Individual chromosomes may be copied from one population into another population of the same species by using genetic markers and backcrossing. At each generation, markers on the chromosome of interest can be checked to determine their population of origin. Without markers, chromosome segments can be transferred less efficiently. Then, phenotypic measurements or progeny testing are needed at each generation. Both of these backcrossing methods are more acceptable to consumers than the use of transgenics. Both methods probably are unacceptable to many purebred livestock breeders.

Genetic selection is already a big business. For example, the yearly increase in U.S. dairy cattle productivity from current selection is worth approximately $200 million. An increase of 2 to 9% in the rate of gain (Brascamp et al., 1993) would justify a multimillion dollar investment by government and industry. A doubling of the rate of gain (Soller, 1994) would justify a stampede toward marker technology. Actual value of MAS will depend on size of QTL effects and on agreement of simulated and real practices. Even the more realistic studies such as that of Meuwissen and van Arendonk (1992) may have overestimated advantages of MAS because the selection among alleles already occurring in purely quantitative models was ignored. Also, the many minor loci will continue to provide genetic progress long after the few major loci have been fixed by MAS.

The first industry breeding programs to adopt MAS may be those that have successfully used other biotechnologies. Few animals will be genotyped in livestock populations that have seen little AI, embryo transfer, or data collection. Nevertheless, these species can benefit if MAS is used in elite seedstock populations and then the genetic gains distributed through natural multiplication. Marker-assisted selection will be the most profitable in dairy cattle because genotyping costs will be offset by greater income per animal and by greater reductions in the generation interval.

If real product prices continue to decline, breeders may actually begin to invest in less data recording, less AI, and less biotechnology rather than more.

Cooperation

To the end user, MAS may be fairly simple. For each animal that is genotyped, the owner receives automatically an improved prediction of breeding value, perhaps on an updated pedigree. Then, selection proceeds normally. Some degree of industry cooperation and coordination will be required before MAS can be that simple. Development of marker maps (Beattie, 1994) was not simple, and maps of QTL will be more difficult to establish and maintain. Before breeders can benefit from MAS, many markers first must be tested in many families.

To discover which markers are useful in which families, DNA samples from many of the most elite members of a breed are needed. Owners of such animals may see little need to contribute to the study unless they also own the progeny and grandprogeny that will benefit from MAS. Poultry and swine breeding companies often own whole families, but family sizes are probably smaller than for elite bulls. Cattle breeding companies usually own just males, and most members of any bull family are owned by someone else.

Thus, cattle breeders must cooperate to obtain DNA from large families. In dairy cattle, paternal half-sib families from the most valuable sires recently have been studied (Cowan et al., 1990; Georges et al., 1994). A Dairy Bull DNA Repository now provides semen from these widely dispersed families to researchers from a central location (Da et al., 1994). Formal agreements for use of the DNA, data, and results already have been established. Researchers will analyze and then pool their data to allow further analysis by other researchers and inspection by the semen donors.

Individual bull names and registration numbers will be suppressed in all research reports. Few researchers have included females in their studies, although blood, milk, or even hair might be obtained as DNA sources.

Obviously, cooperation is not new to farmers. Breed associations currently offer two services that are similar to MAS: recessive testing and blood typing. Progeny testing is performed across much of the United States by AI cooperatives. Collection of performance data also is accomplished by farmer-controlled cooperatives. Private laboratories may need to form new agreements with the existing companies, cooperatives, and breed associations to make MAS work. A few organizations have already formed these new links.

Bloodtyping and recessive testing services are not just similar to DNA genotyping, they may be replaced by DNA techniques as those improve. Currently, bloodtyping costs about $35 to the farmer. DNA genotyping could easily replace bloodtyping if it is less expensive, but with a period of overlap because only bloodtypes rather than DNA types would be available from the previous generation. Recessive carriers may be detected sooner and less expensively with DNA technology than with traditional methods.

Genotyping is like bloodtyping in another way. Information rather than some actual product is produced. Values of the progeny are estimated more accurately with this information but parent values are often unaffected. Value of the whole population is enhanced if the information is made public. A breed association may require by contract that a laboratory report all outcomes, both good and bad, so that buyers will be notified and so that data won't be biased. Marker genotypes from MAS programs will be needed in genetic evaluations, for continuing research, and to match with future progeny.

Breeders have worked together with great success to improve their livestock. For U.S. dairy cattle, genetic progress provided 74% of the increase in milk yield over the past 5 years. (VanRaden et al., 1995). Further investment in genetic research seems justified. Breeders with common goals will invest in biotechnology if rewarded for their individual initiative. Genetic progress will continue and perhaps increase if breeders continue to cooperate locally, nationally, and internationally to improve their populations.

Will American consumers notice the impact of genetic markers? Animal products are already cheap and plentiful in the United States. In much of the rest of the real world, consumers want more animal products. Cooperation could move biotechnology, food, and people from places of high supply to places of high demand. If hundreds of millions of consumers moved to America, the demand for MAS would increase, causing American farmers to apply genetic markers and other technologies to increase production. On the other side of the real world, Chinese farmers with little technology already feed eight times more citizens per acre (four times more people and half as much farmland) than American farmers (Food and Agriculture Organization, 1985). By law, the number of American consumers remains limited, even while the number of genetic markers grows rapidly.

Summary

Genetic selection has been very successful and MAS will increase that success. Contributions of individual loci could be overstated unless estimates of gene effects are regressed. The grandprogeny design is efficient for finding allele differences but less efficient for MAS because current animals and their tested ancestors are one generation further apart. Marker-tested grandsires sired 34% of the elite Holstein cows but only 3% of yearling bulls. Genotypes of an elite bull's first daughters will provide less accurate, but more timely, information than those of his later sons. When individual breeders don't own whole families, MAS must be a cooperative effort. In the past, breeders and breed associations have used similar technologies with goals of verifying ancestry and eliminating harmful recessives. An important goal for animal breeders is to actually feed people, whether we accomplish that with biotechnology or with simpler methods.

Literature Cited

Andersson, L., C. S. Haley, H. Ellegren, S. A. Knott, M. Johansson, K. Andersson, L. Andersson-Eklund, I. Edfors-Lilja, M. Fredholm, I. Hansson, J. Hakansson, and K. Lundstrom. 1994. Genetic mapping of quantitative trait loci for growth and fatness in pigs. Science (Wash DC) 263:1771.

Brascamp, E. W., J.A.M. van Arendonk, and A. F. Groen. 1993. Economic appraisal of the utilization of genetic markers in dairy cattle breeding. J. Dairy Sci. 76:1204.

Beattie, C. W. 1994. Livestock genome maps. Trends Genet. 10:334.

Cowan, C. M., M. R. Dentine, R. L. Ax, and L. A. Schuler. 1990. Structural variation around prolactin gene linked to quantitative traits in an elite Holstein sire family. Theor. Appl. Genet. 79:577.

Da, Y., M. Ron, A. Yanai, M. Band, R. E. Everts, D. W. Heyen, J. I. Weller, G. R. Wiggans, and H. A. Lewin. 1994. The dairy bull DNA repository: A resource for mapping quantitative trait loci. Proc. 5th World Congr. Genet. Appl. Livest. Prod., Guelph, ON, Canada 21:229.

Darvasi, A., A. Weinreb, V. Minke, J. I. Weller, and M. Soller. 1993. Detecting marker-QTL linkage and estimating QTL gene effect and map location using a saturated genetic map. Genetics 134:943.

Dekkers, J.C.M., and M. R. Dentine. 1991. Quantitative genetic variance associated with chromosomal markers in segregating populations. Theor. Appl. Genet. 81:212.

Fernando, R. L., and M. Grossman. 1989. Marker assisted selection using best linear unbiased prediction. Genet. Sel. Evol. 21:467.

Food and Agriculture Organization. 1985. Production Yearbook. United Nations, New York.

Georges, M., D. Nielsen, M. Mackinnon, A. Mishra, R. Okimoto, A. T. Pasquino, L. S. Sargeant, A. Sorenson, M. R. Steele, X. Zhao, J. E. Womack, and I. Hoeschele. 1994. Using a complete microsatellite map and the grand-daughter design to locate polygenes controlling milk production. Proc. 5th World Congr. Genet. Appl. Livest. Prod., Guelph, ON, Canada 21:81.

Goddard, M. E. 1992. A mixed model for analysis of data on multiple genetic markers. Genet. Sel. Evol. 21:467.

Hoeschele, I. 1993. Elimination of quantitative trait loci equations in an animal model incorporating genetic marker data. J. Dairy. Sci. 76:1693.

Hoeschele, I., and E. O. Romano. 1993. On the use of marker information from granddaughter designs. J. Anim. Breed. Genet. 110:429.

Hoeschele, I., and P. M. VanRaden. 1993a. Bayesian analysis of linkage between genetic markers and quantitative trait loci. I. Prior knowledge. Theor. Appl. Genet. 85:953.

Hoeschele, I., and P. M. VanRaden. 1993b. Bayesian analysis of linkage between genetic markers and quantitative trait loci. II. Combining prior knowledge with experimental evidence. Theor. Appl. Genet. 85:946.

Kashi, Y., E. Hallerman, and M. Soller. 1990. Marker-assisted selection of candidate bulls for progeny testing programmes. Anim. Prod. 51:63.

Kennedy, B. W., A. M. Verrinder Gibbins, J. P. Gibson, and C. Smith. 1990. Coalescence of molecular and quantitative genetics for livestock improvement. J. Dairy Sci. 73:2619.

Meuwissen, T.H.E., and J.A.M. van Arendonk. 1992. Potential improvements in rate of genetic gain from marker-assisted selection in dairy cattle breeding schemes. J. Dairy Sci. 75:1651.

Robertson, A. 1949. Crossbreeding experiments with dairy cattle. Anim. Breed. Abst. 17:201.

Ron, M., M. Band, A. Yanai, and J. I. Weller. 1994. Mapping quantitative trait loci with DNA microsatellites in a commercial dairy cattle population. Anim. Genet. 25:259.

Rothschild, M. F., C. Jacobson, D. A. Vaske, C. K. Tuggle, T. H. Short, S. Sasaki, G. R. Eckardt, and D. G. McLaren. 1994. A major gene for litter size in pigs. Proc. 5th World Congr. Genet. Appl. Livest. Prod., Guelph, ON, Canada 21:225.

Rumler, R. H. 1981. Dairy cattle breed organizations in the United States. J. Dairy Sci. 64:1278.

Soller, M. 1994. Marker assisted selection - An overview. Anim. Biotech. 5:193.

VanRaden, P. M., G. R. Wiggans, R. L. Powell, and H. D. Norman. 1995. Changes in USDA-DHIA genetic evaluations (January 1995). AIPL Research Report CH-3. Beltsville, MD.

Weller, J. I., Y. Kashi, and M. Soller. 1990. Power of daughter and granddaughter designs for determining linkage between marker loci and quantitative trait loci in dairy cattle. J. Dairy Sci. 73:2525.

Index